CELLS

HEMODYNAMICS/
CIRCULATION

DEFINITION

DIGESTION

ELECTROPHYSIOLOGY

EXCRETION

HEMOSTASIS

ALLERGY AND
IMMUNOLOGY

ION
CHANNELS

MEMBRANE

METABOLISM

MECHANISM OF
ACTION

MOTILITY

OVERVIEW

SIGNALING/
COMMUNICATION

STRUCTURE/
MORPHOLOGY/
COMPOSITION

SYNTHESIS
AND
SECRETION

TRANSPORT

PHYSIOLOGY OF
REPRODUCTION

HOMEOSTASIS/
REGULATION/
CONTROL

MOSBY'S SUCCESS IN MEDICINE

Physiology

Physiology

MOSBY'S SUCCESS IN MEDICINE™

Uwe Ackermann, Ph.D.
Professor of Physiology,
University of Toronto,
Toronto, Ontario,
Canada

Mosby

St. Louis Baltimore Boston Carlsbad Chicago Naples New York Philadelphia Portland
London Madrid Mexico City Singapore Sydney Tokyo Toronto Wiesbaden

Mosby
Dedicated to Publishing Excellence

A Times Mirror
Company

Vice President and Publisher *Anne S. Patterson*
Editor *Emma D. Underdown*
Developmental Editor *Christy Wells*
Project Manager *Dana Peick*
Production Editor *Jeffrey Patterson*
Manufacturing Supervisor *Tony McAllister*
Book Designer *Amy Buxton*
Cover Designer *Stacy Lanier/AKA Design*

Printed in the United States of America
Composition by Graphic World, Inc.
Printing/binding by R. R. Donnelly

Mosby-Year Book, Inc.
11830 Westline Industrial Drive
St. Louis, Missouri 63146

Published in the United States as Ace the Boards

Library of Congress Cataloging in Publication Data

Ackermann, Uwe.
Mosby's USMLE step 1 reviews—physiology / Uwe Ackermann.
p. cm.—(Ace the boards)
Includes index.
1. Human physiology—Examinations, questions, etc. I. Title.
II. Series.
[DNLM: 1. Physiology—examination questions. QT 18.2 A182m 1996]
QP40.A25 1996
612′.0076—dc20
DNLM/DLC
for Library of Congress 96-8434
CIP

ISBN 0-8151-0054-X (IBM)
ISBN 0-8151-0933-4 (MAC)

96 97 98 99 00 / 9 8 7 6 5 4 3 2 1

Preface

As knowledge expands so do the expectations of instructors and examiners regarding the component that they regard as essential. Students, on the other hand, have too much to learn in too little time.

Success in Physiology was written for those who need a framework for problem-solving, a quick reminder of the important points, and a direct way of preparing for exams. I was thinking primarily of stressed examination candidates but also of senior students and curious practitioners in the life and health sciences.

The content of *Success in Physiology* is similar to that of senior physiology texts that are now on the market but differs from many by the inclusion of chapters on integrative physiology and the pathophysiology of acid-base regulation.

- The information in each unit is presented concisely, in an easily accessible, bulleted, colorized format.
- Medical students have reviewed the book.
- The layout allows students to jot notes while reviewing lecture notes. This will consolidate in one place everything needed for thorough review.
- There are numerous two-color illustrations to enhance understanding.
- Summary tables, lists, and charts are included to help students synthesize material.
- Board-type questions reinforce the clinical utility of the information and are a direct way to prepare for exams.
- Perforated answer sheets are provided to enable students to tear out the answers and place them next to the questions for faster feedback.
- The computer diskette includes hundreds of categorized questions and gives immediate feedback.
- Icons are provided to provide quick access to information.

Although physiology is, above all, an integrating science, each chapter is written as a self-sufficient unit so that extensive cross-referencing is not necessary. The text adheres to the essentials, offering a quick review for the harried student.

Students have helped me decide what must be included and have tested my examination questions. Deborah Atchison, my friend and colleague, provided critical reviews and reality checks in 1524 different ways.

Thank you,

Uwe Ackermann

Test-Taking Strategies

Suzanne F. Kiewit, M.Ed.

Standardized national exams are a part of the medical education process in many countries. To perform well on standardized exams, it is imperative that you begin with a **plan.** Be as efficient and effective as possible by planning well.

MONTHS AHEAD OF THE EXAM

- Sit down with a blank calendar and block in your commitments: classes, final exams, scheduled events.
- Include time for activities of daily life: eating, sleeping, exercising, socializing, banking, maintaining your home, and so forth.
- The remaining time is available for study/review.
- Determine an orderly approach to the material you need to cover that fits your particular set of needs (e.g., subject-by-subject approach, systems approach, pathologic state approach).
- Assign the remaining time to content areas. This is done in various ways: material covered last year first, easiest material first, least comfortable material first, detailed subjects last, whatever. Your plan should reflect your goal: to maximize your score.
- Establish a warm-up, which may consist of breaking the tension in major muscle groups (neck rolls, shoulder rolls, etc.), a quick visualization of you performing successfully, or a brief meditation. Practicing this warm-up routine before each of your study sessions will make it a familiar activity that helps you learn effectively, as well as take exams effectively.
- Designate time at the end of your study period for overall review. Depending on your needs, that might be a week or just several days before the exam.
- Plan for feedback on your efforts. Schedule time for answering questions on the material you are reviewing and for taking at least one mock comprehensive exam.
- Do the comprehensive exam midway through your study period so that you can refine your efforts to reflect the degree of your performance.

DAYS AHEAD OF THE EXAM

- Divide each day into thirds: morning block, afternoon block, and evening block.
- Consider the time of day that is most productive for you and do the most difficult or least favorite material at that time.
- Assign more blocks of study to those areas requiring the most review to reach a comfortable knowledge level.
- A popular way to use blocks is to pair subjects or materials. For instance, pair strong content with weaker content so that you are not always in the position of not knowing material (which would invite negative feelings or ineffectiveness). Or pair a conceptual subject with a detail subject, such as physiology with anatomy, so that you are not always doing the same kind of thinking (this invites positive effort).
- Use your most productive blocks of time for actual study/review. Use the nonpeak times for reinforcement of material covered or feedback by answering questions on material that you have covered.

Planning the Blocks

- Once you determine time allocations for each content area and an orderly approach that fits your needs and goals, you want to specify what you plan to do during each block.
- Be specific as to content area, material to study, and task; for example, MICRO: review chart on viruses; PHYS: answer questions on renal; and so forth.
- Each study block will last approximately 3 to 4 hours. To be most efficient and effective, plan to take a 5- or 10-minute break every hour. If you are having difficulty getting into the study mode, plan to study for 25 minutes, then take a 5-minute break. Reserve longer breaks for switches between subjects. Get up and move around on breaks.

BEFORE THE EXAM

You can maximize your exam performance by

taking control. Adults tend to perform better when they feel that they have a measure of control. With comprehensive exams, it is easy to feel out of control. You are told what time to arrive, where to go, what writing instrument to use, and so on. You want to assume control of as many aspects as possible to maximize your performance:

STUDY Follow the sage advice of planning your work and working your plan to maintain satisfactory preparation with regard to study.

SLEEP Get a good night's rest. Sleep needs vary, but 6 hours is usually minimum. Try to get the appropriate amount of sleep that you require.

NUTRITION Maintain proper nutrition during both study time and exam time. Eat breakfast. Choose foods that help keep you on an even energy level. Eat light lunches on exam days. If you have a favorite food and can take it with you, treat yourself.

MEDICATION It may be cold, flu, or allergy season. Take no medications that may make you drowsy.

CLOTHING Heating and cooling systems are rarely balanced enough to suit everyone. Wear articles of clothing that can be added or removed as necessary. Strive for personal comfort.

READINESS Develop physical, mental, emotional, and psychologic readiness for the exam. Keep your thoughts about the exam and your preparation efforts positive. You must believe that *you can do this!*

ARRIVAL Plan to arrive as close to the designated time as possible and still allow yourself sufficient time to check in. Keep to yourself so that other people's anxieties will not affect you. Find your seat.

ACCLIMATION Settle in and get comfortable. Take several deep breaths. . . RELAX. A relaxed mind thinks better than a tense one. Do your warm-up routine.

ATTENTION Pay attention to the proctor. Complete all identification material as required. Read all instructions carefully. Ask for clarification as needed. Do not begin your test until told to do so. Once you begin, quickly glance through the whole test to see how it is set up and how questions are organized. Again, you want to take control of the situation. A quick review eliminates surprises and allows you to develop a plan.

DURING THE EXAM

Plan Your Approach

There are numerous approaches to answering questions. Answer questions in the order that appeals to you. Doing the easier ones first may give a psychologic boost; however, the ones you skip may stay on your mind and cloud your thinking.

Another approach is to answer each question in sequence. Start with the first one in the section with which you begin and fill in an answer for each question. Do not leave any blanks! The theory behind this is that if you spend any time at all on an item, you should mark your best response at that time and go on. If you are not certain of your choice, mark "R" on the test for review and return to it later as you have time.

There are various formats to questions. The most popular is the one-best-answer multiple choice item. These typically pose a question and ask you to choose between five options. Another popular question type is the matching sets. These may include short leading lists or long leading lists of up to 26 items from which to choose.

Some students plan to do the matching items first. If you prefer matching items, this is a reasonable plan because it helps you get started with items about which you feel confident. It is also reasonable because matching items are not good items on which to guess if you run short of time. **You** must decide the order in which you want to do the questions.

Complete the Bubble Sheet or Answer Card Carefully

There are two schools of thought on this matter. One is to fill in the answer sheet **item by item** as you go. This method minimizes transcription errors. The other method is **block transfer.** Complete a logical chunk of questions in the test booklet (one or two pages) and then transfer responses to the answer sheet. Be sure that the last question number on the page is the last numeral you blacken. This method saves time and offers a mini-mental break at the end of each block. Such minibreaks help decrease fatigue during a long exam. Choose the method that will work for you and *practice* it as you take prep questions.

Budget Your Time

Know how much time is allotted and check the number of questions. A quick calculation will give you the average number of seconds or minutes you can spend on each item and still complete the exam. Obviously, some questions may go more quickly and balance out the ones that take longer. To keep track, you need a pacing strategy. A good strategy is to establish checkpoints at 30-minute intervals. When you overview the booklet, circle the numerals corresponding to where you should be at 30 minutes, 60, 90, 120, and so on. For example, if you have 200 questions on a 3-hour exam, you should be at question number 33 at 30 minutes. As you complete the exam, check your time at the circled items. This technique keeps you from watching the clock too much, yet permits multiple opportunities to adjust your pacing.

If you find yourself spending too much time on any one question, select your best choice at that time, mark an answer, and "R" it for later review. The point is to keep going. Laboring too long on one question limits you from responding to other items you may know well. Remember, controlling your time helps you maximize your points.

ANSWERING THE QUESTIONS

- **Read and *understand* the stems and alternatives.**
  The most frequent error made on exams is misreading or misinterpreting the various aspects of a question. The **stem** is the introductory question or statement. The **alternatives** are the options from which you select the one best response. To encourage reading and understanding, use a process.
- **Follow a process to answer questions.**
  1. Quickly read the stem.
  2. Quickly read the options. (Combined, the first two steps create a preview of the item.)
  3. Carefully read, underline, and mark the stem in a timely fashion.
     - Selectively underline key words and phrases.
     - Pay attention to nouns, verbs, and modifiers.
     - Circle age and gender.
     - Note data in telescopic form (e.g., ↑ BP).
     - Graphically represent material if it helps you to understand (e.g., diagram the renal tubule to answer a question about reabsorption).
  4. Carefully read each alternative. Mark as appropriate.
- **Consider each alternative as one in a series of true, false, or not sure (?) statements.**
  Read each alternative. Rather than slashing out the ones you eliminate, work with each one and designate it as **true, false,** or varying degrees of **true/false/?.** This marking strategy requires you to make judicious decisions about alternatives relative to the stem. It also provides a record of your original thinking, which will save you rethinking time if you need to reconsider a question. Practice this strategy on preparation questions so it becomes second nature.
- **Avoid premature closure.**
  Sometimes you may read a question and anticipate a response. Such a reaction helps focus your attention. However, be sure to read *all* the options so that you are selecting the *best* response. In one-best-answer multiple choice questions, there is one *best* and several *likely* responses. Avoid being misled; consider all the alternatives.
- **Be leery of negative stems.**
  Negative stems require shifting to a negative thinking mode to determine which alternatives are not correct. You can avoid this shift by using this strategy:
  - Circle such words as *except, least, false, incorrect, not true* to raise your awareness of them.
  - Cross out the negative and read the stem as though it were a positive.
  - Mark each option as T/F/?. The F option will then be the appropriate choice.
- **Keep your original answers.**
  To change or not to change answers is a difficult decision. The answer depends on a person's previous history. If you are the kind of student who, if you change answers, changes them from wrong to right, then selectively changing answers may be worthwhile. If, on the other hand, your past experience has been to change right answers to wrong answers, selectively changing answers is probably not a good idea. Good performers change answers, but only if they have reason, such as acquired insight or discovery of misreading or misinterpretation.
- **Stay calm and focused.**
  If a question upsets you, calm yourself. Take several deep breaths. Tell yourself, "I can do this!" Give yourself a mental or physical break. Pay special attention to the next two or three questions after a bout of emotional uneasiness. It is possible to miss items when attention is diffused.

THINKING THROUGH QUESTIONS

- **Use logical reasoning and sound thinking.**
  - Read the item carefully. After careful reading, ask "What is this question really asking?" Restate it so that you know what is being asked.
  - Engage in a mental dialogue with the question. Talk to yourself about what you do know. Always start with what **you** know. Verbalize your thinking.
  - If a diagram or graphic representation is included, orient yourself to it **first** so that the options do not lead your thinking.
- **Use information found within the questions themselves to help you answer others.**
  There will not be obvious answers on a nationally

standardized exam. However, there may be items or graphics that trigger remembrances.

- **Create a diagram, chart, map, or graphic representation of given information.**
  Material that is visually presented usually helps clarify thinking. Use selective, quick sketching as warranted.
- **Reason through information like a detective.**
  - Sift through the details (preview).
  - Determine the relevant information (selectively mark).
  - Put the clues together as in solving a puzzle (reason).
- **Read carefully and note key descriptors.**
  - Note words such as *chronic, acute, greater than, less than, adult, child.*
  - Attend to prefixes such as *hyper-, hypo-, non-, un-, pre-, post-.*
- **Analyze base words and affixes.**
  Studying a question at the word level may help you remember salient information. Look for base words or related words. Determine Latin or Greek word parts and use their meanings to assist you.
- **Consider similar options equally.**
  If you mark one alternative as "false" for a particular reason and another option is false for the same reason, it's probably "false" as well.
- **Trust the questions.**
  The questions are designed to determine if you have a working knowledge of the material. They are not written to confuse you. You need to believe that your medical school curriculum and your study efforts prepared you for most of the questions.
- **Meet the challenge of clinical scenarios.**
  Longer scenarios challenge you to discern the relevant from the irrelevant material. In doing so, you are given multiple clues to consider. To effectively handle the scenario, follow this strategy:
  - Scan the stem and read the first several lines.
  - Skip to the end of the stem and read the last several lines.
  - Check the alternatives to narrow your focus.
  - Now that you know what the question is about, go back to the stem; read and mark what's important to your informed decision making.
  - Make good T/F/? decisions.
- **Reread your underlines and markings when you are down to two choices.**
  By the time you work through a stem and numerous alternatives, it is easy to lose the focus of the question. Check your focus by rereading only the underlines to ensure that you are answering the question being posed.

ANSWERING MATCHING ITEM SETS

Matching items are used to measure your ability to distinguish among closely related items. They require knowledge of specific sets of information. As you study, be alert to potential material that could be tested in this way.

Matching items can be formatted in two ways. **Short leading list matching** items include a set of five, lettered options followed by a lead-in statement and then several numbered stems. **Long leading list matching** items include a set of up to 26 lettered options, followed by a lead-in statement and then several numbered stems.

To efficiently deal with a short leading list item, consider it as an upside-down multiple choice item with the same repeated options. To handle it effectively, do the following:

- Scan the list; determine the topic.
- Read the lead-in statement; determine the focus.
- Quickly read the stem; then read and mark key words.
- In the left margin, create a grid with A, B, C, D, E at the top.
- Make good T/F/? decisions about each stem, marking them in the grid. In this way you can see the pattern of your responses. Similarly, a grid with the item numbers can be drawn beside the leading list and responses marked there.

Handling long leading list matching items effectively requires some modifications in the process. It is not efficient to make T/F/? decisions about each option, so follow this strategy:

- Scan the list; determine the topic.
- Read the lead-in; determine the focus.
- Read a stem and generate your own response.
- Narrow the focus. Put a check mark by those related options in the long list.
- Read and mark specifics in the stem to differentiate among those alternatives you marked.
- Make good T/F/? decisions.

For each stem, mark the narrowed-list options with a different symbol (star, dash, etc.). Items are listed in logical order, alphabetically or numerically. When looking for an option such as "xanthinuria," do not start at the beginning of the list. Looking in the appropriate place saves valuable seconds.

TEST PSYCHOLOGY

How a question is worded can often influence your response to it. Most clues about "test psychology" are a function of the way in which a question is worded—test constructors cannot rename body parts, drugs, diseases, and so forth. Being aware of the psychology behind the wording can often help you answer the test question.

Using techniques of test psychology to arrive at a correct answer has limited value on standardized exams because those who construct the exams are well aware of the use of these techniques. Nonetheless, being aware of these techniques of test psychology may add another point or two to your score, and they can also enhance your sense of control. Knowing these techniques provides additional strategies to employ should the question temporarily confuse you.

The best way to take any exam is to be totally prepared with a strong knowledge base and personal test confidence. The following techniques should be used only if you have exhausted your knowledge base, eliminated all distractors, and cannot come up with the answer even with logical thinking and sound reasoning. Such techniques are **not** a substitute for knowledge, nor are they foolproof.

- **Identify common ideas or themes within the options and between the stem and options.**
  - Circle repeated words in the options.
  - Select the option with the most repeated words or phrases.
  - Circle words repeated in both stem and options.
  - Select the option that contains key words or related words from the stem.
- **Beware of words that narrow the focus or are too extreme because they tend to be incorrect.**
  Circle such words as *all, always, every, exclusively, never, no, not, none.*
- **Options that are look-alikes are good candidates for exclusion.**
- **Note qualifiers that broaden the focus because they may be correct.**
  Circle words such as *generally, probably, most, often, some, usually.*
- **Identify antonyms or two opposing statements as potentially correct options.**
  Test constructors may use pairs of opposites, so this tip may lose its effectiveness.
- **Select the most familiar-looking option.**
  Always go from what you know. Alternatives with unknown terms may be likely distractors.
- **Select the longest, most inclusive answer.**
  This would include "All of the above" as a strong potential response.
- **In numerical items, knock out the high and low alternatives and select one in the middle that seems most plausible.**
- **In negatively stemmed questions, categorize responses; the one that falls out of the category is a likely candidate.**
- **Mark the same alternative consistently throughout the test if you have no best guess and cannot eliminate distractors.**
  Before the test, decide which letter (A, B, C, D, E) will be your choice. In this way, if you have given a question your best effort and cannot decide, mark your favorite response and move to questions that cover more comfortable material.

AFTER THE EXAM

- **Between sessions and overnight:**
  - Take a well-deserved break. Eat nutritionally.
  - If you feel the urge to study, study material that is comfortable, from a source with which you are familiar (e.g., personally developed study cards or your annotated review book).
  - If you discovered a recurring "theme," you might desire to consult that set of information.
  - Do something pleasurable. Relax. Get a good night's rest.
- **After the final booklet:**
  - Recognize that this exam is a measure of what you know on a given day for a given set of information at a given point in time. Keep a reasonable perspective.
  - **Celebrate!**

References

Bushan V, Le T, Amin C: First aid for the USMLE Step 1, ed 5, Norwalk, Conn, 1995, Appleton & Lange.

Contents

MOSBY'S Success IN MEDICINE

Physiology

Chapter 1

Cell Physiology

Cell Structure

Overview The cells of the human body (Fig. 1.1) are eukaryotic cells, which means that they have

- a well-defined membrane surrounding a central nucleus
- a variety of membrane-lined intracellular structures and organelles and
- a number of distinct subcellular compartments in which different chemical reactions, requiring different environments, can occur simultaneously because the membranes surrounding these compartments are selectively permeable and do not permit the ready diffusion of many substances from one side to the other.

Cytoplasmic Organelles, Structures, and Inclusions Figure 1.1 shows organelles and structures that are included in a typical cell

- **Nucleus**
  - Largest organelle (about 5 μm in diameter)
  - Surrounded by the *nuclear membrane*
  - Contains *chromatin* and 1 or 2 *nucleoli*
  - Site of DNA synthesis and repair
  - Site of RNA synthesis
- **Nuclear membrane**
  - A phospholipid bilayer, the outer layer of which is continuous with the *endoplasmic reticulum*
  - The two layers fuse at various points and form *nuclear pores* of 30 to 100 nm in diameter.
  - Acts as a selectively permeable barrier
- **Chromatin**
  - A DNA-protein complex, composed of approximately equal portions of DNA and histone proteins
  - Physically arranged in repeating units of 1 DNA molecule and 8 histone molecules
  - Functionally organized into chromosomes
- **Nucleolus**
  - One or more may be present
  - Made up of ribosomal RNA and not surrounded by a membrane
  - Locus of RNA processing and synthesis of *ribosomes*
- **Ribosomes**
  - Small granules (about 25 nm in diameter)

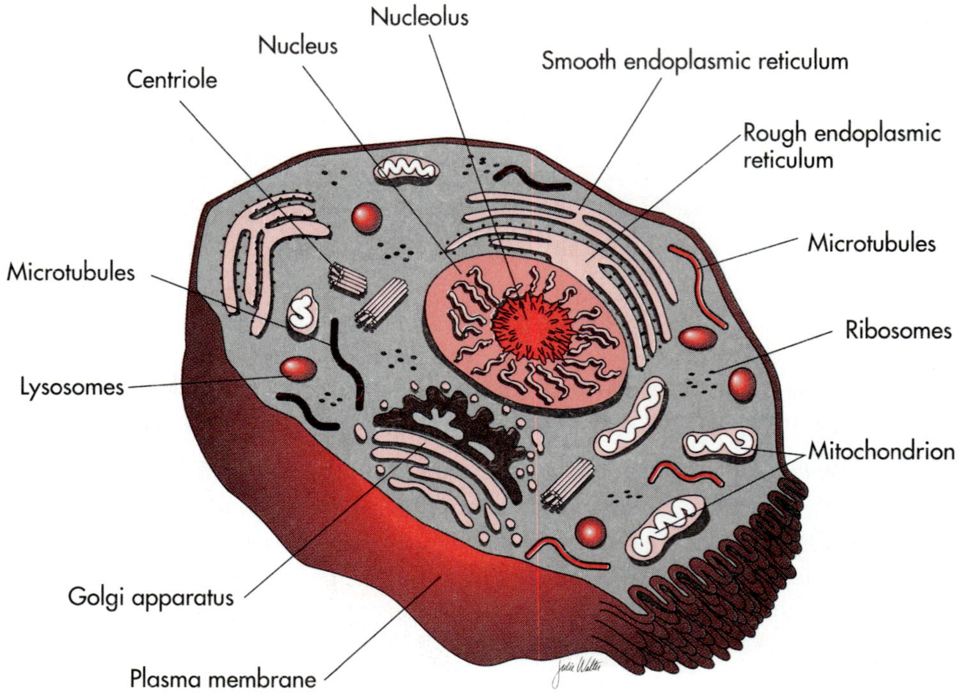

Fig. 1.1 Structure of a typical human cell. (Not every cell will contain all of the organelles or anatomical features that are shown or contain them in the proportion that is shown.)

- Composed of proteins and ribosomal RNA
- May be attached to intracellular membranes or float freely in the cytoplasm
- Sites of protein synthesis

Mitochondrion

- Elongated organelle, surrounded by 2 phospholipid bilayers. The inner bilayer forms *cristae,* which project into the mitochondrial *matrix.*
- The larger the energy needs of a cell, the larger is the number of mitochondria. Hundreds of them may be present in certain cells.
- Produce ATP by aerobic metabolism
- Synthesize urea and heme

Endoplasmic reticulum

- System of interconnected parallel membranes enclosing a fluid-filled space
- Membranes are continuous with the nuclear membrane and the plasma membrane.

— ***Rough ER***

- Studded with ribosomes on the outside of the membrane
- As amino acids are assembled on each ribosome, the growing polypeptide chain is fed into the interior of the ER for further processing.
- Proteins may be retained for functions within the ER, or they may be exported.

- Export from the ER occurs through transport vesicles that form when a portion of the ER membrane encloses a localized volume and pinches itself off.
- Pinched-off transport vesicles move toward the Golgi apparatus.

— ***Smooth ER***

- Contiguous with rough ER
- Receives polypeptides and proteins but does not participate in synthesis
- Synthesizes phospholipids and cholesterol in most cells and performs special tasks in cells of different organs:
    - Synthesizes steroid hormones in some cells; participates in fat metabolism in cells of the GI tract; synthesizes and stores glycogen in cells of liver and skeletal muscle; detoxifies drugs in cells of the liver and kidneys; stores and releases Ca^{++} in cells of striated muscle
- Forms the longitudinal sarcoplasmic reticulum in striated muscle

Golgi apparatus

- A system of stacked, plate-shaped, membrane-lined cisternae, generally located near the nucleus but not attached to the nuclear membrane
- Surrounded by many membrane-enclosed vesicles
- Destination of the transport vesicles that pinched off the rough ER. Transport vesicles from the rough ER migrate toward the Golgi apparatus, fuse with its membrane, and spill their contents into the interior of the Golgi apparatus.
- Golgi modifies and sorts proteins and accumulates them in distinct types of vesicles, destined, respectively, for the plasma membrane, for lysosomes, or for storage granules that are subsequently used in exocytosis.
- The Golgi complex is a major site of membrane formation, and vesicles destined for the plasma membrane are, in fact, portions of new membrane.

Lysosomes

- Spherical, membrane-lined structures containing enzymes that are capable of digesting proteins, carbohydrates, lipids, nucleic acids, and other biologic material.
- The membrane that surrounds lysosomes is normally stable but may rupture as a result of cell injury or hypoxia. This leads to digestion of the cell itself (autolysis).

The Cytoskeleton

- A network of protein filaments extending throughout the cytoplasm
- Controls both the geographic location of intracellular elements and any communication among them
- It is also used as the mechanism for changing and maintaining cell shape and as the "musculature" that causes cell movement.
- The cytoskeleton is composed of 3 types of protein filaments: *microtubules, intermediate filaments, and actin filaments* (Fig. 1.2).

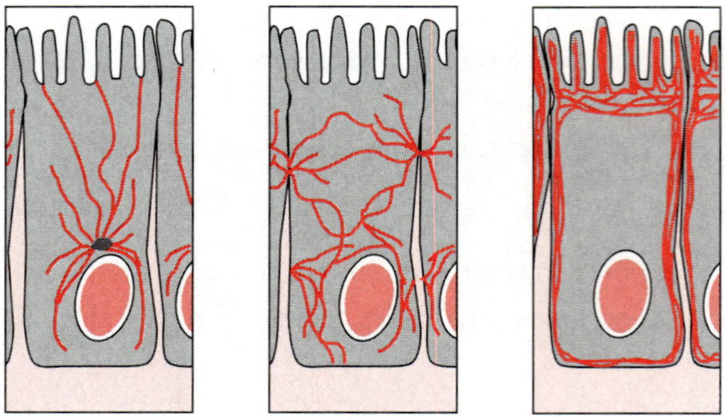

Fig. 1.2 The 3 types of protein filaments that constitute the cytoskeleton. Microtubules generally have one end attached to the centrosome. Intermediate filaments extend across the cytoplasm and give mechanical strength to cells. Actin filaments are most highly concentrated in a narrow region near the plasma membrane.

Microtubules (about 25 nm in diameter)

- Microtubules are hollow, polar cylinders formed by the protein *tubulin.*
  - The *plus end* grows rapidly, and the *minus end* tends to disintegrate unless it is stabilized.
  - In most cells the negative end of most microtubules is anchored and stabilized in the *centrosome,** a region that, in interphase, lies near the nucleus. The centrosome also contains the *centrioles,* a pair of cylindrical bodies, positioned at right angles to each other. During interphase the centrosome duplicates and divides into two equal parts, each containing a centriole pair. When mitosis begins, the two centrosomes move to opposite sides of the nucleus and form the two poles of the *mitotic spindle.*
  - The configuration of microtubules attached to a centrosome is continually changing as new microtubules grow (by tubulin polymerization) and old ones shrink (by depolymerization of tubulin).
  - As long as the plus end of a microtubule is free, the tubule grows along an arbitrary path with the centrosome at its origin. However, the plus end may be stabilized.
- Microtubules are used both as structural supports and as transport paths.
  - Transport occurs along microtubules because they serve as binding sites for *motor proteins* that use energy derived from ATP hydrolysis to carry different cargoes in different directions along the length of microtubules.
    - The *kinesins* are motor proteins that move toward the positive end of the microtubule (away from the centrosome).

*An example of an exception is the nerve cell. In the axon all microtubules are oriented with their plus end pointing away from the cell body. In dendrites, on the other hand, the polarity of the distal end of the microtubules is mixed.

    - The *dyneins* are motor proteins that move toward the negative end of microtubules (toward the centrosome).
    - *Myosin* is a motor protein that is especially abundant in muscle cells but is also found in other cell types.
  - The bending and sweeping motions of cilia are also caused by a motor protein, *ciliary dynein.*
  - Cilia grow from structures called *basal bodies.* Their structure is identical to that of each member of a centriole pair.

- **Intermediate filaments (about 10 nm in diameter)**

  - These are polymers of mechanically stiff polypeptides (keratins, desmin, vimentin, lamins, and others).
  - They occur most abundantly in cells subjected to mechanical stress and are believed to exist for the purpose of giving structural strength to cells and help them resist stretch.
    - The intermediate filaments of different kinds of cells show a distinctive protein preference:
      - Epithelial cells, hair, and nails have intermediate filaments made mostly of *keratins.*
        - Myofibrils in muscle cells are linked together by filaments of *desmin.*
        - Intermediate filaments in nerve cells contain mostly *neurofilament proteins.*
        - The inside surface of the nuclear membrane is lined by a mesh of intermediate filaments (the nuclear lamina), the major protein constituents of which are the *lamins.*

- **Actin filaments (5 to 9 nm in diameter)**

  - They are also called *microfilaments.*
  - They are present throughout the cell, generally in cross-linked bundles, and are concentrated in a narrow band just under the plasma membrane (see Fig. 1.2).
  - They are polar, with a slowly growing minus end and a more rapidly growing plus end. Growth occurs by polymerization and requires the presence of ATP.
  - When cells are moving or changing shape, they often develop surface extensions (*microspikes* and *lamellipodia*) that contain actin filaments. The extensions are believed to result from rapid actin polymerization, and their purpose appears to be to push the cell membrane in the direction of movement.
  - Actin has many functions in cells and they depend on a large variety of actin-binding proteins. Interaction with such proteins can:
    - bind actin filaments into bundles (as in microfilaments)
    - attach actin filaments to the plasma membrane
      - move actin filaments relative to each other (as in muscle)
      - cause movements of the cell surface including phagocytosis

The Plasma Membrane The plasma membrane forms the cell surface. Its major functions are:

- control of the composition of the space it encloses

- cell-to-cell recognition
- cell locomotion

Structure The two major components are lipids and proteins, with their relative contributions varying among different tissues. For example, myelin is at one extreme with only 20% protein, whereas the inner mitochondrial membrane is at the other extreme with 70% protein, most of it resulting from the high enzymatic activity of that membrane.

— ***Lipids*** The major lipid components of the plasma membrane are:

- cholesterol
- phospholipids (which make up more than half of the lipid mass in plasma membranes)
    - phosphoglycerides (Fig. 1.3)
        - *phosphatidylcholine, phosphatidylserine, phosphatidylethanolamine*
    - sphingolipids
        - *sphingomyelin*
        - sphingomyelin resembles phosphatidylcholine in all respects except that it is derived from serine rather than glycerol.
    - inositol phospholipids

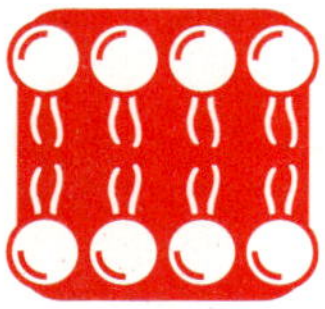

- Every tissue has a distinctive composition and pattern of phospholipids in its cellular membranes.
- The structural characteristics of cell membranes arise from the physicochemical behavior of the phospholipids in an aqueous medium.
    - The presence of electrical charges in the head group allows interaction with water because water is also a polar substance. (Both hydrogen atoms in the water molecule carry a partial positive charge, whereas the oxygen molecule carries a partial negative charge. As a result water molecules interact because the positively charged hydrogen atoms on one molecule are attracted to the negatively charged oxygen atom on another.)
    - Phospholipid molecules arrange themselves in an aqueous medium in such a way that the polar head groups interact with water, but water is excluded from the fatty acid region. This yields 3 possible physical arrangements as shown in Fig. 1.4.
- Because of their cylindrical shape, the phospholipids of plasma membranes arrange themselves in a bilayer.
- The composition of the two halves of the bilayer is different (Fig. 1.5).
    - The inner half, facing the cytosol, contains most of the phospholipids that contain a terminal NH_3 group. This includes phosphatidylserine, the phospholipid with a head group that carries a net negative charge. As a result there is an electrical field across the bilayer and negatively charged phospholipid groups that are required by some cytosolic enzymes

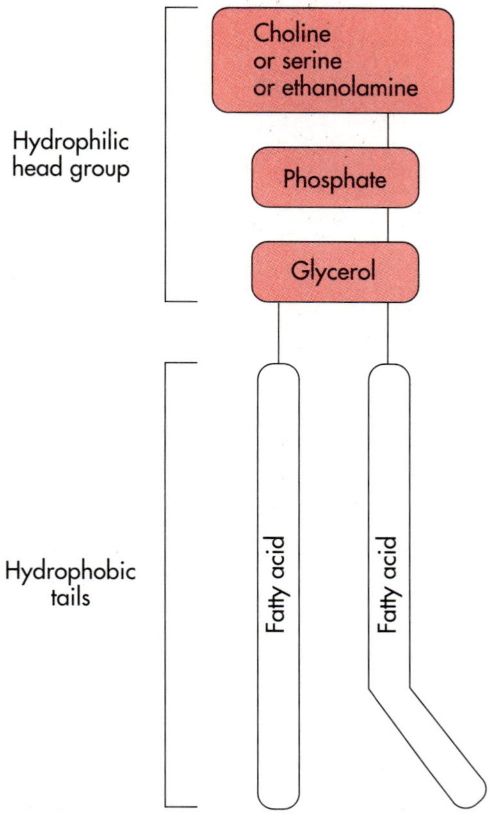

Fig. 1.3 Components of a phospholipid molecule, showing the parts of phosphatidylcholine, phosphatidylserine, and phosphatidylethanolamine.
The tail is nonpolar, but the head group carries electrical charges at the pH of biologic fluids, as follows:
- A minus charge on the singly bonded O of the phosphate group as well as charges on the group that is attached to the phosphate
- A plus charge on the N of choline
- A plus on the N as well as a minus on the carboxyl group in serine
- A plus on the N in ethanolamine

The important feature of the tails is the nature of the chemical bonds. In the phospholipids of cell membranes one tail usually has one or more cis *double bonds, whereas the other tail does not.*
- Each double bond creates a small kink in the tail and thereby gives the molecule an overall cylindrical shape.
- If both tails were straight, the overall shape would be conical, with the head group forming the base of the cone.

(such as protein kinase C) are clustered on the intracellular side.

- The outer half contains most of the choline-containing phospholipids and molecules of glycolipids (lipids with sugar groups attached to them).
    - The glycolipids serve a variety of functions that includes membrane protection, Ca^{++} binding (sialic acid), electri-

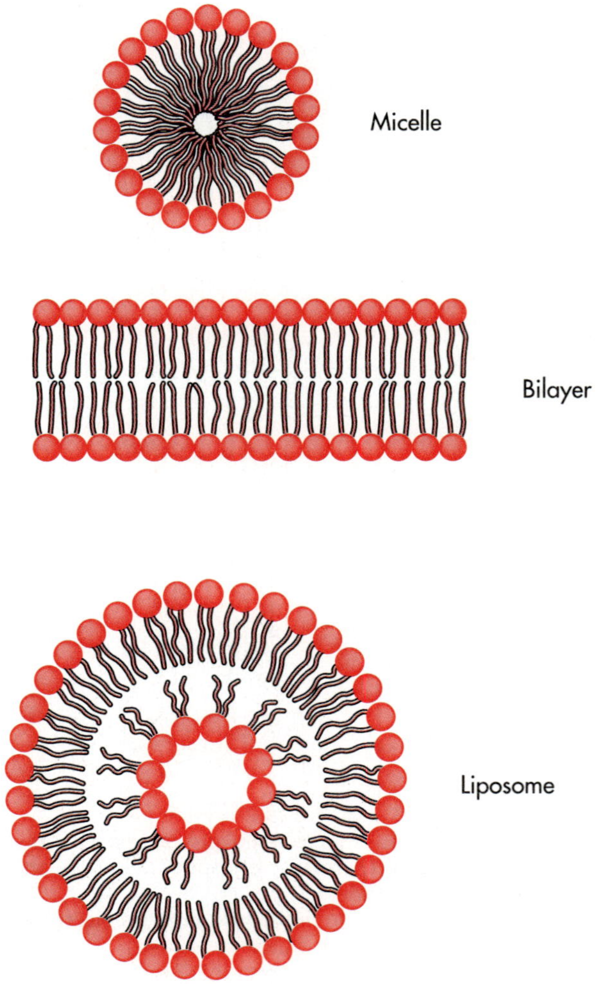

Fig. 1.4 The physicochemical properties of phospholipid molecules determine the patterns of their aggregation in an aqueous solution. Three stable structures are possible.
- Conical molecules (those with straight lipid tails) form micelles or liposomes.
- Cylindrical molecules (those with one or more kinks in a tail) form bilayers.

Each structure provides a central, water-free lipid region.

cal insulation, cell recognition, and interactions with the extracellular matrix.

- Plasma membranes also contain *cholesterol* (see Fig. 1.5). The physical rigidity imparted to this molecule by the steroid ring structure enhances the impenetrability of the plasma membrane because it makes the lipid bilayer stiffer in the tail region so that water-soluble molecules cannot "wiggle" their way into and through the bilayer.

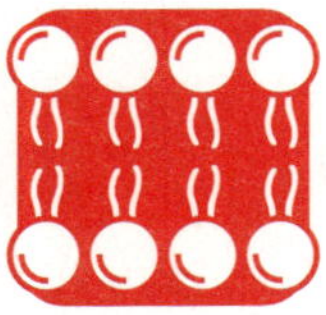

Membrane proteins The lipid bilayer also contains a variety of proteins that either penetrate the bilayer or attach to one of its sides (Fig. 1.6). These proteins are responsible for most of the biologic functions of cell membranes.

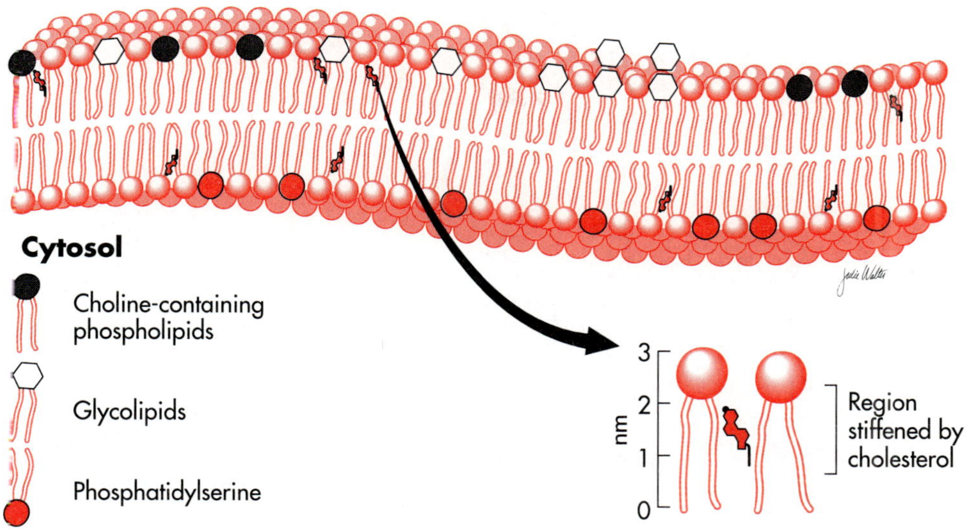

Fig. 1.5 The bilayer forming the plasma membrane is not symmetrical. The side facing the extracellular fluid contains all the glycolipids and most of the phospholipids that have choline in their head groups. On the other hand, most of the phospholipids that have a terminal amino group in their head portion (phosphatidylethanolamine and phosphatidylserine) are in the cytosolic half of the membrane. Cholesterol is equally distributed in the two halves of the bilayer.

— *Membrane carbohydrates*

- The surface of all cells is covered by the *glycocalyx,* a cell coat that is rich in carbohydrates (see Fig. 1.6).
- Membrane carbohydrates serve many functions:
  - The variety of their covalent linkages permits intricate coding for cell-to-cell recognition processes.
  - The density of the glycocalyx protects the cell surface from random mechanical and chemical interactions or damage.
  - The precision of their three-dimensional shapes allows specific cell-to-cell adhesion to occur (such as sperm-egg interactions).

Cell Nourishment

Living cells contain systems of energy production and energy utilization. Energy production involves the breakdown of macromolecular fuels such as carbohydrates, fats, or proteins into smaller molecules and yields storable energy. The proportion of each of these fuels that is used differs depending on the tissues and is further influenced by dietary and hormonal factors. Energy utilization is necessary to maintain cellular functions.

Energy Metabolism

Metabolism of proteins, carbohydrates, and fats Fig. 1.7 shows the 3 stages in the generation of energy from the oxidation of foodstuffs.

- In the first stage no useful energy is generated. Large molecules are broken down into smaller units: Proteins are hydrolyzed to amino acids; carbohydrates are hydrolyzed to simple sugars such as glucose; fats are hydrolyzed to glycerol and fatty acids.

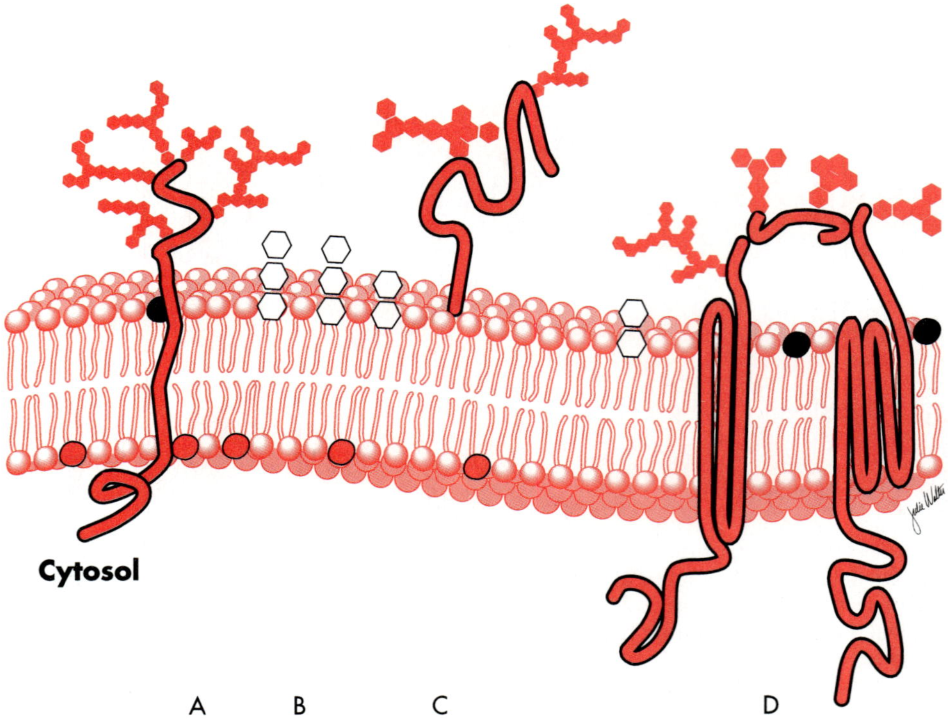

Fig. 1.6 Membrane proteins and elements of the glycocalix. *A,* Most of the carbohydrate is attached to intrinsic proteins. In addition, the glycocalyx contains: *B,* glyco lipids; *C,* peripheral glycoproteins; and *D,* glycoproteins and proteoglycans that are adsorbed to the cell surface by means of membrane proteins. All carbohydrate residues face the extracellular space.

- In the second stage some ATP is generated, whereas most of the molecules produced in stage 1 are converted into the acetyl unit of *acetyl CoA.*
- In the third stage most of the ATP is generated as acetyl CoA brings acetyl units into the Krebs cycle where they are completely oxidized to CO_2 and ATP is generated by oxidative phosphorylation.

ATP and other high-energy phosphates

- The molecule *adenosine 5′-triphosphate (ATP)* is the main link between energy production and utilization in mammalian cells. Energy production involves the formation of the terminal phosphate group of the ATP molecule, and energy utilization involves its hydrolysis or transfer.
- In addition to the adenine triphosphate, ATP, various nucleoside triphosphates are involved in transferring energy during cellular processes. They include *guanosine triphosphate (GTP),* the energy source used in gluconeogenesis and protein synthesis, *uracil (UTP),* used in glycogen synthesis, and *cytosine (CTP),* used in lipid synthesis. In mammalian cells there exists a variety of enzymes that permit transfer of the terminal energy-rich phosphate bond from ATP to these other nucleotides.

Cell Growth

The Cell Cycle Growth and replication of cells is a cyclic event (Fig. 1.8) in which a newly formed cell becomes a parent cell by doubling its content during *interphase.*

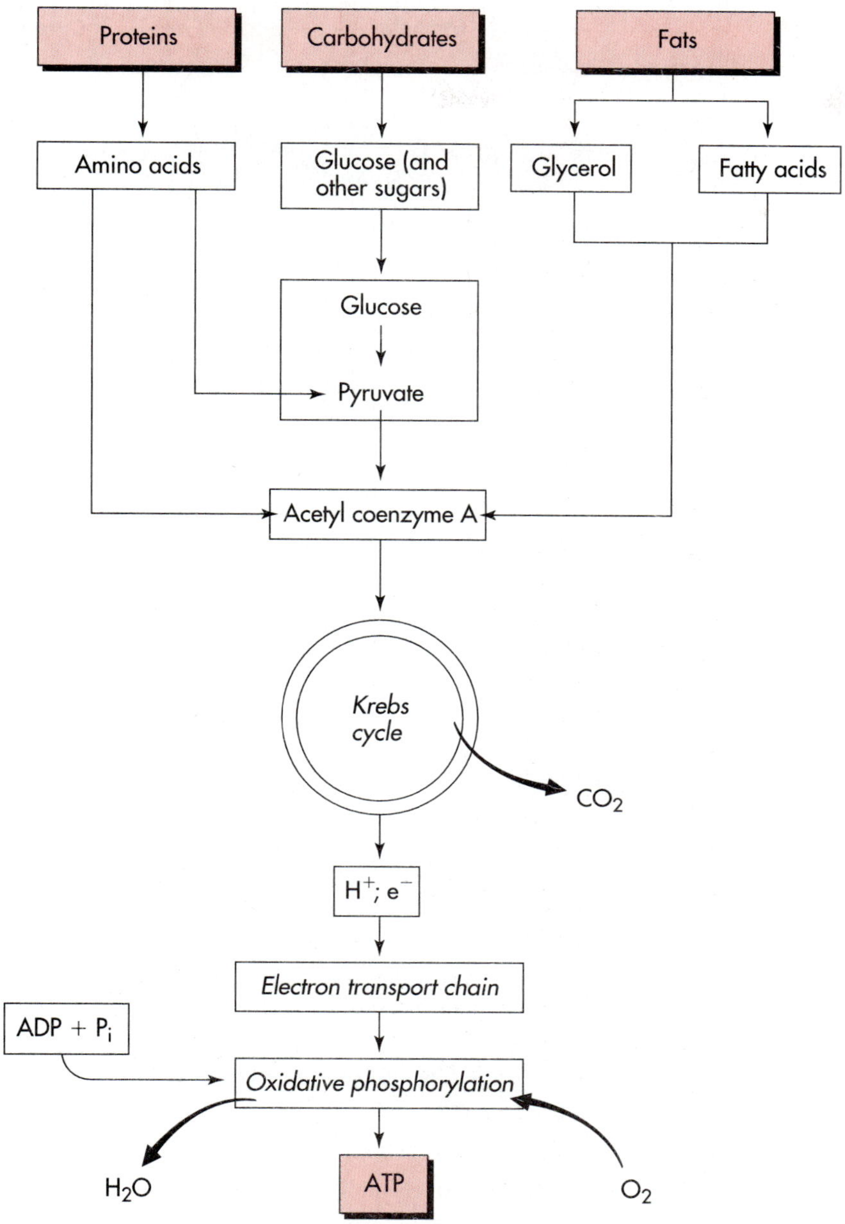

Fig. 1.7 The formation of ATP from food.

• Late in interphase the centrosome duplicates, but all the components of the duplicated complex remain on one side of the nucleus. Mitosis begins when the centrosome complex splits in two, allowing each centriole pair to become part of a separate microtubule organizing center, radiating an array of microtubules called an *aster.* The 2 asters move to opposite sides of the nucleus to form the 2 poles of a mitotic spindle.

• The chromatin material of the nucleus condenses from long, loosely coiled strands into well-defined chromosomes.

• The cell enters *prophase* at the point when condensed chromosomes first become visible. The nuclear envelope begins to disintegrate in late prophase, and such breakdown defines the beginning of *prometaphase,* the portion of the cell cycle during which microtubules of each aster begin to capture randomly mov-

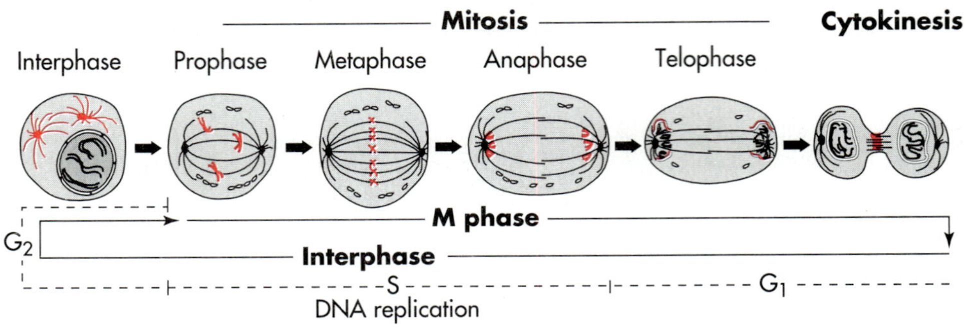

Fig. 1.8 The phases and events of the cell cycle. Mitosis is the small portion of the cycle when the nucleus divides. Mitosis together with cytokinesis (the process of cytoplasmic division) constitute the M phase of the cycle. Successive M phases are separated by interphase. Interphase begins with a quiescent gap (G_1), is followed by a period in which DNA replicates, or synthesizes (S), and ends with another quiescent gap (G_2).

ing chromosomes. Prometaphase lasts about 10 minutes and is followed by *metaphase.*

- Metaphase is a period of about 30 minutes during which all the chromosomes become attached at their centromere to the microtubules of the spindle and become aligned across the middle of the spindle, with each pair of sister chromatids being held by oppositely directed microtubules. The point of attachment is the *kinetochore,* a multiprotein complex that forms on each chromatid and binds 30 to 40 microtubules.
- The paired chromatids are held together along their length. This binding is suddenly disrupted when the degradation of *cyclin* causes inactivation of *M-phase–promoting factor (MPF).* These events signal the beginning of *anaphase,* when sister chromatids separate in unison and begin to move toward the spindle poles. They complete the migration to the poles within about 5 minutes.
- At *telophase* the chromosomes at each pole fade, new nuclear membranes form, and the parent cell begins the processes of *cytokinesis.* An early event in cytokinesis is formation of a constriction ring of actin filaments and myosin around the midbody of the elongated cell. The cytoplasm then divides by a process called *cleavage.*

Regulation of the Cell Cycle and Cell Growth

- Two families of proteins control the cell cycle:
  - The *cyclins* are activating proteins that bind to cyclin-dependent protein kinases (CDK) and thereby control CDK ability to phosphorylate relevant target proteins. Intracellular cyclin concentration undergoes spontaneous fluctuations.
  - The *cyclin-dependent protein kinases (CDK)* induce cell cycle events by phosphorylating participating proteins.
- A cell is driven into mitosis at the end of a chain that begins in G_2 with accumulation of mitotic cyclin. The complex of mitotic cyclin bound to CDK forms MPF, the *M-phase–promoting factor.* MPF is inactive when it is formed but becomes activated by enzyme-driven phosphorylations and dephosphorylations.
- Mitosis begins when a critical level of active MPF has been reached.
- Mitosis wanes when cyclin is suddenly destroyed by proteolysis.
- Cell growth is regulated by extracellular protein growth factors by means of receptor-mediated intracellular cascades that regulate gene transcription and

Table 1.1 ***Examples of Protein Growth Factors and their Actions***

Factor	Typical Actions
Platelet-derived growth factor (PDGF)	Stimulates growth of connective tissue cells
Interleukin-2 (IL-2)	Stimulates proliferation of T-lymphocytes
Interleukin-3 (IL-3)	Stimulates proliferation of various blood cell precursors
Insulin-like growth factors I and II (IGF-I, IGF-II)	Stimulate cell metabolism
Transforming growth factor β (TGF-β)	Modulates cellular responses to other growth factors
Nerve growth factor (NGF)	Promotes dendritic growth in and survival of specific neurons
Erythropoietin	Promotes proliferation and differentiation of erythrocyte precursors

cell cycle control systems. These growth factors are normally present at low concentration, and cell number is limited by competition for these factors. Table 1.1 lists examples of growth factors and some of their actions.

The Cell Environment

Each cell in the body is embedded in the *extracellular matrix.* Its major components, which are secreted by cells called *fibroblasts* are:

- *glycosaminoglycans,* unbranched polysaccharide chains
- *proteoglycans,* a group of molecules formed by a core protein to which linear polysaccharide chains of the class glycosaminoglycans are bonded and
- *fibrous proteins* that serve either:
    - structural functions (such as *collagen* and *elastin*) or
    - adhesive functions (such as the glycoproteins *fibronectin, laminin, entactin, thrombospondin,* and *tenascin*)

Glycosaminoglycans (GAG) These molecules tend to form gels. Their high density of negative charges binds clouds of ions (such as Na^+), the osmotic activity of which attracts water into the extracellular matrix. The resulting hydrated gel provides mechanical support to tissues while permitting rapid diffusion of water and water-soluble molecules. Four main groups of glycosaminoglycans (GAG) have been identified:

1. Hyaluronic acid (hyaluronan)
2. Chondroitan sulfate; dermatan sulfate
3. Heparan sulfate; heparin
4. Keratan sulfate

Except for hyaluronic acid, they all attach themselves to a core protein to form proteoglycans (Fig. 1.9).

Proteoglycans Figure 1.9 shows the typical arrangement of glycosaminoglycan (GAG) residues around a protein core. Individual proteoglycan subunits, with the help of linking proteins, may form large aggregates when their core proteins bind to a backbone of hyaluronic acid. Smaller proteoglycans are associated with the cell surface glycocalyx, and some have core proteins that span the plasma membrane. The wide variety of GAG assemblies and proteoglycan structures implies a variety of functions. The main functions of proteoglycans are:

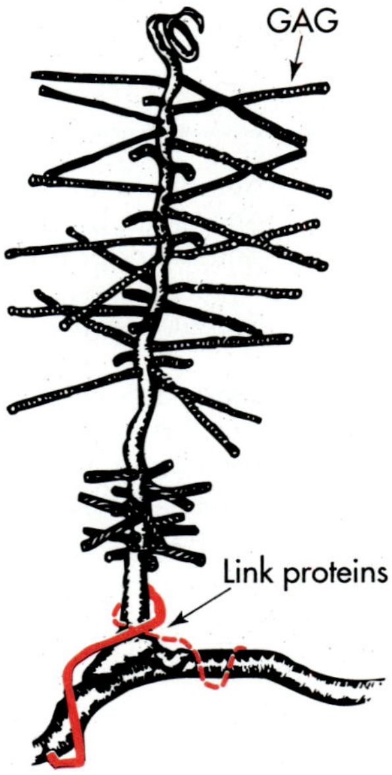

Fig. 1.9 Schema of a proteoglycan molecule. *GAG,* Glycosaminoglycans.

- providing hydrated space around cells
- serving as interstitial sieves by forming gels of different pore sizes and charge densities
- binding cells to the extracellular matrix
- binding and modulating the activities of enzymes and growth factors
- providing transmembrane signaling pathways

Collagen Collagen forms 30% of the proteins in the human body, and 15 types have been identified so far. Their distinctive molecular feature is a long, firm structure in which 3 polypeptide chains are wound around one another in a superhelix. Formation of the triple helix involves *prolyl hydroxylase,* an enzyme with an absolute requirement for vitamin C (ascorbic acid) as a cofactor. Hence, inadequate supplies of this vitamin lead to the complex of tissue disorders known as *scurvy.*

The collagens are classified into three major groupings:

- Fibril-forming collagens (types I, II, III, V, and XI)
  - Type I is the predominant collagen. It forms the interstitial supporting structures of all tissues.
  - Type III dominates in blood vessel walls.
- Fibril-associated collagens (types IX and XII). This group is found attached to the surface of collagen fibrils and is believed to link fibril to fibril or fibril to extracellular matrix.
- Mesh-forming collagens (types IV and VII)
  - Type IV is found mostly in basement membrane.
  - Type VII forms the anchoring filaments in epithelia.

Elastin The amino acid sequence of elastin differs from that of collagen in such a way that the elastin molecule is not able to form a regular helical structure. Instead it exists as an unordered, extensively cross-linked coiled structure and is therefore capable of storing elastic energy when it is deformed. It enables tissues to stretch without tearing.

Glycoproteins Glycoproteins participate in the functions of the extracellular matrix by means of their role in cell adhesion. Four significant members of this class of matrix components are *fibronectin, laminin, thrombospondin,* and *tenascin.*

- Fibronectin and laminin molecules contain binding sites for cell surface integrin receptors and for fibrin, collagen, and other constituents of the extracellular space.
- Thrombospondin and tenascin contain amino acid sequences that promote cell adhesion as well as other sequences that promote antiadhesion and cell migration.

Physiology of the Cell Surface Membrane

Membrane Transport Mechanisms Transport of substances across the cell wall occurs by diffusion through the lipid bilayer or by means of two classes of membrane-embedded proteins, *carrier proteins* or *channel proteins.*

Permeability of the plasma membrane

- If there were no proteins in the plasma membrane, the ease with which molecules diffuse across the phospholipid bilayer down a concentration gradient would depend partly on molecule size, partly on the solubility of the molecule in lipid, and partly on whether the molecule is electrically charged.
  - Lipid-soluble (a.k.a. *hydrophobic,* or *nonpolar*) molecules diffuse rapidly. Examples are O_2, CO_2, N_2.
  - Lipid-insoluble (a.k.a. *polar*) molecules diffuse rapidly provided that they are small and uncharged. Examples are H_2O and urea. Large uncharged polar molecules such as glucose and sucrose barely diffuse.
  - Charged molecules (ions), even when they are small (Na^+, K^+, Cl^-) do not significantly permeate phospholipid bilayers unless special transport mechanisms are present.

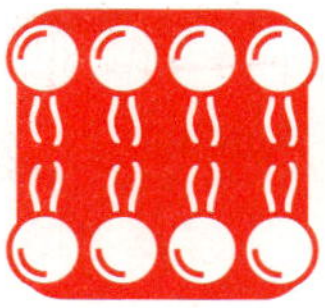

- Tranport of ions and large polar molecules is accomplished by specific transmembrane proteins.
  - *Carrier proteins* transfer substances by physical movement of one part of the protein molecule relative to the others. Transport by carrier proteins may be passive or may be active (coupled to a source of metabolic energy).
  - *Channel proteins* are arranged so as to form a narrow pore for the transport of substances down their electrochemical gradient. Such transport requires no direct expenditure of metabolic energy and is therefore called *passive transport.*
- Transport of macromolecules such as proteins is accomplished by enclosure of the substance that is to be transported within a membranous sac.
  - If the substance is being transported out of the cell, the process is called *exocytosis.*
  - Transport into the cell is called *endocytosis.*

- **Passive transport**
  - The driving force is provided by:
    - the difference in electrical potential across the membrane (if the transported substance carries a charge) or
    - the difference in concentration (activity) of transported substance or in osmolality on the two sides of the membrane. Note that transport of water is commonly ascribed to differences in osmolality rather than to differences in the concentration of water. There is no conceptual difference between the two.
  - The permeability (resistance, conductance) of the membrane modulates the ability of a driving force to transport solutes across the membrane.
  - Passive transport may occur directly through the phospholipid bilayer, or through *carrier proteins,* or *channel proteins.*
- **Active transport**
  - Active transport always occurs by means of carrier proteins called *transporters.* The pumping activity of the protein:

    (1) is tightly coupled to a source of metabolic energy,
    - *Primary transporters* require the direct utilization of ATP (such as active transport mechanisms for Na^+, K^+, and Ca^{++}).
    - *Secondary transporters* are proteins whose transporting function has an absolute requirement for the simultaneous movement of an ion (generally Na^+) down a concentration gradient that is maintained by primary transporters (such as transport mechanisms for glucose and amino acids).

    (2) transports substances against their electrochemical gradient, and

    (3) occurs in only one direction across the plasma membrane.
  - Carrier proteins that transfer a single solute across the membrane are called *uniporters.*
  - More complex carrier proteins function to transport two solutes in a coupled fashion so that the transfer of one depends on the coupled transfer of the other either in the same direction *(symport)* or in the opposite direction *(antiport).*

Ion Channels Most channel proteins in cell membranes form narrow, highly selective pores that are concerned specifically with ion transport. Hence they are called *ion channels.*

- **Structure**
  - Channel proteins consist of one or more subunits, with each subunit further consisting of membrane-spanning domains that are linked in series by amino acid chains looping into the cytosol or into the extracellular space (Fig. 1.10, *A*).
  - Several subunits may be arranged around a central "pore" (Fig. 1.10, *B*) through which a selected ion species may pass when the channel is activated.
- **Function** Transport through channel proteins is always passive, and its

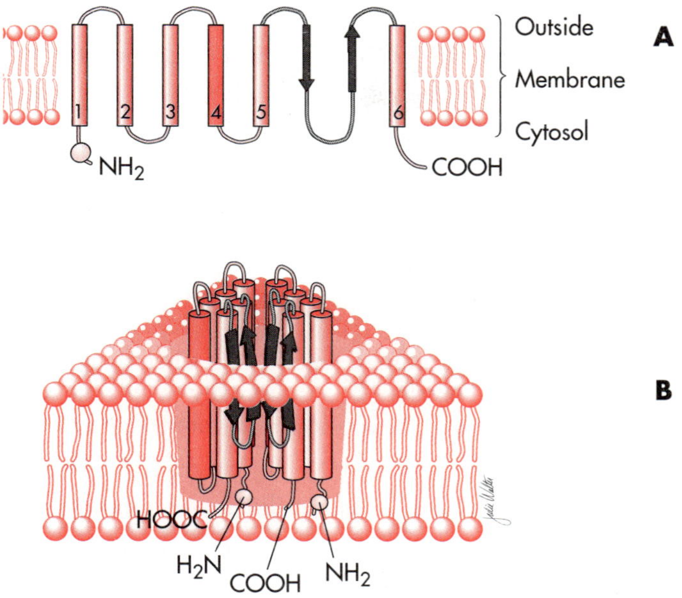

Fig. 1.10 Schema of a typical channel protein. **A,** Arrangement of six membrane-spanning domains. In voltage-gated channels one of the domains (usually domain 4) is believed to contain the voltage-sensing element. **B,** Two identical subunits are shown in an arrangement that is suggestive of how a central "pore" might be formed by the subunits.

maximum rate is about 1000 times greater than that of carrier proteins. Ion channels differ from aqueous pores in two respects:

- Ion channels are highly selective for specific ions.
- Ion channels are not continuously open but alternate between an open and a closed state.
  - They are opened in response to a specific stimulus that can be electrical *(voltage-gated channels),* chemical *(ligand-gated channels),* or mechanical. Voltage-gated cation channels, conveying Na^+, K^+, or Ca^{++}, are of special importance for the function of excitable cells such as nerve or muscle. Ligand-gated channels are ubiquitous and influence the behavior of cells in response to a variety of chemicals.

Membrane Potentials

Balance of forces across cell surface membranes The plasma membrane represents a leaky barrier interposed between regions that have different concentrations for any one of many different ion species. Transmembrane concentration differences for any one species represent a force that might cause ion concentrations to equilibrate across the plasma membrane. However, when ions move down their respective concentration gradient, they create an opposing gradient in electrical potential.

- As a result, passive ion (net) transport stops when the force arising from the remaining concentration gradient is balanced by the opposing force arising from the gradient in electrical potential.
- The outcome is a steady state in which ion species are maintained at a concentration difference across the cell surface membrane by an equal and opposite electrical force.

- **Ion equilibrium potential**
  - — ***Definition*** The *ion equilibrium potential* (E_{ion}), or the *Nernst potential* of an ion, is the electrical driving force that would
    - be equal in magnitude but opposite in direction to a concentration gradient driving force
    - prevent net passive transport of that ion species
    - lead to a steady state with respect to the concentration gradient for that ion species
  - — ***Determination***

• E_{ion} can be measured directly only when there is but one ion species present. Therefore E_{ion} is normally calculated from the existing concentrations of the ion species of interest and the valence of the ion:

$$E_{ion} = -\frac{61}{Z} \log \frac{\text{Intracellular concentration of ion}}{\text{Extracellular concentration of ion}}$$

(Z is the valence of the ion.)

- — ***Significance***

• E_{ion} is a fictitious number. Its magnitude and polarity are equal to the electrical potential that would have to be applied to the inside of the cell if the existing concentration difference for that ion is to be maintained by an opposing electrical force alone.

• If the calculated equilibrium potential for an ion is equal to the resting membrane potential (the voltage that can be measured across the surface membrane of the resting cell), it is likely that the steady-state distribution of the ion on the two sides of the cell membrane is determined by passive transport mechanisms only.

• If the calculated equilibrium potential for an ion across a cell surface membrane is different from the resting membrane potential of the cell, active transport mechanisms are involved in maintaining the distribution of the ion across the surface membrane.

- **Resting membrane potential**
  - — ***Definition***

• The resting membrane potential of a cell is the voltage that can be measured across the surface membrane of the resting cell. It is *not* simply the algebraic sum of all ion equilibrium potentials because that sum does not account for losses in potential resulting from the passage of each ion through the resistance of the membrane.

- — ***Determination*** The resting membrane potential of a cell is usually determined by direct voltage measurement.

However, it can be calculated with the help of the *Goldman-Hodgkin-Katz equation:*

$$E_{rest} = 61 \log \frac{P_K [K_o] + P_{Na} [Na_o] + P_{Cl} [Cl_i] + \ldots}{P_K [K_i] + P_{Na} [Na_i] + P_{Cl} [Cl_o] + \ldots}$$

where E_{rest} = Resting membrane potential

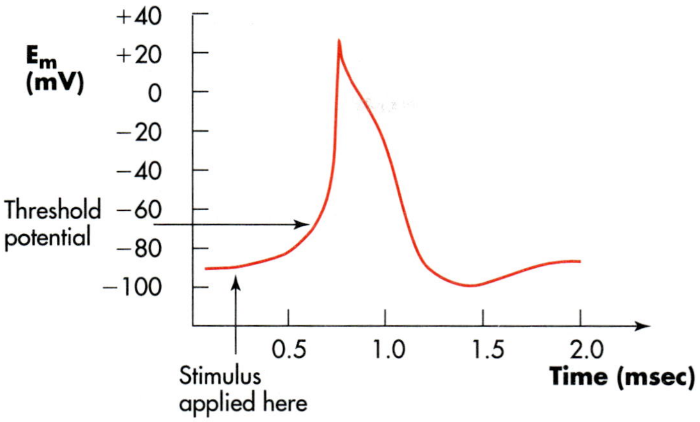

Fig. 1.11 After an adequate stimulus the membrane potential (E_m) of an excitable cell typically displays the electrical behavior shown here. It is called an action potential.

P_x = Membrane permeability coefficient for ion species x
K = Potassium ion
Na = Sodium ion
Cl = Chloride ion
o = Extracellular concentration
i = Intracellular concentration

- **Action potential** Some cells, called *excitable cells,* have membrane properties that allow their membrane potential to change temporarily from the stable resting level to a less negative (perhaps even a slightly positive) value.

— ***Definition***

- An action potential (Fig. 1.11) is a stimulus-triggered, momentary excursion in membrane potential from the resting membrane potential to a peak value that is more positive than threshold and back to resting membrane potential.

— ***Ion movements across the cell surface membrane*** Action potentials arise from instabilities in voltage-gated ion channels within the membrane. They are initiated only by stimuli that are strong enough to raise the membrane potential to threshold. At threshold the membrane permeability to Na^+ increases greatly and the following sequence is triggered:

1. Na^+ ions enter the cell rapidly, driven by the large concentration gradient between intracellular and extracellular space. This produces the upstroke of the action potential.
2. The rapid Na^+ influx would stop once the membrane potential reaches E_{Na} (the Na^+ equilibrium potential, about + 40 mV), but even before the membrane potential reaches this value, open Na^+ channels spontaneously revert to the inactive state, and Na^+ permeability begins to return toward normal.
3. K^+ channels activate more slowly than Na^+ channels after the

threshold value is reached. A sufficient number of them are open only by the time Na^+ channels are already being inactivated. As a result, K^+ permeability of the membrane reaches a peak when Na^+ permeability is decreasing. At that point:

4. K^+ ions leave the cell rapidly, driven by the K^+ gradient. This produces the downstroke of the action potential. The outflow of K^+ ions stops when the membrane potential reaches the K^+ equilibrium potential (about − 80 mV).

At the end of the action potential,

- the membrane potential is back at its normal resting value;
- the inside of the cell has a slight excess of Na^+ ions and a slight deficit of K^+ ions. (These ionic imbalances are so small that several hundred thousand action potentials could be generated before the cell would run low on K^+.)
- The Na^+/K^+ pump restores ionic balances in the intervals between action potentials.

Cell-to-Cell Communication

Junctions Cells make extensive contacts with other cells and with the extracellular matrix. Such contacts occur at structurally and functionally distinct regions, called *cell junctions.*

Junctions are classified, according to their function, into 3 groups: *occluding, anchoring,* or *communicating* (Fig. 1.12).

Occluding junctions also known as tight junctions Occluding junctions are prominent in transporting epithelial cells where they are located near the apical end of the cells and serve 2 purposes.

- They restrict leakage of solutes through the space between two neighboring cells. The degree of restriction varies among substances and, for any one substance, among tissues.
- They confine lateral migration of membrane proteins to a region that is bounded by two neighboring tight junctions. This prevents, for example, transport proteins located on the luminal side of the nephron from migrating to the basolateral side.

Anchoring junctions Anchoring junctions are a feature of tissues that regularly undergo mechanical stress.

- They serve as anchoring points for filaments of the cytoskeleton, allow a group of cells to function as one geometric unit, and are grouped according to their attaching filaments.
- They consist of 2 classes of proteins:
  - Intracellular bonding proteins, which form a characteristic *plaque* on the intracellular side of the plasma membrane (as shown in the desmosome in Fig. 1.12).
  - Transmembrane linking proteins, the intracellular domain of which binds to an intracellular site while the extracellular domain binds to the extracellular matrix or to the transmembrane protein of a neighboring cell. The *integrins* are a family of membrane proteins that anchor cells to the extracellular matrix. The *cadherins* (molecules responsible for calcium-dependent cell adhesion) are

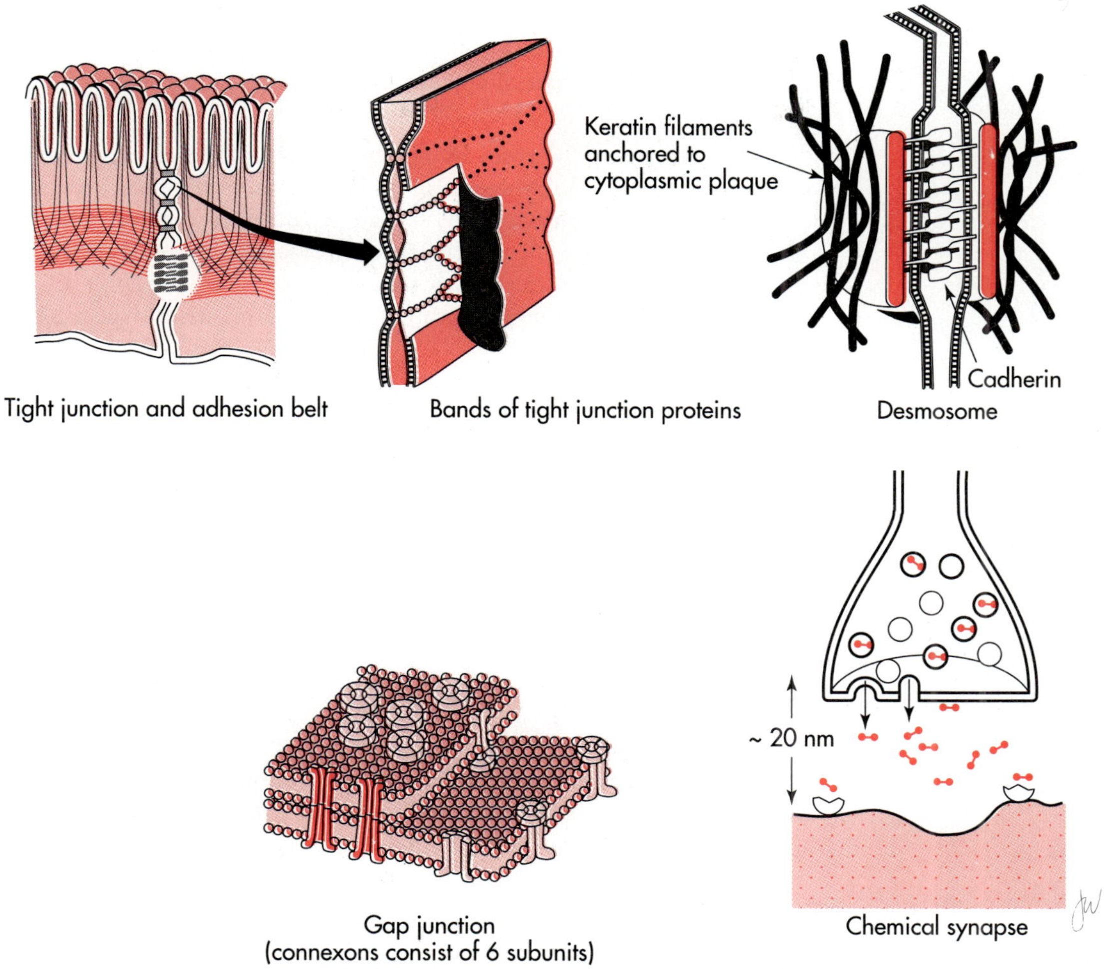

Fig. 1.12 Junctions found between mammalian cells. The upper row shows a tight junction and a desmosome containing two types of anchoring junctions. The lower row shows two types of communicating junctions.

membrane proteins that connect a cell to the cadherins of an adjacent cell.

— ***Actin filament attachment loci*** The *zonula adherens* is a prominent example of such a junction. It forms either localized attachment points or a continuous *adhesion belt* (see Fig. 1.12).

— ***Intermediate filament attachment loci*** *Desmosomes* and *hemidesmosomes* are the two most abundant attachment sites for intermediate filaments. Desmosomes function to couple adjacent epithelial cells together (see Fig. 1.12), whereas hemidesmosomes couple epithelial cells to the basal lamina. The transmembrane linking proteins belong to the cadherin family.

- **Communicating junctions**

— ***Gap junctions*** Gap junctions are regions where the membranes of two neighboring cells are separated by a uniform, narrow gap of 2 to 4 nm. The gap is bridged by channel proteins *(connexons)* that align themselves (see Fig. 1.12) such that ions and water-soluble molecules can pass from cell to cell.

- In this way these junctions allow both electrical and chemical communication.
- The state of openness of gap junction channels is regulated by Ca^{++} or H^{+}.

— ***Synapses*** Synapses are points of contact for the purpose of information transfer between a nerve and another cell. The two synapsing cells do not touch physically but are separated by a narrow cleft. Most synapses are chemical in that information is transmitted by means of neurotransmitter agents that elicit their biologic effect after coupling to a specific receptor protein in the plasma membrane of the postsynaptic cell.

— ***Receptors*** Receptors are protein molecules that specifically bind a signaling molecule (called a *ligand*) and then initiate cascades that result in a biologic response of the target cell.

- Most classes of receptors are located in the plasma membrane and interact with extracellular ligands.
- Some receptor classes are located within the cell, either in the cytosol or in the nucleus. The most prominent among these are receptors for steroid hormones, thyroid hormones, or vitamin D.

Transmembrane Signaling

Membrane Receptors and Signal Transduction Receptors in the plasma membrane act as signal transducers in that they first bind a signaling ligand and then produce intracellular "signals" that cause responses in the target cell. Membrane receptors are grouped according to the transduction mechanisms into *ion channel–linked receptors, G protein (= GTP-binding protein)–linked receptors,* or *enzyme-linked receptors.*

G protein–linked receptors

- The family of G protein–linked receptors is the largest family of cell surface membrane receptors. This receptor class:
  - responds to a large variety of ligands
  - consists of a single polypeptide chain that threads back and forth across the lipid bilayer
  - has an extracellular ligand-binding domain and an intracellular domain for G-protein binding
- G proteins are GTPases that exist in either a resting or an active state.
  - Resting G protein consists of 3 subunits and has a molecule of GDP bound to one of them.
  - When a G protein is activated by receptor-ligand interaction, it releases its GDP, replaces it with GTP, and splits off the GTP-carrying subunit, allowing it to interact with an effector. The interaction can be excitatory or inhibitory.
  - Activated G protein spontaneously reverts to its resting state when it has hydrolyzed the bound GTP into GDP and the GDP-carrying subunit has reunited with the other 2 subunits.
  - Activated G proteins can directly modulate ion-channel behavior, but in most instances the activation of G proteins changes the in-

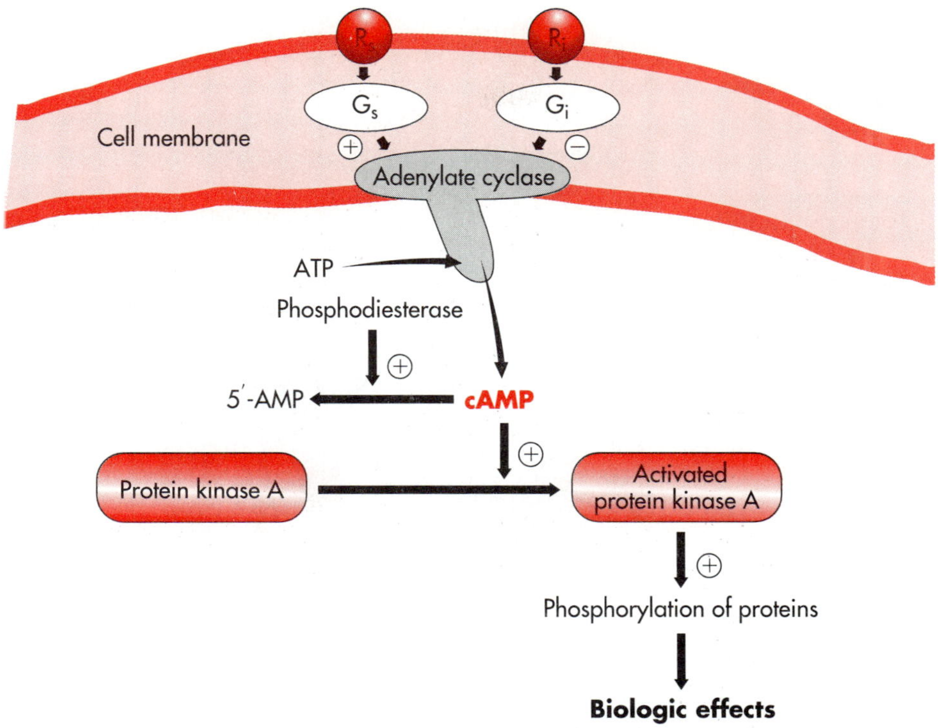

Fig. 1.13 The cAMP second-messenger system.

tracellular concentration of mediating chemicals called, *second messengers.* cAMP and Ca^{++} are the most common second messengers. *Diacyl glycerol* and *cGMP* are used as well.

The adenylate cyclase system

- The second messenger, cAMP is synthesized from ATP by the plasma membrane-bound enzyme, *adenylate cyclase* and is rapidly destroyed by one or more phosphodiesterases (Fig. 1.13).
- Extracellular signaling ligands control intracellular cAMP levels by changing the activity of adenylate cyclase. Both stimulatory (R_s) and inhibitory ligands (R_i) occur and operate by means of stimulatory (G_s) or inhibitory G proteins (G_i) respectively (see Fig. 1.13).
- The biologic effects of cAMP arise from its ability to activate the enzyme *protein kinase A* (see Fig. 1.13).
- The most prominent example of a receptor system operating by means of adenylate cyclase is the adrenoceptors.

The phospholipase-C system

- Ca^{++} acts as an intracellular second messenger in several cellular responses.
- Intracellular Ca^{++} concentration is normally low (near 10^{-7} M). On the other hand, $[Ca^{++}]$ in the extracellular fluid or the endoplasmic reticulum is high (about 2×10^{-3} M). As a result Ca^{++} will enter cells quickly if Ca^{++} channels are opened.
- Release of Ca^{++} from the endoplasmic reticulum is triggered by inositol trisphosphate (IP_3). IP_3 is produced by hydrolysis of phosphatidylinositol bisphosphate (PIP_2), a minor phospholipid component of the cell mem-

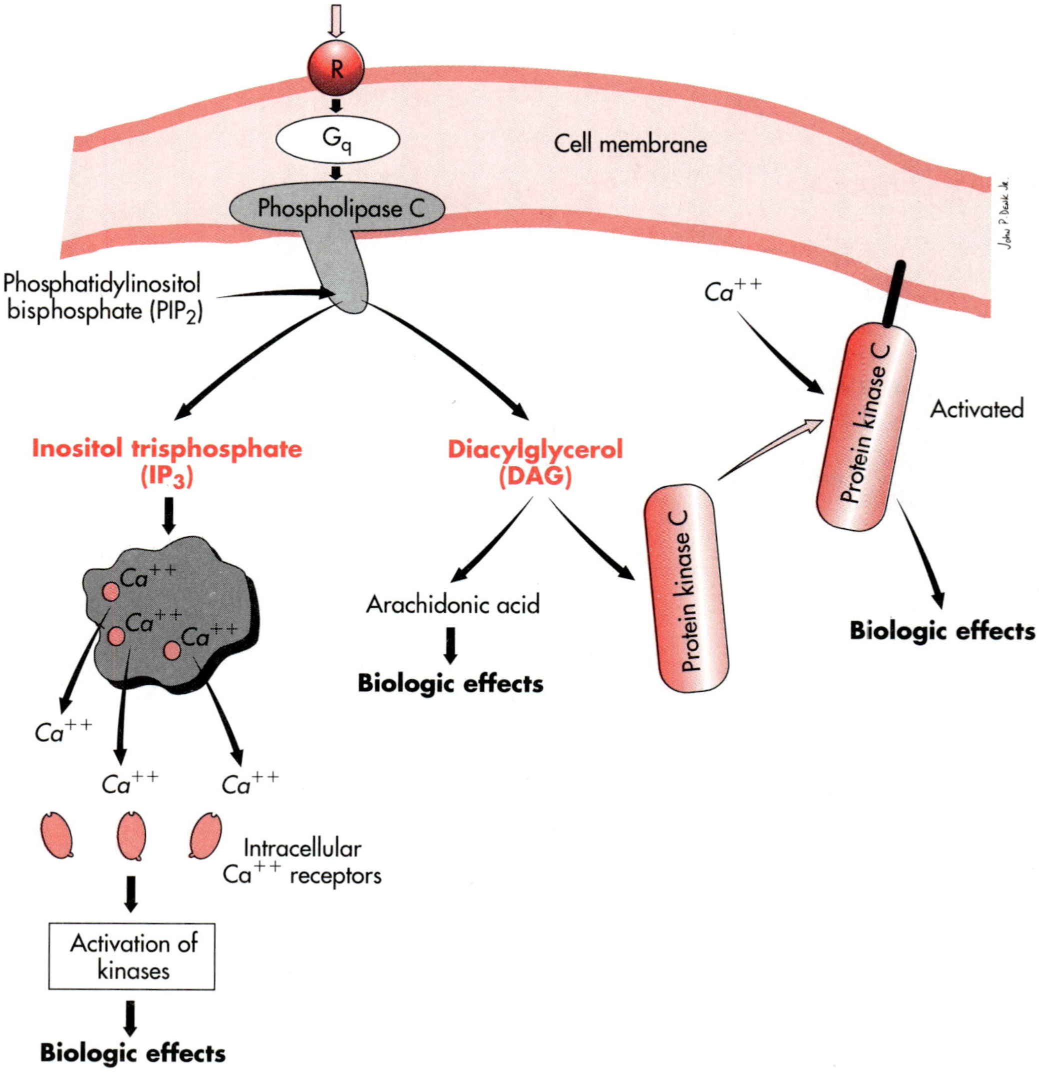

Fig. 1.14 The phospholipase C system of transmembrane signaling.

brane. The cascade begins with the coupling of a ligand to a membrane receptor, activating a specific G protein (G_q), which then activates *phospholipase C* (Fig. 1.14).

- Activated phospholipase C hydrolyzes PIP_2 and produces IP_3 as well as *diacylglycerol (DAG)*.
- Ca^{++} that has been released into the cytosol binds to intracellular receptor proteins such as *calmodulin* (most mammalian cells) or *troponin C* (striated muscle cells).
    - The Ca^{++}/Ca^{++}-receptor complex binds to a variety of kinases and activates them so that they can exert their biologic effects by protein phosphorylation.
- DAG has two significant functions:
    - It can be further cleaved to produce *arachidonic acid,* which may act as a second messenger.
    - It causes *kinase C* to move from the cytosol to the intracellular face

of the plasma membrane. Once attached to the plasma membrane, protein kinase C becomes activated by a cluster of factors that includes DAG, Ca^{++}, negative membrane charges, and membrane phospholipids.

Cyclic GMP

- cGMP is an intracellular second messenger for extracellular ligands like *nitric oxide* (NO) and *natriuretic peptides* (ANP, BNP, and CNP).
- It exerts its biologic effects by interacting with 3 types of intracellular proteins. They are:
  - cGMP-sensitive ion channels (the cation channel in rod photoreceptor cells of the retina and the amiloride-sensitive Na^+ channel of the inner medullary collecting duct of the nephron)
  - cGMP-dependent protein kinases (mostly in smooth muscle, platelets, and cerebellum)
  - cGMP-regulated phosphodiesterases

Multiple Choice Review Questions

1. Which one of the following statements concerning cell metabolism is correct?
   a. Oxidative phosphorylation occurs in mitochondria.
   b. The end product of cell metabolism is stored in the form of reduced equivalents of nicotinamide adenine dinucleotide (NADH).
   c. Acetyl CoA is a product of the citric acid cycle (Krebs cycle).
   d. Fatty acids are converted to glucose by the formation of pyruvate from acetyl CoA.
   e. Oxygen has no influence on the formation of acetyl CoA from pyruvate.

2. In mammalian cells the ion equilibrium potential for Ca^{++} is +80 mV. Which one of the following statements best describes the significance of this?
   a. It means that the intracellular Ca^{++} concentration is lower than the extracellular Ca^{++} concentration.
   b. It means that no active transport mechanism is required to maintain the existing Ca^{++} concentration difference between the inside and the outside of the cells.
   c. It means that the electrical driving force for Ca^{++} entry into cells is increased during an action potential.
   d. It means that Ca^{++} would be in equilibrium across the cell membrane if 80 mV were added to the resting membrane potential.

Questions 3 and 4 refer to a fictitious cell with a resting membrane potential of +280 mV. Its surface membrane is freely permeable to the ions, Rh^- and Sp^+ but quite impermeable to the ion, Fd^{++}. The relative intracellular concentrations of the 3 ion species as well as their equilibrium potentials are shown in Table 1.2.

3. Which one of the following statements regarding the concentration difference for Sp^+ across the cell membrane is true?
   a. The concentration difference for Sp^+ is being maintained by an active pumping mechanism that pumps Sp^+ into the cell.
   b. The concentration difference for Sp^+ is being maintained by an active pumping mechanism that pumps Sp^+ out of the cell.
   c. The concentration difference for Sp^+ is being maintained by the opposite concentration difference for Rh^-.
   d. The high equilibrium potential for Sp^+ is attributable to the concentration difference of Fd^{++} across the cell membrane.

4. Suppose that the membrane of the cell described above were damaged so that the concentration difference for Fd^{++} across the membrane became smaller. Which one of the following consequences would be true?
   a. The equilibrium potential for Sp^+ would change.
   b. The equilibrium potential for Sp^+ would become more positive.
   c. The concentration difference for Sp^+ across the cell membrane would not change.
   d. The equilibrium potential for Rh^- would not change.

5. Each the following substances directly participates in signal transduction across the plasma membrane *except:*
   a. Ion channels
   b. G proteins (guanine triphosphate–binding proteins)
   c. Guanylate cyclase
   d. Cholesterol
   e. Adrenoceptors

Table 1.2 ***Electrophysiology Measurements in a Fictitious Cell***

Ion Species	Intracellular Concentration	Extracellular Concentration	Equilibrium Potential
Fd^{++}	High	Low	+280 mV
Rh^-	High	Low	+280 mV
Sp^+	Low	High	+350 mV

6. Each of the following statements relating to cell-extracellular matrix interactions is correct *except:*
   a. A major function of proteoglycans is the modulation of gene transcription.
   b. Fibronectin fragments play a significant role in wound healing by promoting cell adhesion to the extracellular matrix.
   c. Collagen synthesis requires vitamin C.
   d. Laminin induces the replication of endothelial cells.

7. The functions of the plasma membrane depend on each of the following component properties *except:*
   a. Proteins may penetrate the bilayer.
   b. Phospholipids are amphipathic.
   c. Phospholipids maintain osmotic equilibrium across the plasma membrane by permitting free diffusion of ions and small water-soluble molecules.
   d. Proteins permit selective transport of ions.

8. A swimmer is desperate to improve his time in the 100-meter free-style event and uses blood doping to increase the oxygen-carrying capacity of the blood. An additional effect of blood doping is hyperkalemia. Which one of the following changes would *not* result from an increase in extracellular potassium?
   a. Less negative resting membrane potential in excitable cells
   b. Increased excitability of nerves and muscle
   c. Decreased amplitude of action potentials in excitable cells
   d. More positive threshold potential in excitable cells

9. Each of the following statements about the phospholipase C signal transduction pathway is true *except:*
   a. Protein kinase C requires Ca^{++} for full activation.
   b. Protein kinase C is activated in part by diacylglycerol (DAG).
   c. Inositol trisphosphate (IP_3) can be enzymatically cleaved to produce arachidonic acid.
   d. Protein kinase C requires membrane lipids for full activation.
   e. The biologic effects of inositol trisphosphate (IP_3) require the presence of intracellular Ca^{++} receptors.

Chapter 2

Blood

Overview

Blood, a suspension of cells in plasma, is a portion of the extracellular fluid volume. It is confined to special transport channels, the blood vessels, and its 3 major functions are *transport, communication,* and *organism preservation.*

- **Transport** The cargo of blood is gases, nutrients, metabolic waste, heat, defense agents, and specialized chemicals such as buffers, enzymes, and hormones.
- **Communication** The communications functions of blood are performed by a variety of chemicals including *hormones.*
- **Organism Preservation** Blood operates at 3 levels to preserve the normal function of the organism:
  - **At the organ level** Clotting factors operate to prevent blood loss.
  - **At the microscopic level**
    - *Phagocytosis* defends the organism against foreign cells or cellular debris.
    - A special relationship among the capillary endothelium, plasma proteins, interstitial fluid, and lymph maintains blood in a fluid state and maintains an optimal blood volume.
  - **At the molecular level** *Immunoglobulins* protect the organism against molecular debris or invaders.

Composition of Blood

Blood is a suspension of *formed elements* (erythrocytes, leukocytes, and platelets) in *plasma.* Their relative abundance is shown in Table 2.1.

- **Formed Elements**
  - **Erythrocytes** These cells are the primary oxygen transport system of the body.
  - **Leukocytes** These cells are broadly classified on the basis of histological appearance or affinity for certain dyes:
    - *Granulocytes*
      - *Neutrophils* are the primary defense against bacterial infection and the primary mediator of inflammatory responses.
      - *Eosinophils*, *Basophils* } participate in allergic reactions
    - *Monocytes*
      - Modulate immune responses
      - Scavenge dead cells or denatured protein

Table 2.1 ***Composition of Human Blood***

Component	Composition (% by volume)	Differential Count (% of all leukocytes)
Erythrocytes	42-48	
Leukocytes	<1	
Granulocytes		
Neutrophils		60-70
Eosinophils		2-4
Basophils		0.5-1
Monocytes		3-6
Lymphocytes		20-25
Platelets	<1	
Plasma	52-58	

— ***Lymphocytes*** These cells are mediators of immune responses.

- **Platelets** The major function of these cells is the formation of hemostatic plugs after blood vessel injury.

Plasma

- Plasma is the fluid component of blood. It consists of water (90%), minerals (3%), and proteins (7%).
- The major minerals are Na^+, $C1^-$, and HCO_3^-. Na^+ and $C1^-$ provide osmotic equilibrium across the cell membrane, whereas HCO_3^- is a major regulator of the H^+ concentration in body fluids.

- **Plasma proteins** The plasma protein constituents are *albumin* (4.5 g/dL), several *globulins* (2.5 g/dL), and *fibrinogen* (0.3 g/dL). Most are synthesized by the liver and have 5 major functions:
  - *Carriers* for hormones, trace metals, or drugs
  - *Proteolytic agents* in the cleavage of various hormonal or enzymatic precursors
  - *Inhibitors* for a variety of interstitial proteases or plasma proteases
  - Provide the *colloid osmotic pressure* in plasma
  - Provide the *humoral immunity* portion of the immune system

Erythrocytes

The function of erythrocytes is to transport respiratory gases between the lung and other tissues. This function dictates their physical attributes:

- They are a 3 g/dL solution of *hemoglobin* (the major O^2-transporting protein) and some *carbonic anhydrase* (a catalyst for CO_2 carriage).
- Their exterior membrane is pliable. They have no mitochondria, no structural elements, and no nucleus (in humans). These aspects give them remarkable ductility and allow them to recover again and again from deformations that occur as they are squeezed through capillaries.
- Their biconcave shape is conducive to surface exchange phenomena because it gives them maximum surface area for a given volume.

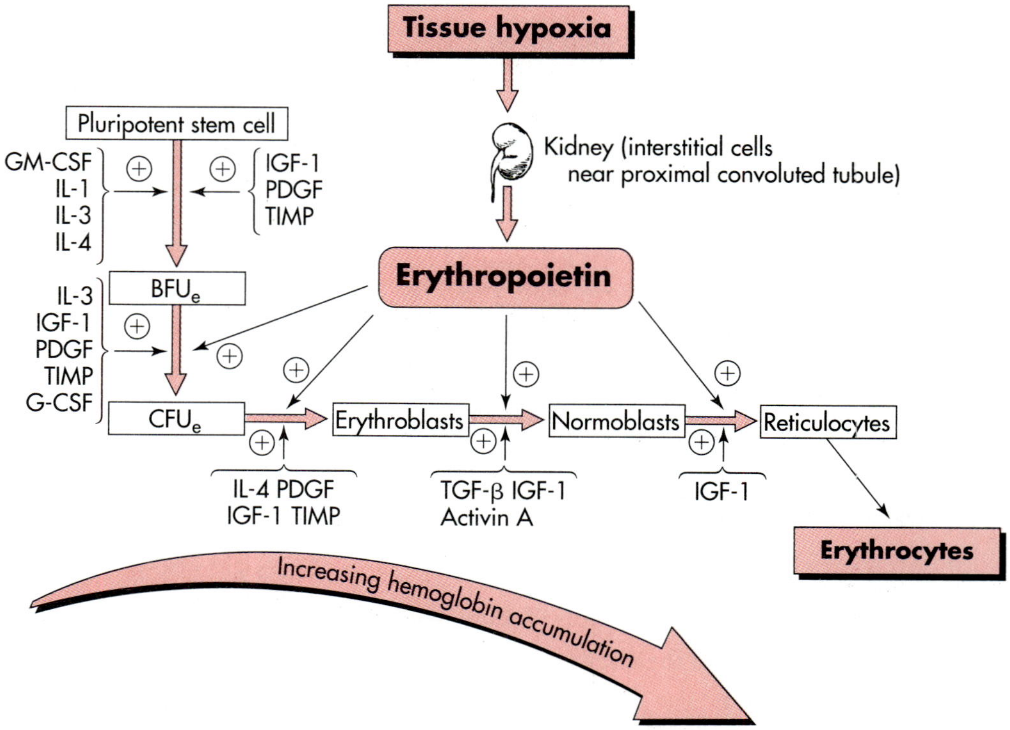

Fig. 2.1 Formation of erythrocytes in response to tissue hypoxia. About 2^{11} mature erythrocytes are formed from one pluripotent stem cell by the action of several growth factors. *BFU_e,* erythrocyte burst-forming unit; *CFU_e,* erythrocyte colony-forming unit; *G-CSF,* granulocyte colony-stimulating factor; *GM-CSF,* granulocyte-macrophage colony-stimulating factor; *IGF-I,* insulin-like growth factor–1; *IL-1* to *IL-4,* interleukin 1 to interleukin 4; *PDGF,* platelet-derived growth factor; *TGF-β,* transforming growth factor–β; *TIMP,* tissue inhibitor of metalloproteinase.

■ **Erythropoiesis** In the adult human about 200×10^9 red blood cells are formed each day in the marrow of the ribs, sternum, vertebrae, and pelvis. They, as well as the platelets and most leukocytes, originate from a single type of primitive cell, the *pluripotent stem cell.*

- **Prerequisites** In addition to the general need for vitamin B_{12} and folic acid to permit DNA synthesis for cell division, red blood cell formation (erythropoiesis) requires 5 important factors: (1) a stimulus, (2) the presence of a normal, growth-promoting environment in the bone marrow, (3) the presence of several growth factors, (4) erythropoietin, and (5) iron.

- **Stem cell conversion** Growth factor influences from local bone marrow cells guide the conversion of pluripotential stem cells to cells that are committed to the erythroid series and beyond that to erythrocytes (Fig. 2.1).

  - The earliest recognizable erythroid precursor cell is called an *erythroid burst-forming unit* (BFU_e). Survival and proliferation of these cells requires erythropoietin and at least one other factor (such as IL-3) to form the next stage, called an *erythroid colony-forming unit* (CFU_e). CFU_e cells represent the earliest committed cells in the erythroid series.

  - Conversion of CFU_e cells to hemoglobin-synthesizing erythroblasts and eventually to erythrocytes requires the presence of *erythropoietin.* Other growth factors generally participate, but only erythropoietin is obligatory.

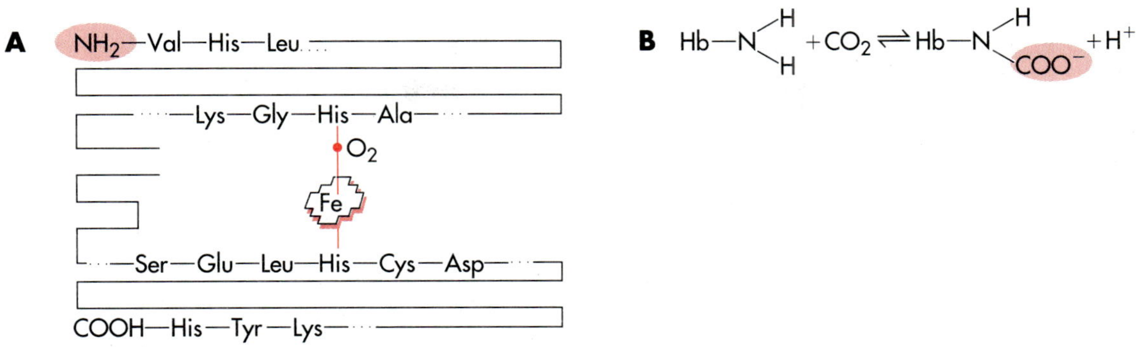

Fig. 2.2 Transport of oxygen or carbon dioxide by hemoglobin. **A,** O_2 is carried between Fe in the heme group and a His of an adjacent globulin chain. **B,** CO_2 is carried at the NH_2 terminal.

Erythropoietin

- Erythropoietin is produced to some extent in the liver, but mostly by certain peritubular renal interstitial cells. The hormone is produced constitutively, and hypoxia causes recruitment of additional synthesizing cells.
- Its synthesis is regulated, presumably as a result of conformational changes associated with the presence or absence of oxygen, by a heme protein with rapid turnover.
- Erythropoietin interacts with its target cells through specific surface membrane receptors.

Hemoglobin (Hb)

- Hb synthesis is the most significant intracellular event during erythrocyte formation. It begins in CFU_e cells, and its accumulation increases steadily, reaching a plateau once the reticulocyte stage is reached.
  - Hb synthesis begins in mitochondria, using *succinate* as a substrate, continues in the cytoplasm, where *porphyrinogen* is formed, and returns to the mitochondrion for the formation of *heme,* an iron-containing pigment, and finally to the cytoplasm for the combination of heme with *globin* subunits.
  - The final Hb molecule consists of a heme ring suspended between neighboring globin loops.
- Hb binds O_2 and CO_2 reversibly, being responsible for most O_2 that is carried by blood and for about 25% of CO_2. (The remaining 75% is carried in the form of dissolved H_2CO_3/HCO_3^-.)
  - O_2 is carried between the iron of the heme moiety and a neighboring histidine residue (Fig. 2.2, *A*).
  - CO_2, on the other hand, is carried as a carbamino group at the NH_2 terminal of the molecule (Fig. 2.2, *B*).

Life Cycle of Erythrocytes Erythrocytes have a life-span of about 120 days, all of them spent within the lumen of blood vessels. Aging cells undergo membrane changes that allow mononuclear phagocytes in the marrow, liver, and spleen to recognize and remove the deteriorating cells. In the course of these processes:

- Heme is dissociated from the globin portion and is oxidized. This separates Fe from the pigment portion.
- *Transferrin,* the iron-transporting protein, carries the iron back to the CFU_e cells for incorporation into new erythrocytes.

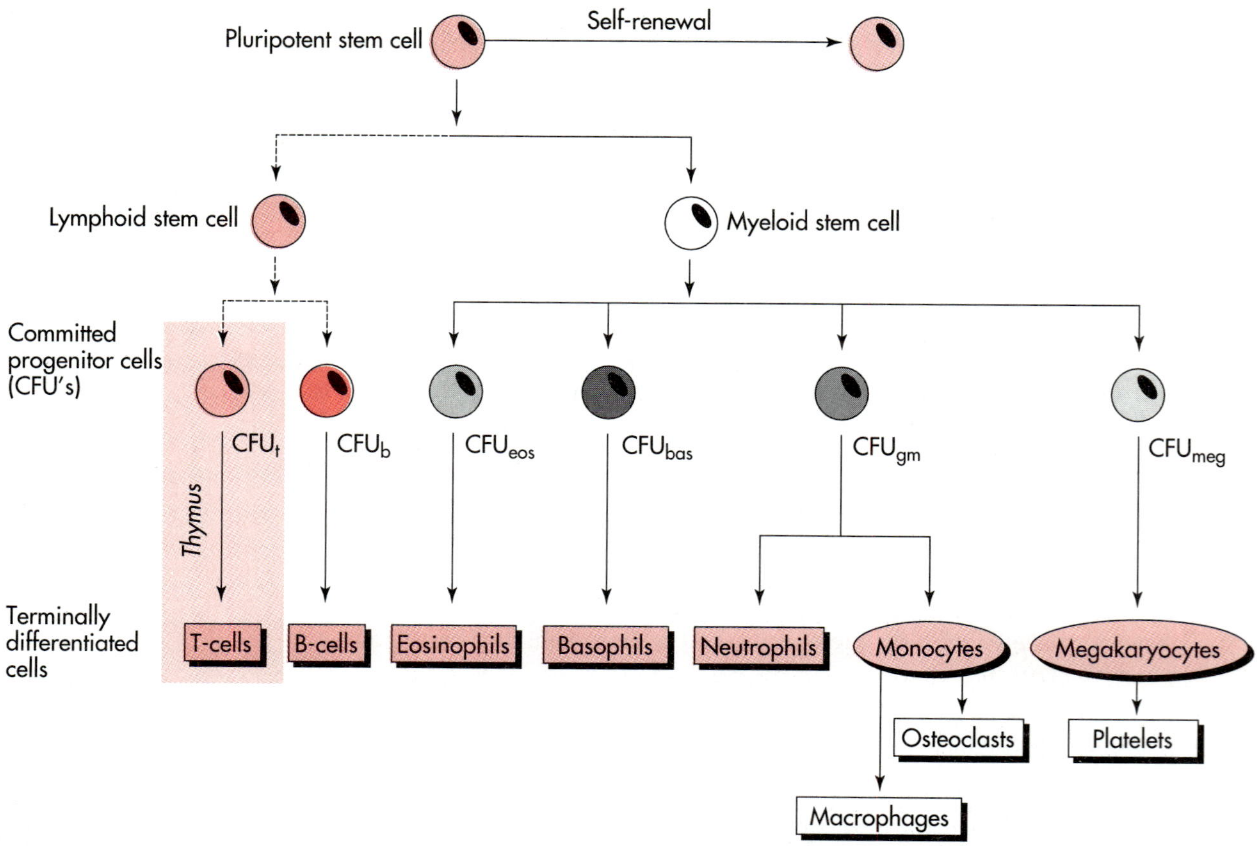

Fig. 2.3 Formation of leukocytes. Except for T-lymphocytes, which form in the thymus, leukocytes are formed in the bone marrow from stem cells. Stem cells either can produce daughter stem cells in a process of self-renewal or can differentiate during a few cell divisions into committed progenitor cells (colony-forming units, CFU) that yield blast cells, which are the earliest morphologically recognizable precursors of specific blood cells. In view of uncertainties regarding precursors for T-lymphocytes and B-lymphocytes, their paths are shown as dashed lines. Neutrophils and macrophages develop from a common progenitor cell, called the granulocyte-macrophage progenitor cell (CFU$_{gm}$).

- The pigment portion of heme is reduced to *bilirubin* and is excreted in bile into the GI tract, giving stool its characteristic brown color.
- The globin chains are broken down into their amino acids and released to the body pool of amino acids.

Leukocytes

Overview Groups of leukocytes (white blood cells) differ by their histological properties (microscopic appearance or affinity for certain stains), but they all have one major function: They are a rapid and specific defense mechanism against infectious molecular agents or microorganisms.

Formation Under the influence of locally produced or circulating growth factors, *pluripotential stem cells* can be diverted from self-renewal toward a committed line of *progenitor cells* (Fig. 2.3). These differ from stem cells in two ways:

1. Progenitor cells have lost the capacity for self-renewal.
2. Progenitor cells are not pluripotential but are committed to produce (under the proper growth conditions) daughter cells of a particular type.

During maturation each cell line acquires distinctive properties.

Table 2.2 *Comparison of Lymphocyte Responses*

B-CELLS	T-CELLS
Responsible for:	
• humoral immune responses	• cell-mediated immune responses
Cell Surface Recognition Sites:	
• are complete immunoglobulin molecules	• resemble the Fc region of a typical immunoglobulin
Can Recognize:	
• freely circulating antigen	• only antigens that are presented on the surface of another cell
Respond to Receptor Activation:	
• by synthesizing a highly specific antibody	• in various ways, depending on the nature of the protein that couples the antigen to the surface of the presenting cell

Lymphocytes B-lymphocytes and T-lymphocytes respond differently to antigens and have different immunological tasks. These are summarized in Table 2.2.

- **B-cells**

  • Antigens can stimulate B-cells to form clones that synthesize a specific immunoglobulin antibody different from the antibodies synthesized by all other B-cell clones.

  • The specificity of antibody synthesis is directed by sites of recognition and binding that are located within the B-cell surface membrane.

- **T-cells** The surface recognition or binding sites on T-cells respond to appropriately presented antigen as well as to the glycoproteins that couple antigen to the surface of the carrier cell. These coupling glycoproteins are antigenic surface markers that are present on all cells. They belong to a single family of molecules whose synthesis is directed by a tight cluster of genes, generally named the *major histocompatibility complex* (MHC) or, specifically in humans, named the *human leukocyte antigen* (HLA).

  The nature of the HLA product on the surface of the carrier cell determines the nature of the T-cell response to the antigen. Two types of proliferative responses occur (Fig. 2.4):

  - One response yields a population of T-cells that modulates the processes by which B-cell clones are formed.
  - The other response yields a population of cytotoxic T-cells that can directly deactivate the antigen.

- **Immunoglobulins** These proteins are of paramount importance in the function of lymphocytes.

  — ***Structure*** The basic structural units are a pair of "heavy" amino acid chains and a pair of "light" amino acid chains (Fig. 2.5).

  • The variable domains in the heavy chains designate the immunoglobulin as belonging to one of five types: IgG, IgA, IgM, IgD, or IgE.

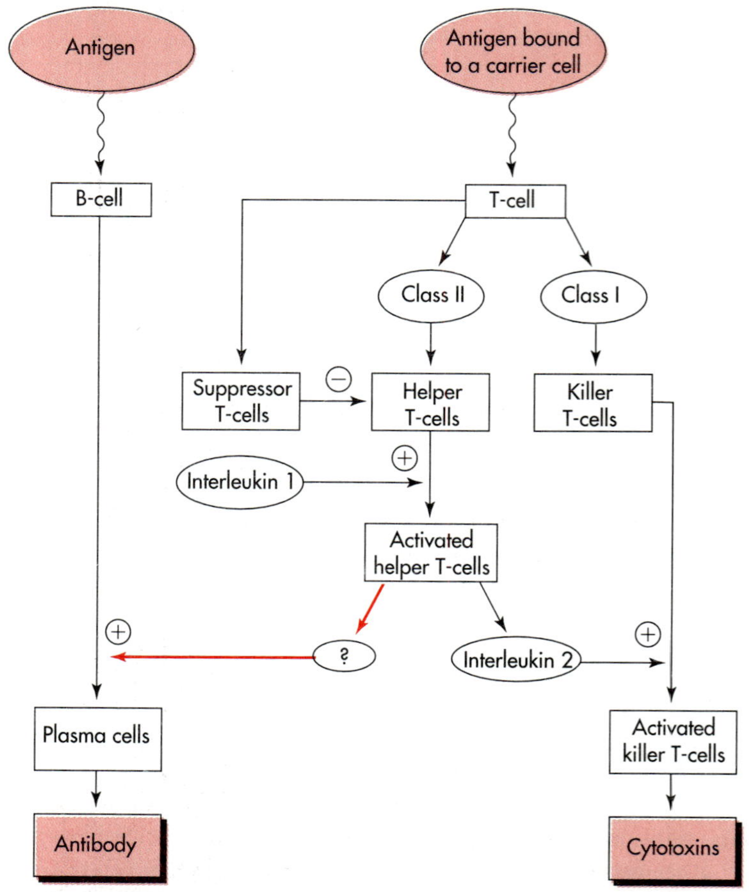

Fig. 2.4 Outline of typical immune reactions. B-cell activation and cloning is under continuous control by T-cells. T-cell behavior is determined by the interaction between the T-cell and the HLA-determined glycoprotein-binding sites on the antigen-presenting cell: Class I refers to those glycoproteins of the HLA complex that will, upon activation, lead to cytotoxic T-cell responses. Class II refers to HLA glycoproteins whose activation will lead to proliferation of helper T-cells.

• The hypervariable regions within the variable domains on neighboring light and heavy chains form specific antigen recognition or binding sites.

— ***Function*** The unique function of immunoglobulins is to recognize and bind to other proteins. The purpose of forming such antigen-antibody complexes is either

• to precipitate antigen from solution or

• to attach antigen to phagocytic and cytotoxic cells for subsequent destruction

• Precipitation requires that several molecules of antigen and antibody form a physical aggregate. Attachment of complexes to other cells requires that the receiving cell have surface receptors for the Fc portion of the antibody.

■ **Immunity** The immune system is a defense mechanism characterized by recognition of nonself, specificity, and memory. It has 2 basic components: *natural immunity* and *acquired immunity.*

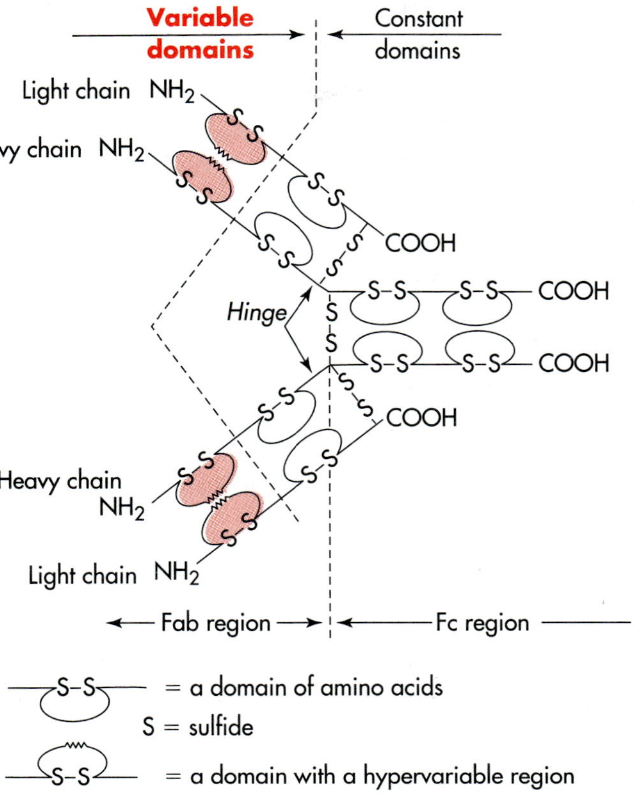

Fig. 2.5 Structure of a typical immunoglobulin molecule. Each chain consists of several domains linked to neighboring domains by amino acids. Some domains occur in all immunoglobulins (constant domains) and some occur only in certain classes of immunoglobulins (variable domains). Within the variable domains, regions of hypervariability exist. Papain cleaves immunoglobulin molecules at a point called the "hinge." The region to one side of the hinge is named the Fc region; that on the other side of the hinge is named the Fab region.

- **Natural immunity** Natural immunity is bestowed by immunological components that are capable of acting directly and immediately on foreign matter (examples are *interferon* and *properdin*).
- **Acquired immunity** Acquired immunity is a normally dormant component of the immune system. It can be activated in response to specific stimuli. Passive activation (by the injection of previously activated components) is possible, but the essence of the immune system is *active acquired immunity,* derived from circulating lymphocytes.
- **Immune reactions** Fully developed immune reactions involve antibody synthesis by B-cells (humoral response) as well as direct, cytotoxic T-cell responses (see Fig. 2.4).
  - ***Humoral responses*** Humoral responses begin when a B-cell recognizes and binds antigen to its surface binding sites (see Fig. 2.4). However,
    - B-lymphocyte cloning occurs only after the B-cell–antigen complex has received a signal from an activated helper T-cell. Formation of activated helper T-cells is a two-step process:

1. Presentation of the antigen, bound to the surface of a presenting macrophage in conjunction with class II HLA products. (The presence of class II HLA products directs the formation of a helper T-cell.)
2. Activation and proliferation of the helper T-cell. This stage requires interleukin-1, a soluble factor derived from macrophages.

- Activated helper T-cells are able to recognize antigen bound to the surface of B-cells in association with class II HLA products and are stimulated by this complex to secrete the factor or factors required for B-cell proliferation and antibody production.

— ***Cytotoxic responses*** Cytotoxic responses are initiated when precursors of killer T-cells recognize and bind class I HLA surface markers on a foreign cell (see Fig. 2.4). Subsequent differentiation and proliferation occurs only after a signal from activated helper T-cells:

- Recognition of coexisting class I incompatibilities (by precursors of killer T-cells) and class II incompatibilities (by activated helper T-cells) on the same cell causes release of interleukin-2 from activated helper T-cells. This release leads to the formation and replication of activated killer T-cells that are specific for the class I HLA surface incompatibility that initiated the response.
- A variety of cytotoxic agents will then destroy the foreign cell.

Granulocytes and Monocytes

Granulocytes Granulocytes mature in the bone marrow and arrive in the tissues fully differentiated as *eosinophils, neutrophils,* or *basophils* for specific phagocytic tasks (see Fig. 2-3). They contain sophisticated mechanisms by which they can progress rapidly from a harmless circulating intravascular cell to a specific phagocytic cell and killer of bacteria. These mechanisms include:

- an internal microtubular skeleton against which contractions can occur
- a system of cytoplasmic contractile proteins
- a system of membrane receptors for chemoattractive substances, for activated complement, or for immunoglobulins
- systems of *primary granules* and *secondary (specific) granules* containing phagocytic substances

Monocytes After they leave the bone marrow, monocytes undergo further differentiation (see Fig. 2.3).

- In bone they form *osteoclasts,* large multinucleated cells that erode the bone matrix.
- In other tissues they form a pool of stationary or mobile phagocytic cells called *macrophages.*

Functions The major functions of granulocytes and monocytes relate to *phagocytosis* and *inflammation*

— ***Phagocytosis*** Phagocytosis is a process of immobilization, ingestion and digestion of foreign agents by granulocytes and monocytes. A vital first step in phagocytosis is the *activation of the complement system.*

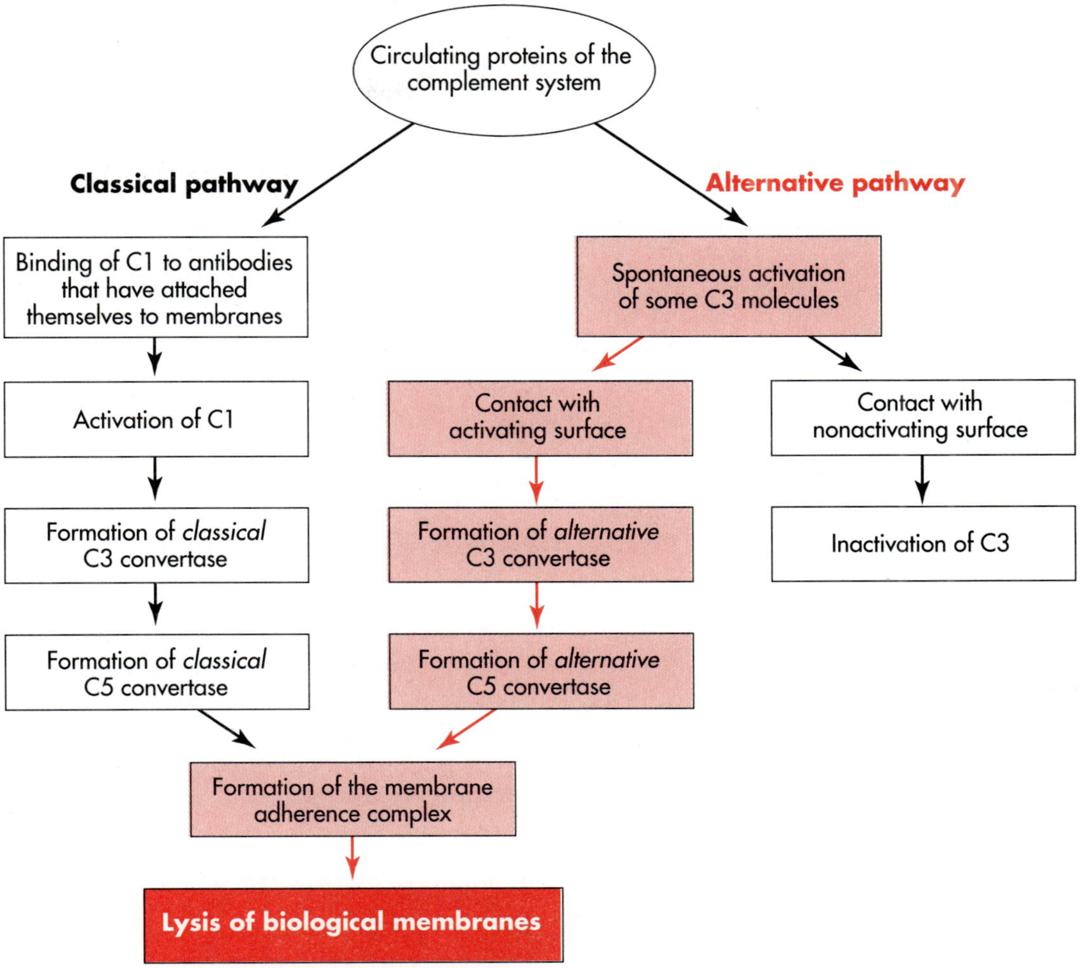

Fig. 2.6 Complement activation, a vital first step in phagocytosis, can proceed along one of two paths: The *classical pathway* is precipitated by physical attachment of C1 molecular filaments to "bridges" that have formed between antibodies and a foreign membrane. The *alternative pathway* follows C3 activation. C3 molecules are sufficiently labile that a few are always spontaneously activated. If the activated few encounter a surface that has particular structural or immunological features, the full cascade proceeds.

Complement activation. About 20, normally dormant, proteolytic plasma proteins make up the complement system. Two among them, named C1 and C3, can, by separate pathways, activate the system to full, cascading activity. Fig. 2.6 summarizes the steps by which either path leads to formation of a *membrane adherence complex.* This complex may either cause membrane lysis directly or may act as a coupler between phagocytic cells and the membrane to be attacked. Such coupling is called *opsonization.*

— ***Inflammation*** In acute inflammation granulocytes, accumulating locally in response to *chemoattractants,* form the initial defense. Macrophages follow later.

Chemoattraction. Chemoattractive agents interact with surface receptors on granulocytes to cause 4 effects:

1. Granulocyte accumulation in and egress from the microcirculation of the affected region

2. Emigration of granulocytes from blood vessels toward the noxious agent
3. Attachment of granulocytes to the opsonized surface of the agent (if it is large) or else engulfing of the agent within a phagocytic vacuole
4. Fusion of granules with the opsonized membrane and release of granule contents

Lysis. Granulocytes release lysosomal proteases. They also cause the release of arachidonic acid (AA). AA oxidation yields *leukotrienes, prostaglandins,* and *superoxide radicals.*

LEUKOTRIENES

- They result from the oxidation of AA by the lypoxygenase pathway.
- They are potent inducers of leukocyte aggregation and chemotaxis.

PROSTAGLANDINS AND SUPEROXIDE RADICALS ($O_2^\bullet$)

- Prostaglandins result from the oxidation of AA by the cyclooxygenase pathway. They may act as local vasodilators, but their chief role may be to act as a local brake on the release of proteases.
- $O_2^\bullet$ quickly forms 2 metabolites, hydrogen peroxide (H_2O_2) and hydroxyl radical (OH^*). These have little inflammatory or bactericidal activity by themselves but can react with other substances to form effective destroyers of bacteria.

PLATELETS

Platelets are disk-shaped cells, 2 to 4 µm in diameter.

- They are without nucleus but contain all other organelles that are typically found in cells.
- Their surface membrane is covered with glycoproteins and receptors for a variety of agents.
- They are formed in bone marrow by budding from the cytoplasm of megakaryocytes and spend their lifetime (estimated to be about 10 days) within the intravascular compartment, many of them being held in the red pulp of the spleen.
- Their major function is in hemostasis.

Function

- Platelets normally circulate freely.
- Their function is to aggregate in localized clumps at sites of vessel injury. Whether platelets clump together depends entirely on surface forces governing interactions among platelets and between platelets and vascular endothelium. These surface forces are ruled by products of arachidonic acid. As a result:
- Phospholipid metabolism is a major influence on the microenvironment that surrounds platelets.

Metabolism of Membrane Phospholipids Phospholipases are released and activated when surface membrane receptors interact with their specific ligand. Ligands that have been shown to influence platelet aggregation include collagen, serotonin, immune complexes, and others (Fig. 2.7).

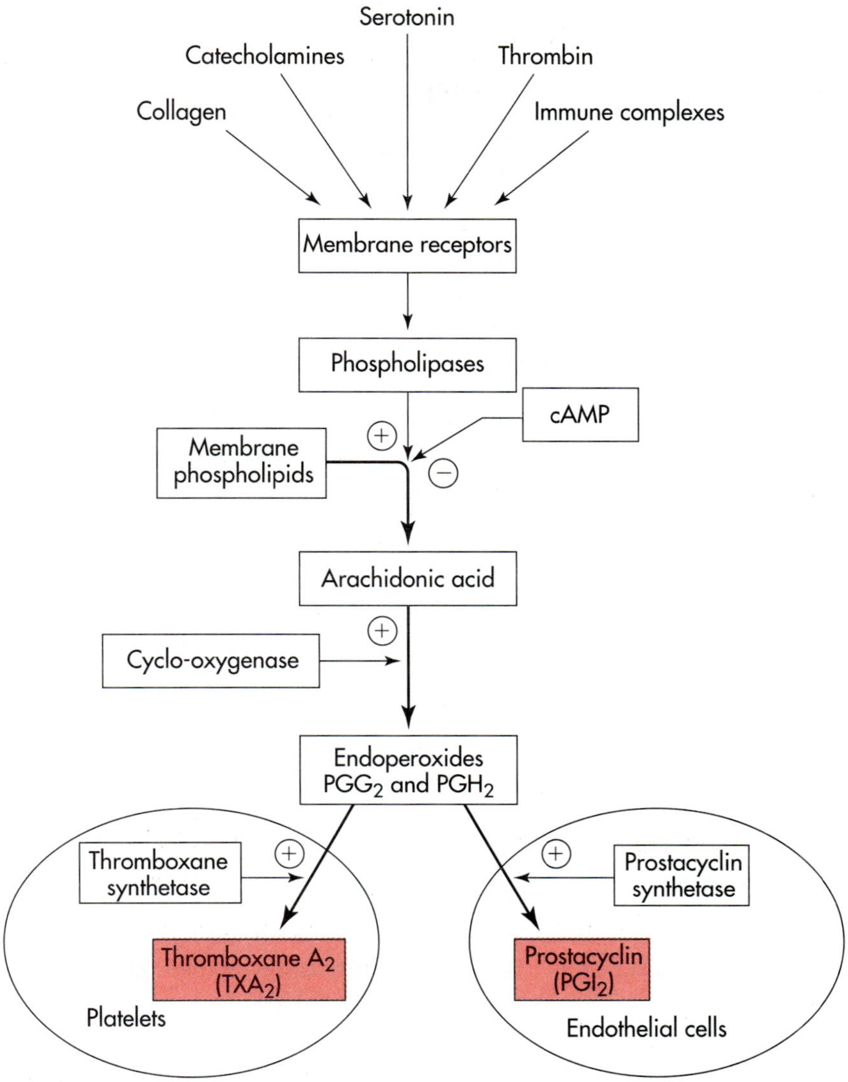

Fig. 2.7 Summary of phospholipid interactions among platelets and between platelets and vascular endothelium. Phospholipases, produced and released subsequent to activation of surface membrane receptors, act to hydrolyze phosphatidylcholine, phosphatidylinositol, and other membrane phospholipids. This yields prostaglandin endoperoxides (PGG_2 and PGH_2). The chemical nature of subsequent metabolic products is determined by local availability of appropriate enzymes.

- Platelets contain thromboxane synthetase. As a result they produce thromboxane A_2 (TXA_2), a powerful mediator of and specific stimulus for platelet aggregation and adhesion.
- Endothelial cells contain prostacyclin synthetase; therefore they produce prostacyclin (PGI_2), an inhibitor of platelet aggregation.

- Under normal circumstances platelet clumping is prevented by an antagonistic equilibrium between platelet-derived *thromboxane* A_2 (TXA_2), a mediator of platelet aggregation and endothelium-derived *prostacyclin* (PGI_2), an inhibitor of platelet aggregation.
- Plug formation occurs when the balance is shifted toward production of TXA_2.

- **Formation of Hemostatic Plugs** When the endothelial lining is damaged, the equilibrium between TXA_2 and PGI_2 is shifted toward TXA_2. Two factors are involved:

  - Local loss of endothelial cells and
  - contact between platelets and underlying extracellular matrix components including collagen. Such contact leads to platelet activation, increased production of TXA_2, and eventually the *release reaction.* The release reaction includes
    - contraction of platelets and consequent release of granule contents into the surrounding medium
    - release of enzymes that are capable of converting fibrinogen to fibrin
  - Platelet adhesion is facilitated by *von Willebrand factor,* an adhesive glycoprotein that is synthesized in vascular endothelial cells and forms linkages between the platelet and nonplatelet surfaces.
  - Platelets will aggregate to form a hemostatic plug if the surrounding medium supplies fibrinogen so that fibrin can be formed on the platelet surface to hold neighboring platelets together.

Hemostasis

- **Overview** Prevention of blood loss after vessel injury involves 3 physiologic mechanisms:
  1. Vasoconstriction in the affected area
  2. Formation of a platelet plug
  3. Formation of a firm blood clot

  - **Vasoconstriction** Two aspects are involved:
    - Nervous factors (initiated by pain or by the anticipation of pain)
    - Chemical factors (tissue chemicals released by the injury as well as serotonin released from activated platelets)
  - **Platelet plug** Contact of circulating platelets with collagen fibers in the intima of the injured vessel leads to platelet activation and consequent formation of a hemostatic plug. It also activates the enzymatic cascade that leads to the formation of a blood clot.

- **Clot Formation** Blood vessel injury or contact between blood and an inappropriate surface will, within seconds and in circumscribed areas, activate normally present plasma proteins and transform fluid, flowing blood, into a gel.

  - All clotting proteins, except factor VIII and lipoprotein-associated clotting inhibitor (LACI), are synthesized in the liver.
    - Many of them are proteolytic enzymes of the *serine protease* class, meaning that they have a highly reactive serine residue in their active site.
    - Four of them can be provided with vital Ca^{++}-binding sites only if vitamin K is present. These four are prothrombin and factors VII, IX, and X.
  - Activation of the clotting cascade by vessel trauma follows the *extrinsic pathway.*
  - Activation of the cascade by contact with a wettable surface follows the *intrinsic pathway.*

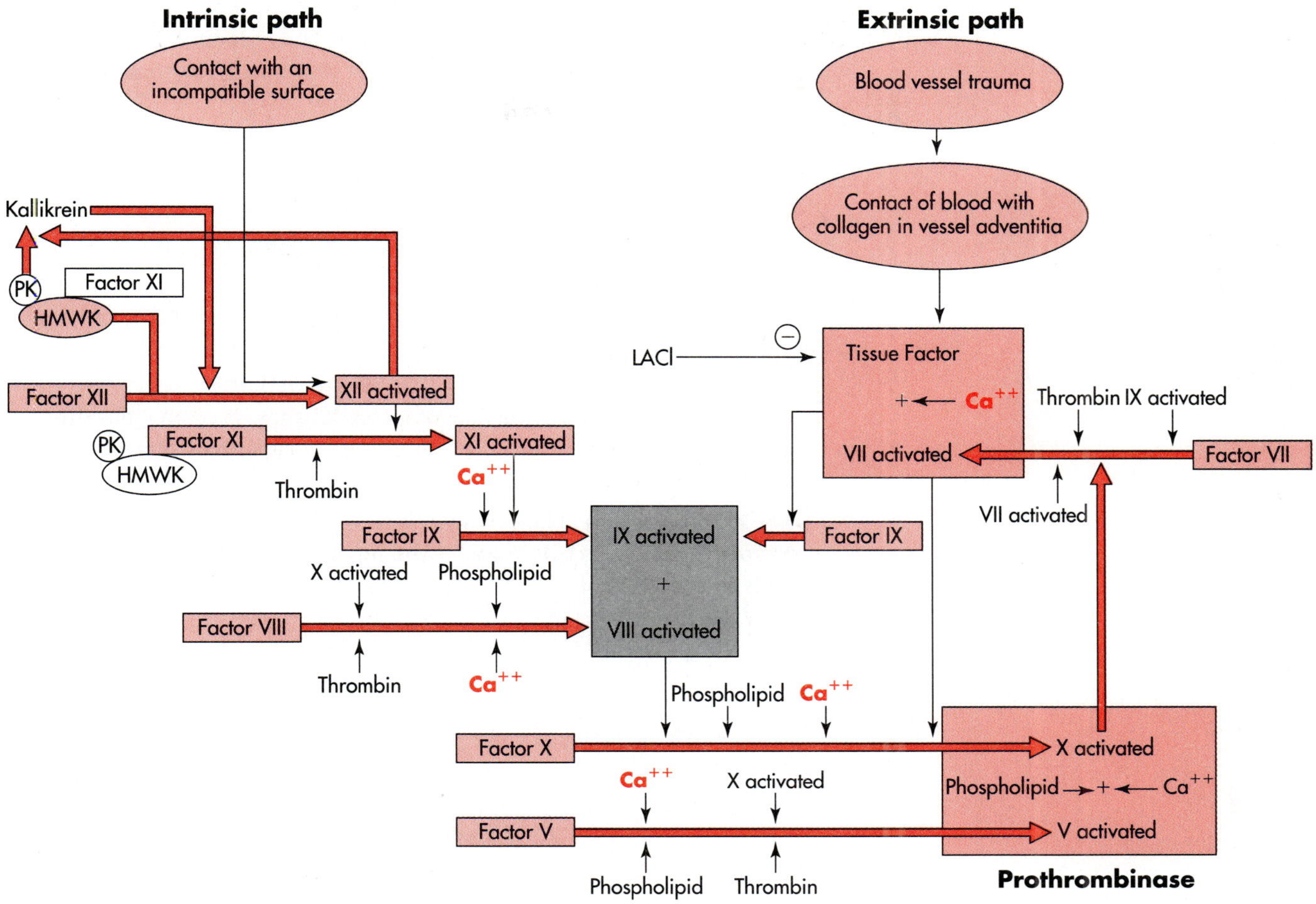

Fig. 2.8 Generation of prothrombinase by the intrinsic or extrinsic paths. Except for the contact-activation reactions leading to the formation of factor XIa, all reactions require the presence of Ca^{++}. Three essential enzymes in the cascade, factors VIIa, IXa, and Xa, are themselves physiologically inert but become catalytically effective when they are bound to cofactors. As shown in the illustration, these are tissue factor, factor VIIIa, and factor Xa, respectively. *HMWK,* high-molecular-weight kininogen; *LACI,* lipoprotein-associated coagulation inhibitor; *PK,* prekallikrein.

Once initiated, the clotting process can be divided into 3 stages:

1. Generation of *prothrombinase*
2. Activation of prothrombin
3. Conversion of fibrinogen to fibrin

- **Generation of prothrombinase** Prothrombinase is activated factor X complexed with activated factor V in the presence of Ca^{++} and platelet-derived phospholipids (Fig. 2.8). Its formation can follow either of two pathways, depending on the nature of the triggering event.

— ***The intrinsic path*** This path is initiated when blood contacts a wettable surface such as glass.

- Four molecular species are involved: factor XII (Hageman factor), high-molecular-weight kininogen (HMWK), prekallikrein (PK), and factor XI.
  - PK and factor XI circulate bound to HMWK.
  - The intrinsic path cascade is initiated when HMWK and factor XII contact the wettable surface. Such contact leads, pre-

sumably by a conformational change, to autoactivation of a small quantity of factor XII. XIIa thus produced activates PK to kallikrein, and kallikrein causes full-blown activation of factor XII.

- Sufficient activation of factor XII then rapidly activates factor XI. This step is further accelerated in the presence of thrombin.

• Activated factor XI participates in the subsequent clotting cascade (see Fig. 2.8).

• It should be noted that HMWK, PK, and factor XII are essential for triggering coagulation in a test tube or in extracorporeal bypasses; they play no known role in normal hemostasis.

• Patients with a deficiency of factor XI show excessive bleeding after surgical interventions, but only some of them will show abnormal bleeding after tissue trauma.

— ***The extrinsic path*** The extrinsic path is initiated by blood vessel trauma. It short-circuits many of the intrinsic activation steps and leads to prothrombinase after the activation of only one factor, factor VII.

• In this pathway the initial reaction is formation of a complex between factor VII, Ca^{++}, and *tissue factor* (see Fig. 2.8).

- Tissue factor is a transmembrane glycoprotein found in certain cells of blood vessel adventitia, though not in cells to which circulating blood is normally exposed. It is tightly associated with phospholipid and has high affinity for activated factor VII. In the presence of Ca^{++} the two will form a one-to-one complex after vascular injury.
- The VIIa–tissue factor complex is, in isolation, a weak activator of either of its biologic substrates, factor IX and factor X. However, activation of small amounts of factor X will initiate a positive feedback mechanism by which factor Xa preferentially activates factor VII that is complexed to tissue factor.

• The extrinsic pathway is of short duration because of the presence in blood of lipoprotein-associated coagulation inhibitor (LACI). It quickly inactivates the VIIa–tissue factor complex.

• Other inhibitory factors are antithrombin III and activated protein C.

- Antithrombin III is the primary inhibitor of factors IXa and Xa.
- Activated protein C inhibits the cofactor activities of factors VIIIa and Va.

● **Prothrombin activation** Prothrombin is a circulating protein. Its synthesis (in the liver) requires vitamin K and can be inhibited by substances that compete with vitamin K (such as warfarin).

• Prothrombinase is the only known physiologic activator of prothrombin (Fig. 2.9). The process involves:

- binding of factor Va to the surface of activated platelets where it serves as a binding site for factor Xa

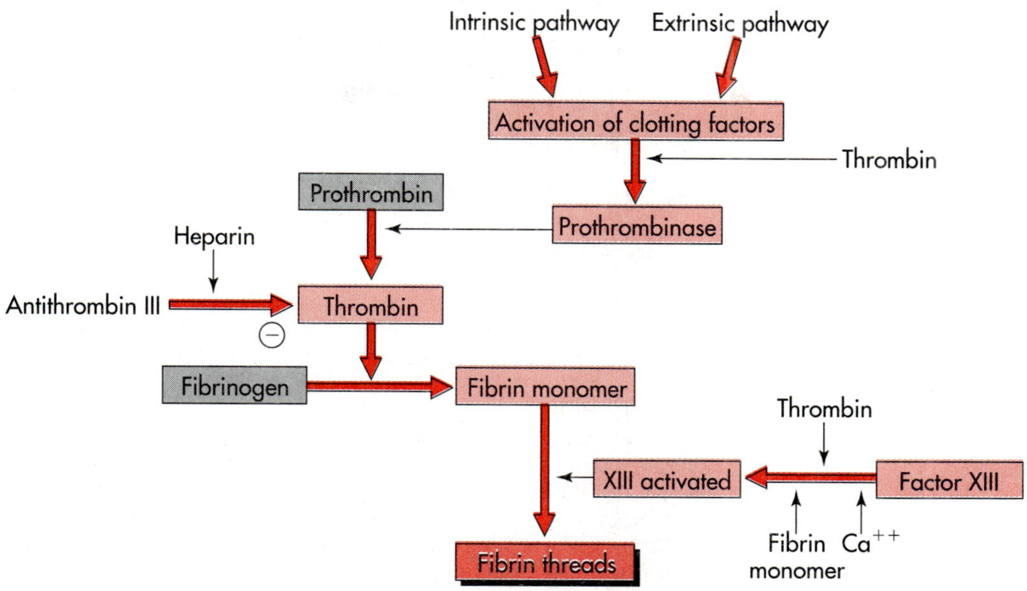

Fig. 2.9 The formation of fibrin threads by the enzymatic activity of thrombin.

- formation of a prothrombinase complex in which phosphatidylserine is transposed from the inner leaflet of the phospholipid bilayer of the platelet surface to the outer leaflet to provide the phospholipid coagulant activity there

• Cleavage of prothrombin at two sites yields *thrombin,* a molecule that is capable of moving off the platelet surface to catalyze conversion of fibrinogen to fibrin in the surrounding plasma and extracellular fluid (see Fig. 2.9).

• Prothrombin cleavage can be inhibited by antithrombin III, and this inhibition is promoted by the anticoagulant *heparin* (see Fig. 2.9).

Conversion of fibrinogen to fibrin The objective of blood coagulation is to form stabilized fibrin threads from fibrin monomer.

• Fibrin monomer derives from plasma fibrinogen as a result of the enzymatic activity of thrombin (see Fig. 2.9).

• Once fibrin has been formed it is further stabilized by strong cross-linking. That process requires Ca^{++} and is promoted by activated factor XIII (fibrin stabilizing factor). Thrombin promotes activation of factor XIII.

Fibrinolysis

• With time, fibrin is removed, and the clot is resorbed by the proteolytic action of *plasmin.*

• Plasmin results from the activation of plasminogen, a freely circulating, inactive proenzyme. Its conversion requires cleavage of only a single peptide bond and is accomplished by *plasminogen activators.* Two such activators are released from cells:

- Tissue-type plasminogen activator (t-PA)
- Urokinase-type plasminogen activator (u-PA)

Endothelial cells secrete both, and u-PA is also secreted by monocytes and fibroblasts.

Multiple Choice Review Questions

1. Which one of the following statements is correct?
   a. Antibodies are produced by lymphocytes.
   b. The major phagocytic leukocyte is the basophil.
   c. The average half-life of a neutrophil in the circulation is 120 days.
   d. Erythropoietin is mostly produced by the liver.

2. Which one of the following statements regarding activation of the clotting cascade via the Extrinsic Path is correct?
   a. It can occur only *ex vivo* in the test tube.
   b. It does not require phospholipid to form prothrombinase.
   c. It will lead to the formation of fibrin that is structurally different from that formed after intrinsic activation.
   d. It differs from the intrinsic pathway in that it requires the presence of factor VII.

3. Which one of the following statements correctly pairs a commonly used anticoagulant with its mechanism of action?
   a. Heparin inactivates factor V and thereby inactivates prothrombinase.
   b. Warfarin competes with vitamin K in the hepatic synthesis of vitamin K–dependent clotting factors.
   c. Aspirin (acetylsalicylic acid) competes with fibrinogen for the active site on thrombin.
   d. Tissue plasminogen activator (t-PA) acts by blocking the serine residues of thrombin.

4. Which one of the following statements correctly describes the sequelae to platelet contact with subendothelial collagen after vascular trauma?
   a. The platelets activate tissue factor, and the clotting sequence begins.
   b. Heparin is released from the extracellular matrix, inactivates thrombin, and thereby limits the coagulation process to the site of vascular injury.
   c. The collagen fibrils are rapidly degraded by platelet-derived thrombin.
   d. PGI_2 (prostacyclin) is released from the collagen fibrils and further promotes platelet aggregation.
   e. The platelets are activated, and the release reaction is initiated.

5. Which one of the following statements concerning vitamin K is correct?
   a. Thrombin requires vitamin K as a cofactor.
   b. Vitamin K is necessary for incorporation of Ca^{++}-binding residues into some clotting factors.
   c. Vitamin K is necessary for normal platelet function.
   d. Vitamin K is necessary for normal differentiation of platelets in the bone marrow.
   e. There is no known dietary source of vitamin K.

6. An individual has a blood volume of 5 L and a plasma volume of 3 L. Which one of the following statements is correct?
   a. She has a hematocrit of 0.60.
   b. She has a hematocrit of 0.66.
   c. She has a hematocrit of 0.40.
   d. She has a hematocrit of 1.67.

7. All the following statements concerning the mechanism of clot formation are correct ***except:***
   a. Thrombin, like many of the clotting enzymes, is a serine protease.
   b. Vitamin K is required for the formation and insertion of Ca^{++}-binding, paired residues of gamma-carboxyglutamic acid (GLA) into four essential clotting factors.
   c. Conversion of soluble fibrinogen to insoluble fibrin requires the proteolytic action of thrombin.
   d. Antithrombin III acts by cross-linking fibrin monomers into large meshworks of fibrin threads in which platelets become lodged.

8. Which one of the following statements is ***incorrect?***
   a. The duration of the effect of aspirin on blood coagulation is directly related to platelet lifespan.
   b. Heparin is an effective anticoagulant in blood drawn into a test tube as well as when injected into the bloodstream.
   c. Warfarin is equally effective as an anticoagulant whether it is taken orally or added to blood in a test tube.
   d. A patient lacking antithrombin III would tend to have inappropriate clotting more readily than a normal person would.

Chapter 3

The Autonomic Nervous System

The Nervous System and Regulation of Physiologic Function

Categories of Nervous Function There are 3 categories of physiologic function (Fig. 3.1):

- Mental processes
- Actions on the internal environment (autonomic reflexes)
- Actions on the external environment (sensory-motor integration)

Mental processes Voluntary actions on the external environment are guided by thought. On the other hand, higher mental processes are not required for maintenance of the internal environment or for involuntary reactions to external stimuli. However, processes at the highest cortical level provide the cultural and ethical boundaries within which autonomic responses are permitted to express themselves.

Autonomic reflexes The autonomic nervous system governs actions on the internal environment. It is a system of central nervous nuclei and efferent nerves that integrates and coordinates all reflex mechanisms that maintain a living person in a steady state with the changing environment.

Classification of Nerve Fibers Nerve fibers are classified according to their conduction velocities. Two classification systems are in use, the *Gasser system* and the *Lloyd system.* Fig. 3.2 shows the designations given to different fiber classes in the two systems. It also shows the approximate diameters and conduction velocities of some peripheral autonomic axons.

The Gasser system The Gasser system is the most commonly used to describe autonomic nerves.

- Rapidly conducting, large myelinated fibers are designated as *group A.*
- Within the A group, subgroupings α, β, γ, and δ exist according to axon diameter.
- Autonomic myelinated fibers are designated as *group B.*
- Unmyelinated fibers are designated as *group C.*

The Lloyd system This system is generally used to designate afferent fibers from muscle. It ranges from group I (large, fast, myelinated fibers with subgroupings a and b) to group IV (unmyelinated, slow fibers)

Patterns of Physiologic Activity Involving the Nervous System The common pattern of physiologic activity is that:

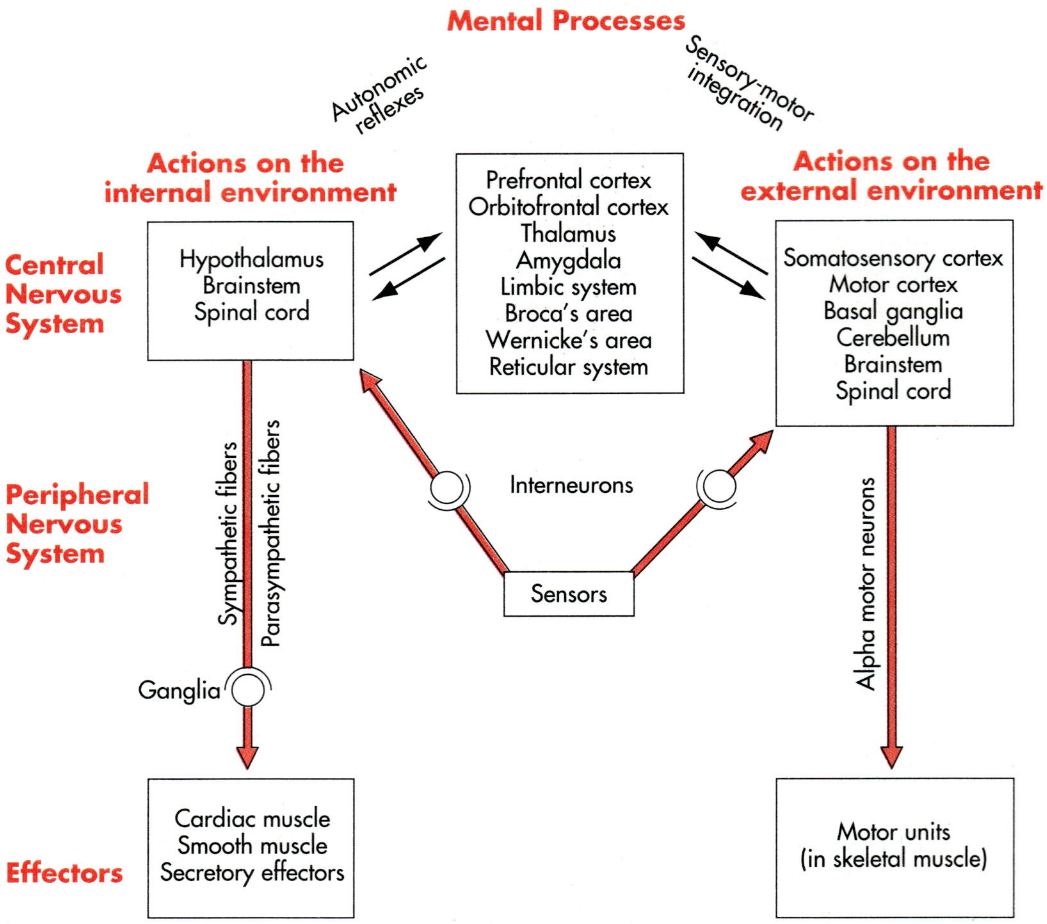

Fig. 3.1 The divisions of human nervous function. There are 3 categories of nervous function: mental processes, actions on the internal environment, and actions on the external environment. Both the central and peripheral nervous systems participate in all categories of nervous function.

- Central nervous patterns of action potentials cause a mechanical effect in peripheral effector mechanisms.
- The mechanical effect is sensed in the periphery and is transformed and transmitted as action potential patterns to the central nervous system.
- The success of the mechanical effect is evaluated in the central nervous system, relative to a goal.
- If necessary, a modified central nervous command pattern is issued to the effector mechanisms.

Functional Components of the Nervous System The typical pattern of activity described above requires 5 functional components.

These are also shown in Fig. 3.1:

1. A peripheral nervous *sensing mechanism* that transforms a specific stimulus modality (stretch, touch, smell, pH, light, sound, taste, temperature, or pain) into trains of action potentials.
2. *Afferent nerves* and *interneurons* that conduct the action potentials toward the central nervous system.
3. A *center* that evaluates all incoming information, determines whether a deviation from the normal set point has occurred, and generates an appropriate outgoing program.

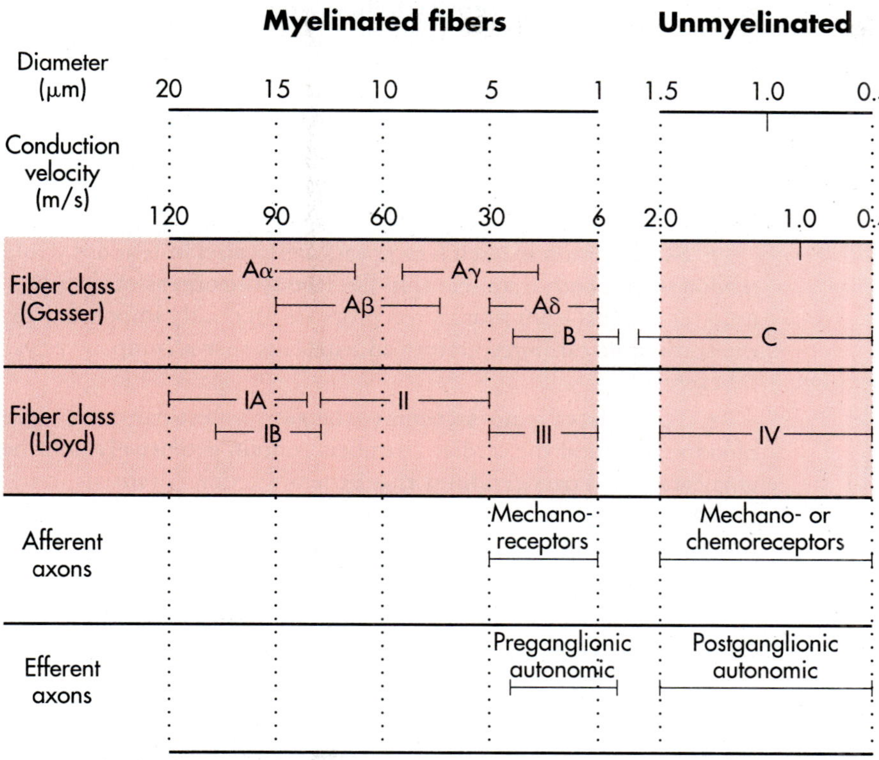

Fig. 3.2 Nerve fibers are classified according to their conduction velocities and are named by either the *Gasser* system or the *Lloyd* system.

4. *Efferent paths* (which may be nerves) that convey the outgoing program to the periphery.
5. *Effector mechanisms* that are capable of converting the outgoing program into an appropriate somatic or autonomic response.

- **Neurons** These cells generate action potentials and conduct them through their axons to synapses with other cells. Axons may be surrounded by myelin, a good electrical insulator that prevents current leakage. In myelinated fibers an action potential generated at one point is conducted rapidly to the nearest interruption in the myelin sheath, called a *node of Ranvier,* and there generates another action potential that is conducted to the next node. This arrangement is called *saltatory conduction.* It allows high transmission speed.

— ***Afferent axons*** The autonomic nervous system is mostly concerned with regulation of body temperature, blood pressure, blood gases, blood flow distribution, local distensions, and sphincter diameters. Accordingly, its major sensors are *thermoreceptors; mechano-*(or *stretch-) receptors,* and *chemoreceptors.*

Thermoreceptors. Although all receptors change firing rate with temperature, thermoreceptors are nerve endings, located in the hypothalamus, skin, and mucous membranes, that display far greater temperature sensitivity than all other receptors do. Thermoreceptors feature prominently in the autonomic adjustments required for body temperature regulation. Thermosensitive cells of the hypothalamus are most important for this.

Mechanoreceptors. Mechanoreceptors are nerve terminals that respond to changes in physical deformation of their environment such as touch, vibration, acceleration, or stretch. Receptors relevant to the function of the autonomic nervous system are found in skin, skeletal muscle, and blood vessel walls.

- Skin mechanoreceptors are present in the form of branched free nerve endings, hair follicle endings, pacinian corpuscles, Meissner's corpuscles and Merkel's disks. They provide information guiding autonomic responses to the external environment.
- Skeletal muscle stretch receptors are located in tendons (Golgi tendon organ) and in muscle spindles (the sense organs of skeletal muscle). Signals from muscle receptors are especially important for the autonomic cardiorespiratory adjustments that accompany physical exercise.
- The walls of many sections of the gastrointestinal tract, some blood vessels, and the cardiac chambers contain extensively branched, unencapsulated nerve endings that generate action potentials in response to wall stretch. They provide feedback information for the autonomic adjustment of GI tract peristalsis, arterial blood pressure, and extracellular fluid volume.

Chemoreceptors. Chemosensitive cells are found in a variety of anatomical locations. Those germane to autonomic adjustments are:

- located in the central nervous system and in the cardiovascular system (carotid and aortic bodies as well as the cardiac ventricles). They are specialized neurons that produce generator potentials in response to intracellular H^+ concentration. Those in the aortic and carotid bodies also are able to respond to changes in ambient PCO_2 or PO_2 These units feature prominently in the regulation of respiration.
- A special group of chemosensitive receptors is located on the ventricular epicardial surface, in the lung, and in the wall of the aortic arch. They do not respond to H^+, O_2, or CO_2 but are activated by serotonin. Activation of these receptors depresses cardiovascular function, and their true physiologic role is not known yet.

— ***Efferent axons***

- Efferent axons of the *somatic nervous system* directly innervate motor units of skeletal muscle (see Fig. 3.1).
- Efferent axons of the *autonomic nervous system* synapse in a ganglion before they proceed to an effector organ (see Fig. 3.1).

● **Synapses** Synapses in the autonomic nervous system are chemical. Their presynaptic side, but not their postsynaptic side, contains many intracellular vesicles (20 to 80 nm in diameter) filled with *neurotransmitter.*

— ***Neurotransmitters*** Many different neurotransmitters are involved in the central portions of the autonomic nervous system. In the peripheral portions the major neurotransmitters are *acetylcholine* and *norepinephrine.* They are synthesized locally. In settings where the releasing stimulus is strong enough, significant amounts of *neuropeptide Y* may be co-released with norepinephrine from sympathetic varicosities.

Biologic effects of neurotransmitters result from interaction with a postsynaptic *receptor.* Receptor activation leads to an electrical or chemical change in the postsynaptic cell.

- In nerve-to-neuron synapses the effect of interest is changes

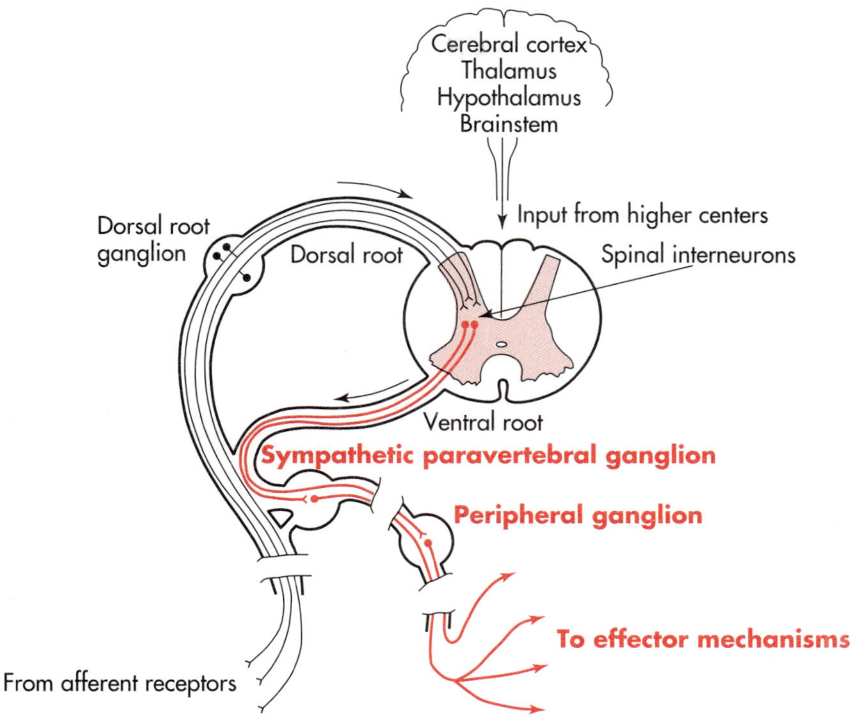

Fig. 3.3 Anatomical organization of the nervous system at the level of a spinal segment. Higher autonomic centers modulate segmental circuits through synapses between descending spinal tracts and spinal interneurons.

in transmembrane ion flux and subsequent generation of either an *excitatory postsynaptic potential (EPSP)* or an *inhibitory postsynaptic potential (IPSP).*

- In nerve-to-effector organ synapses the effect of interest is often a change in postsynaptic cytosolic $[Ca^{++}]$ because that ion governs many mechanical or secretory responses.

The Peripheral Autonomic Nervous System

General Structure Fig. 3.3 shows the arrangement of afferent and efferent nerve endings at the level of the spinal cord.

- At each spinal segment 2 spinal nerves connect with the cord, one on the left and the other on the right.
- Each spinal nerve branches into a dorsal and a ventral root (see Fig. 3.3).
  - The dorsal root contains afferents from a specific region of the body.
  - The ventral root contains somatic and sympathetic efferent nerves for a specific region of the body. Contrary to somatic efferents, sympathetic efferents do not innervate peripheral effectors directly but form a synapse in a peripheral *ganglion.*

Divisions of the Autonomic Nervous Systems The two branches of the autonomic nervous system are named the *sympathetic* and *parasympathetic* divisions respectively. As summarized in Table 3.1, the two divisions differ from each other in the following respects:

- Location of spinal outflow

Table 3.1 *Comparison of Features of the Two Sections of the Autonomic Nervous System*

Sympathetic System	Parasympathetic System
Location of Spinal Outflow	
Preganglionic fibers originate from neurons lying in the intermediolateral column of the spinal cord and leave through ventral roots in the thoracolumbar region.	Preganglionic fibers originate from neurons lying in the cranial nerve nuclei of the brainstem (cranial outflow) or from neurons lying in the lateral columns of the spinal cord in the sacral region (sacral outflow).
Location of Ganglia	
Ganglia form separate, discrete structures.	Ganglia are formed by *ganglion cells* within the walls of the effector organ.
Anatomy of Terminal Nerves	
Synapses with target organ are located in varicosities spaced at 3 to 10 μm intervals.	Terminal *boutons* synapse with target organ.
Postganglionic neurotransmitter	
Norepinephrine in most cases.	Acetylcholine in most cases.

- Location of ganglia
- Anatomy of terminal nerves
- Nature of neurotransmitter released at the effector organ synapse

The effect of increased activity in either system can be excitatory in some target organs and inhibitory in others. In target organs that are innervated by both systems, sympathetic and parasympathetic effects often oppose each other. In addition to such additive postsynaptic effects, there is also considerable presynaptic cross-talk between the systems, leading to such complex biologic effects as facilitated antagonism.

Ganglia Sympathetic ganglia are found at 2 sites:

- In the sympathetic trunks alongside the vertebral column.
- In the viscera (cervical, or stellate, ganglion; mesenteric ganglion). Preganglionic fibers that synapse in a visceral ganglion pass through the ganglia of the sympathetic trunk without synapsing there.
- The fibers of the parasympathetic division are not collected in ganglia but are distributed in the walls of the effector organs.
- Ganglia are not simple relay stations but function as sites of *convergence* or *divergence.*
  - Convergence occurs when the number of preganglionic fibers is larger than the number of postganglionic fibers.
  - Divergence occurs when the number of postganglionic fibers is larger than the number of preganglionic fibers.

Effector Organs The effector organs of the autonomic nervous system are secretory units, cardiac muscle, and smooth muscle.

- **Secretory effectors** Two types of secretory effectors are controlled by the autonomic nervous system: Exocrine glands (most prominently those modulating gastrointestinal or sexual function) and endocrine glands, most prominently the endocrine pancreas and the adrenals.

- **Cardiac muscle** The autonomic nervous system is a dominant influence in the short-term control of the rate and vigor of cardiac contraction.
- **Smooth muscle** Smooth muscle, like cardiac muscle, is supplied by the autonomic nervous system. Like the nerves supplying cardiac muscle, smooth muscle innervation is used to modulate contractile behavior, not to initiate it.

Receptors Autonomic nerves are abundantly supplied with both presynaptic and postsynaptic receptors.

- Presynaptic receptors modulate neurotransmitter release when they are activated.
- Postsynaptic receptors are responsible for the biologic effects that are associated with neurotransmitter release.
- Autonomic receptors are broadly divided into *cholinergic* and *adrenergic* types. Both are proteins with 7 membrane-spanning domains. Several receptor subtypes exist within each division.

Cholinoceptors

- Postsynaptic receptors in ganglia are *nicotinic cholinergic.*
- Receptors in effector organs are, for the most part, *muscarinic cholinergic* in the parasympathetic system.
- So far 5 subtypes of muscarinic receptors have been identified. They are identified as M_1 to M_5. They are distinguished from one another only by the relative potency of antagonists because no selective agonists and no highly specific antagonists have been found. All are found in the central nervous system. In addition,
  - The M_2 subtype is expressed at high density in the heart.
  - M_1 and M_3 are found in the peripheral ganglia and exocrine glands.
  - M_3 is found in vascular endothelium.
- The predominant effector pathway of receptor activation is activation of the phosphoinositide pathway by means of a G protein (see Chapter 1).

Adrenoceptors Receptors in effector organs are *adrenergic* in the sympathetic system.

- There are two types of adrenergic receptors, designated α and β. Both α and β receptors have several functional subtypes.
- Within the α grouping:
  - α_1 receptors predominate postsynaptically on the effector organ membrane and are further subdivided into α_{1A}, α_{1B}, α_{1C}, α_{1D} on the basis of the relative affinity of antagonists.
  - α_2 receptors dominate presynaptically, on adrenergic nerve terminals themselves. There they modulate norepinephrine release. They are subdivided into 3 groups, α_{2A}, α_{2B}, α_{2C}.
- The predominant effector pathway of α-adrenoceptors is activation of the phosphoinositide pathway by means of a G protein.
- Within the β grouping:

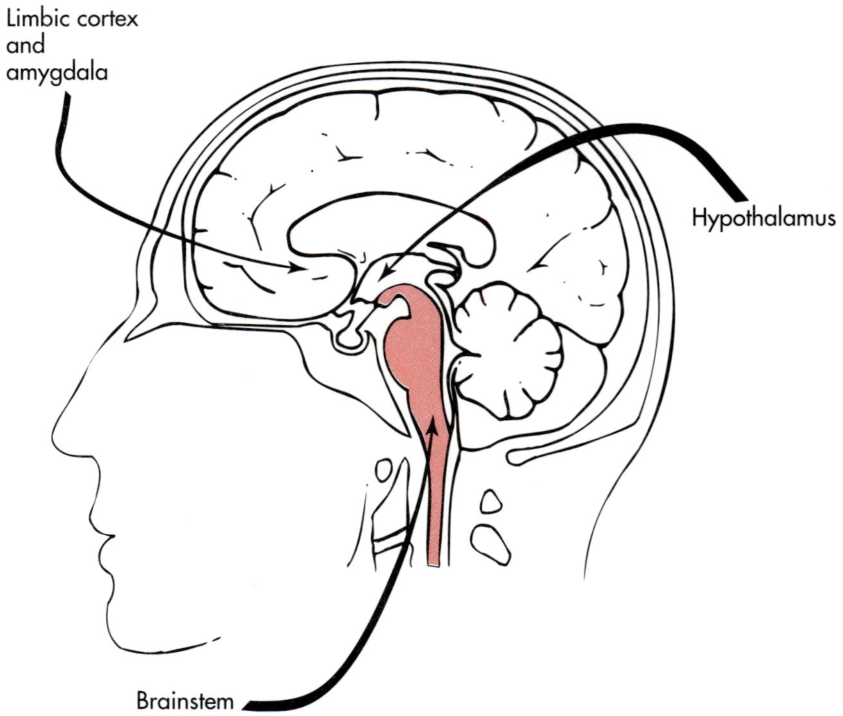

Fig. 3.4 Location of the major reflex centers governing autonomic reflexes.

- β_1 receptors are found mostly in cardiac myocytes.
- β_2 receptors are found in smooth muscle and in secretory effectors.
- β_3 receptors have limited distribution and are present in low levels in adipose tissue (where they stimulate lipolysis) and the GI tract (where they stimulate gut motility).

- The predominant effector pathway of β-adrenoceptors is activation of adenylate cyclase through a G protein.

- **Neuropeptide Y receptors** Subtypes Y_1 and Y_2 are present in the periphery. They operatee predominantly through inhibition of adenylate cyclase by means of a G protein.

The Central Autonomic Nervous System and Reflex Centers

Patients with high cord transections are capable of reflex responses to mild stimuli. This capability indicates that there are spinal segmental reflex loops. However, true autonomic integration requires the presence of higher nervous centers. These centers are shown in Fig. 3.4.

- **Brainstem** The brainstem area consists of 3 anatomically distinct regions: *midbrain, pons,* and *medulla.* They are linked at the core by the *reticular formation.*
  - **Midbrain** The midbrain acts as a conduit for ascending and descending fibers. It contains nuclei associated with complex neurological patterns that are not normally controlled by autonomic activity but do have autonomic correlates:
    - *Periaqueductal gray* organizes motor responses to external stimuli.

- *Superior* and *inferior colliculi* map visual and auditory space and map body movements within them.
- *Red nuclei* coordinate somatic motor functions.
- The *substantia nigra* modulates limb movements.

Pons

- Contains nuclei for cranial nerves V, VI, VII, and VIII.
- Contains reflex centers for cardiovascular and respiratory control.

Medulla

- Contains nuclei for cranial nerves IX, X, XI, and XII.
- Contains nuclei that act as reflex centers for autonomic control or as intermediaries in somatic control.
- The pons-medulla region is an autonomous center for reflex responses to afferent signals from respiratory, cardiovascular, or gastrointestinal receptors. The physiologic responses to activation of neurons in these areas are physically complete and system specific.

Reticular formation The reticular formation is a collection of both ascending and descending neurons having as major functions:

- the determination of the state of consciousness
- balancing autonomic and somatic activity with the level of consciousness

Hypothalamus The hypothalamus contains nuclei that provide:

- an interface between higher nervous centers and the endocrine system
- coordination of whole-body autonomic responses to behavioral drives (such as fear) or to input from autonomic sensors (such as baroreceptors)

The hypothalamus coordinates all autonomic systems. This involves:

- integration of responses to hunger, thirst, and sexual drives
- integration of thermoregulation
- integration of defense reactions
- control of several endocrine secretions
  - adrenal medulla
  - posterior pituitary
  - anterior pituitary
- Responses that are coordinated by the hypothalamus are physically complete and involve the whole body.

Limbic Cortex and Amygdala These regions of the central nervous system are the gateway to domains of mental processing.

- Their activity governs the extent of autonomic correlates that may accompany emotional states such as fear, rage, embarrassment, or sexual desire.
- They also represent a level of conscious control over aspects of autonomic function. The range of this control over resting autonomic functions is demonstrated during states of deep meditation. In this state metabolic rate, heart rate, arterial blood pressure, and distribution of blood flow can be modified by application of conscious mental effort.

Multiple Choice Review Questions

1. The sympathetic division of the autonomic nervous system is characterized by:
   a. Inhibition of target-organ neuroeffector junctions by muscarinic antagonists.
   b. Adrenergic preganglionic fibers.
   c. Short postganglionic fibers.
   d. Thoracolumbar outflow from the spinal cord.
   e. Efferent fibers that are predominantly located in the vagus nerves.
2. During an operation all the nerve fibers leading into the right and left paravertebral sympathetic chains are cut. What would be the immediate consequence of this operation?
   a. Inability of arterioles in arm muscles to constrict.
   b. Inability of arterioles in arm muscles to constrict in response to unloading of cardiovascular mechanoreceptors.
   c. Inability of the heart to increase its rate in response to unloading of cardiovascular mechanoreceptors.
   d. Inability of the heart to increase its performance.
3. *Atropine* is a muscarinic antagonist and prevents the action of acetylcholine on smooth muscles, glands, and the heart. Which one of the following effects would you expect after administration of a physiologic dose of atropine to a patient?
   a. Loss of control over the diaphragm
   b. Cardiac asystole
   c. Pupillary constriction
   d. Loss of control over micturition
   e. Failure of skeletal muscle arterioles to constrict in response to stimulation of sympathetic efferent neurons
4. Drug X is classified as a ***nonspecific** beta-adrenergic antagonist.* It prevents the vasodilator and cardioaccelerator actions of epinephrine but does not prevent the vasoconstrictor action of either epinephrine or norepinephrine. These data are consistent with one of the following conclusions. Which one?
   a. Epinephrine produces vasodilatation by stimulating β_2-receptors.
   b. Epinephrine produces vasodilatation by stimulating alpha-receptors.
   c. Epinephrine produces vasoconstriction by stimulating alpha-receptors.
   d. Epinephrine produces bradycardia by stimulating beta-receptors.
   e. Norepinephrine produces vasodilatation by stimulating alpha-receptors.

The following items, lettered a to e, apply to questions 5, 6, and 7. For each of the questions choose the lettered item that best applies. Each lettered item may be used once, more than once, or not at all.

a. Autonomic nervous system
b. Paravertebral sympathetic ganglia
c. Hypothalamus
d. Nicotine
e. Norepinephrine

5. Initiates contraction of the urinary bladder but not of the diaphragm.
6. Acts as a reflex center for autonomic function.
7. Blocks biologic effects in both the sympathetic and parasympathetic divisions of the autonomic nervous system.

Chapter 4

Muscle Physiology

Overview

Muscle is excitable, contractile tissue. It is classified, on the basis of its microscopic appearance, as *striated muscle* or as *smooth muscle.*

- The difference in appearance between striated and smooth muscle is attributable to a difference in the physical arrangement of the contractile proteins.
- Striated muscle is further divided into *skeletal muscle* and *cardiac muscle.*
- Smooth muscle occurs in two arrangements:
    - In *unitary smooth muscle,* cells are arranged in sheets with multiple gap junctions providing extensive electrical and metabolic contact with neighboring smooth muscle cells.
    - In *multiunit smooth muscle,* cells are arranged as discrete muscle fibers, with each fiber being innervated by an individual nerve. Electrical conductance between neighboring fibers is poor.

Ultrastructure

Striated Muscle

Morphology Striated muscle, viewed through a light microscope, consists of repeating motifs of light and dark bands, bounded at intervals of about 2 µm, by narrow, dark bands called the *Z line* (or *Z band*). A typical cell incorporates several Z lines and is bounded by an external membrane, called the *sarcolemma.*

— ***Sarcolemma*** The sarcolemma penetrates deeply into each cell by means of the system of transverse tubules (*T-tubules*). These tubular invaginations allow extracellular fluid to be close to the cell interior.

— ***Sarcoplasmic reticulum*** The endoplasmic reticulum of striated muscle cells is called the *sarcoplasmic reticulum* (SR). It is a system of membrane-lined intracellular sacs that is arranged in contiguity with the system of T-tubules.

— ***Contractile elements*** The microscopic appearance of striated muscle arises from the physical arrangement of its constituent proteins. Fig. 4.1 shows the proteins *myosin* and *actin,* arranged in a repeating, interdigitating pattern to form the structural skeleton of thin and thick filaments. In the region where thin and thick filaments overlap, a portion of the myosin molecule projects toward the actin molecule. *The myosin projection would spontaneously bind to actin if the proteins* tropomyosin *and* troponin-T, -C, *and* -I *did not provide a physical shield. The obstruction provided by them is called* steric hindrance.

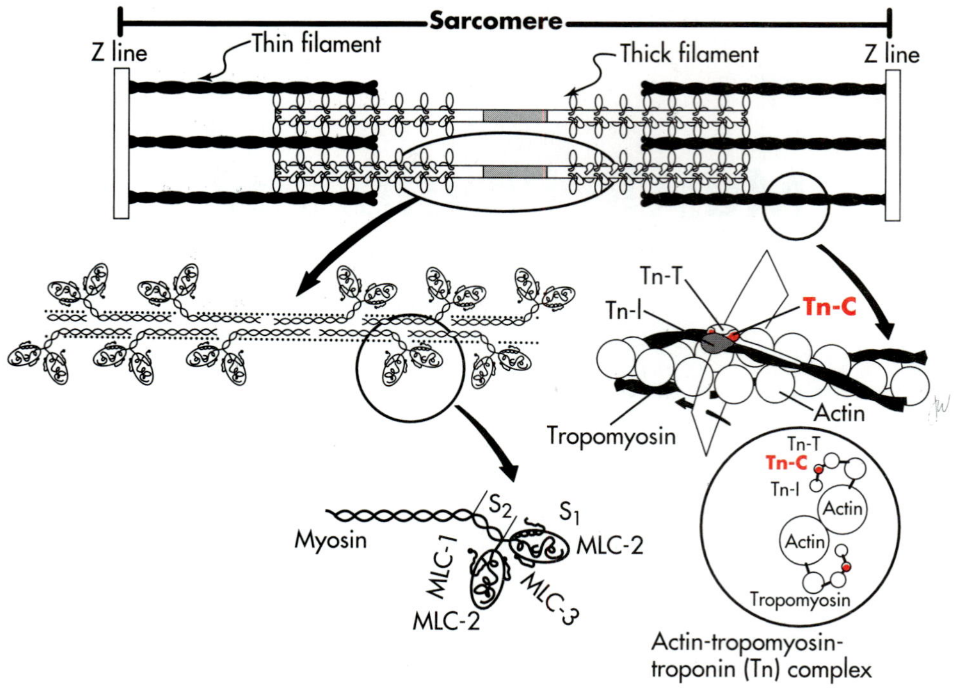

Fig. 4.1 Ultrastructure of striated muscle. The *thick* filaments are composed of the protein myosin. Myosin is a molecule that forms long filaments, with a short projection, the *head,* at one end of it. The *thin* filaments are composed of the proteins actin, tropomyosin, and troponin. Troponin exists as an assembly of 3 subunits, troponin-T, -C and -I. In the resting state actin, tropomyosin, and the troponin subunits are arranged so as to prevent myosin heads from binding to actin. *MLC–i,* Myosin light chain–i; *S1,* subfragment 1 portion of the myosin head; *S2,* subfragment 2 portion of the myosin head; *Tn-C,* troponin-C; *Tn-I,* troponin-I; *Tn-T,* troponin-T.

Myosin

- Myosin forms the thick filament
  - It is composed of two intertwined heavy chains.
  - The heavy chains can be cleaved by trypsin into a tail portion and a portion incorporating *subfragments 1* and *2* (S1, S2) (see Fig. 4.1).
  - Each heavy chain is associated with a "head," called *subfragment 1* (S1) (see Fig. 4.1), which can be cleaved from S2 by papain.
  - S1 is formed by three dissimilar light chains. Of these, MLC-2 is attached to the heavy chain (see Fig. 4.1) and is essential for contractile function because it contains the binding site for ATP and one for actin.

• A major role of the tail portion is to allow polymerization of myosin molecules so as to form a symmetrical aggregate around the M-line in the sarcomere with the tail portion of the molecules arranging themselves in both parallel and antiparallel fashion. In each section of the thick filament the head groups of the molecule are polarized away from the center. Such polymerization and polarization is crucial for the function of myosin, which is to move actin filaments toward the M-line.

Actin. Actin forms the thin filament.

- It is synthesized as an almost globular molecule named G-actin.
- In the presence of ATPase and Mg^{++}, G-actin aggregates to form a filamentous polymer named F-actin, the major protein of the thin filament.

Tropomyosin. Tropomyosin is a rod-shaped protein that lies near the grooves of the F-actin filament (see Fig. 4.1). Its function is to strengthen the actin filaments.

Troponin. Troponin exists in muscle as a complex of three dissimilar subunits, designated *Tn-C, Tn-T,* and *Tn-I* (see Fig. 4.1). Tn-T binds to tropomyosin and holds the complex in place. Tn-I inhibits the actin-myosin interaction at resting conditions. Tn-C is a Ca^{++}-binding protein.

Electrophysiology Resting and active behavior of striated muscle depends critically on ion-selective channels for Na^+, K^+, Cl^- and Ca^{++}.

— ***The inactive state***

- In or near the resting state, membrane currents are largely carried passively by K^+, Cl^- and Ca^{++}. The active Na^+, K^+ pump also participates because its stoichiometry (3 Na^+ out for 2 K^+ in) provides net outflow of positive charge.
  - The dominant K^+ channel at rest is the K_1 channel. It normally conducts K^+ out of the cell, but can carry K^+ into the cell whenever the membrane potential is more negative than the potassium equilibrium potential. K_1 is called "inwardly rectifying" because its conductance in the K^+-inward mode is greater than that in the K^+-outward mode.
  - Cl^- flux is small and directed out of the resting cell. Chloride ions protect the surface membrane from the spontaneous depolarization that could result from K^+ buildup, particularly in the T-tubules.
  - Resting Ca^{++} flux occurs mostly through the calcium-leakage channel, Ca^{++}_{leak}. It provides a slow inward Ca^{++} flow and thereby helps to maintain a stable resting membrane potential in the face of outward current flows provided by the K_1 potassium channel and the active Na^+, K^+ pump.
- The surface membrane of cardiac muscle also contains a sodium-calcium exchange mechanism. It uses the sodium electrochemical gradient as the source of energy for moving Ca^{++} and Na^+ across the cell membrane. It is described in more detail in Chapter 6.

— ***Action potentials*** The most significant ion currents involved in action potentials of striated muscle are Na^+, Ca^{++}, and K^+. Sodium and calcium channels regulate the upstroke and, in cardiac muscle, the plateau phase of the action potential. Potassium channels determine the return of the membrane potential to its resting state.

Sodium channels. Na^+ ions provide the principal current during the action potential upstroke.

Calcium channels. Most excitable cells have many types of Ca^{++} channels, but the L and T types are prominent in striated muscle.

- The T type:
  - is activated at a relatively low membrane potential of about −70 to −50 mV.
  - is inactivated within 20 to 50 ms.
- The L type:
  - is activated at a membrane potential that is 10 to 25 mV more positive than the activation potential for T-channels.
  - is inactivated slowly, within 200 to 300 ms.

• The T-tubule network of skeletal muscle is a rich source of L type Ca^{++} channels. They are localized in those regions of the T-tubule membrane that directly face the terminal sacs of the sarcoplasmic reticulum (SR) and are arranged in a regular pattern, facing Ca^{++}-release channels (ryanodine receptors), which are localized to the SR membrane. The release channels are also called *ryanodine receptors* because they will specifically bind the plant alkaloid, ryanodine. Low concentrations of ryanodine hold the channel permanently open. High concentrations of ryanodine will block the channel.

• The L type Ca^{++} channels in the T-tubule membrane and apposing ryanodine receptors in the SR cisternae interact directly to couple T-tubule events to the sarcoplasmic reticulum. The nature of the coupling differs between skeletal muscle and cardiac muscle and the difference arises from the activation kinetics of the channel:

- The intramembrane domains of the channel contain highly mobile, charged gating domains that respond rapidly to changes in membrane potential. Their early, fast movement is followed by a slower conformational change of the whole molecule and leads, eventually, to channel opening and Ca^{++} flux through the channel.
- A skeletal muscle twitch begins less than 10 ms after the start of an action potential. Hence the importance of the L type Ca^{++} channels in skeletal muscle arises from their ability to sense changes in membrane voltage. The absolute value of the potential difference across the T-tubule membrane controls coupling to the ryanodine receptor, not the influx of Ca^{++} ions.
- A cardiac muscle twitch, on the other hand, is delayed by almost 200 ms from the start of an action potential. Hence cardiac-muscle L-type Ca^{++} channels mediate long-lasting Ca^{++} currents and permit the translation of membrane depolarization into an influx of Ca^{++} that is capable of triggering intracellular responses such as "calcium-mediated calcium release."

Potassium channels. Several functional subtypes have been identified:

• Channels carrying the current, I_{K1}. These channels are the major carrier of basal potassium current in striated muscle cells.

• Channels carrying the delayed rectifier current, I_K. The role of this channel in all excitable membranes is to provide outward current for repolarization of the membrane toward its diastolic potential.

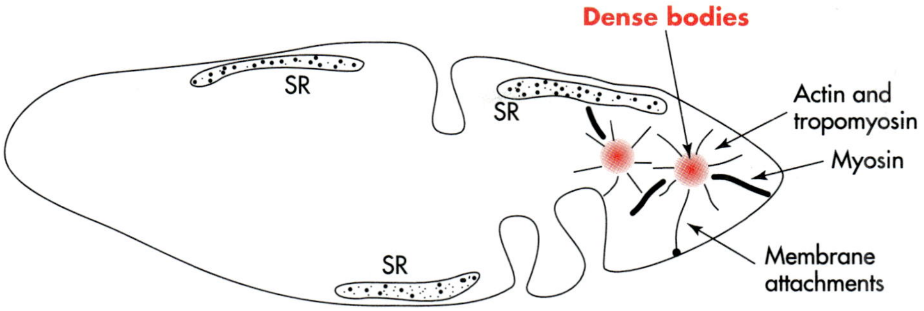

Fig. 4.2 Ultrastructure of smooth muscle. Thick and thin filaments are arranged in rosettes radiating from *dense bodies*. Several dense bodies have thin filament attachments to the cell membrane. This allows the cell to shorten when thick and thin filaments interact. *SR*, Sarcoplasmic reticulum.

— ***End-plate potentials*** Cardiac muscle displays automaticity because it is functionally and structurally coupled to spontaneously active pacemaker cells.

- Skeletal muscle fibers are not spontaneously active but require a stimulus. This stimulus is normally provided by somatic nerves using acetylcholine as a neurotransmitter.
- Acetylcholine activates postsynaptic nicotinic receptors in the end-plate region. Nicotinic receptors form an intrinsic cation channel. Their activation causes increased flow of Na^+, K^+, and Ca^{++} across the membrane. Inward flow of Na^+ and Ca^{++} dominates and depolarizes the cell.
- When the release of neurotransmitter is insufficient to depolarize the postsynaptic membrane to threshold, there results a localized subthreshold change in postsynaptic potential, called an *end-plate potential.*
  - A propagated action potential is generated when an end-plate potential is sufficiently large to depolarize the postsynaptic membrane to threshold.
  - Spontaneous release of minute quanta of neurotransmitter causes *miniature end-plate potentials (MEPPs).*

Smooth Muscle

Morphology

— ***Smooth muscle cells*** Smooth muscle cells are spindle-shaped or branched. Their plasma membrane has several distinct regions, and their cytosol contains actin and myosin as contractile proteins.

Plasma membrane. In addition to a gap junction region, the plasma membrane shows histologically and, presumably, functionally different portions (Fig. 4.2).

- One portion has surface invaginations (*caveolae*).
- Another portion has closely apposed sarcoplasmic reticulum.
- A third portion has attachments to intracellular dense bodies.
  - Caveolae greatly increase the surface area of smooth muscle cells but appear to have no other function.

- The membrane portion near sarcoplasmic reticulum is probably a major site of signal transduction.

Sarcoplasmic reticulum. The appearance and function of smooth-muscle SR are like those of SR in striated muscle. In the place of the special arrangement that is seen in striated muscle between SR and T-tubules, smooth muscle shows electron-dense structures that appear to couple the sarcoplasmic reticulum to the plasma membrane.

Contractile elements. The contractile machinery of smooth muscle is not as well organized as that of striated muscle, but it resembles the latter in all the important features. There are thin filaments, composed mostly of G-actin and thick filaments, composed of myosin. Their physical arrangement is suggestive of a sliding-filament mechanism of contraction.

- *Dense bodies,* rich in the actin-binding protein α-actinin, are found throughout the cytoplasm and as attachments to the intracellular face of the plasma membrane. They form insertion structures for the thin filaments that radiate from each dense body (see Fig. 4.2).
- Thin filaments are composed mostly of G-actin and tropomyosin. They also contain *filamin, caldesmon,* some *calmodulin,* and a variety of kinases and phosphatases. Filamin is an actin-binding protein of no known function. Caldesmon is a calmodulin-binding protein. Calmodulin binds Ca^{++}.
- Thick filaments consist of smooth muscle myosin, which is immunologically different from striated muscle myosin, but has the same ultrastructural appearance. The thick filaments are interspersed among the thin filaments in such a way that each of them is surrounded by a rosette of about 15 thin filaments.

Electrophysiology

— ***Ion channels*** Smooth muscle has a variety of voltage-gated and ligand-gated channels.

Voltage-gated channels

- Ca^{++} channels are the most important contributors to the upstroke of smooth-muscle action potentials and to excitation-contraction coupling. L-channels dominate because of their higher conductance and more prolonged "on" time.
- Voltage-gated K^+ channels conduct the rectifier K^+ current that repolarizes the cell at the end of an action potential.
- Voltage-gated Na^+ channels have also been identified in some smooth muscle. When they are present, they contribute to the change in membrane potential during the upstroke of the action potential.

Ligand-gated channels. Such channels are influenced by chemical rather than electrical changes. The influence of the ligand is direct in some cases (as it is, for example, in channels that are sensitive to ATP or in channels that are activated by increases in intracellular [Ca^{++}]), and it is indirect in others where interaction of the ligand with its membrane receptor produces a second messenger that can gate ion channels.

— ***Resting membrane potential*** Smooth muscle cells have a more positive resting membrane potential than striated muscle cells have. It is

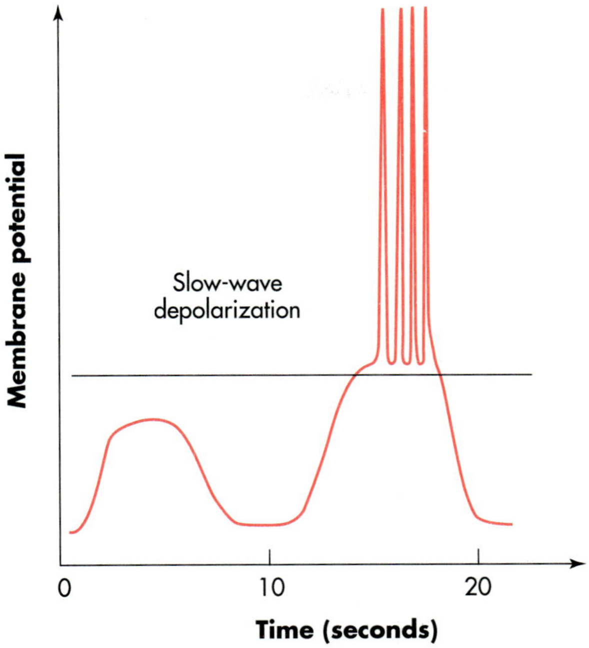

Fig. 4.3 Typical membrane potentials in smooth muscle cells. Slow-wave activity is shown. Occasionally a slow-wave potential reaches threshold, and one or more action potentials are then generated and propagated.

between −40 and −60 mV, which is more positive than the potassium equilibrium potential, E_K.

- The most likely explanation for this is greater contributions from Na^+ and Cl^- currents.
- Most vascular smooth muscle has, in addition to the ubiquitous Na^+,K^+ pump, an active mechanism that transports Cl^- into the cell.

— ***Action potentials*** Three electrically distinct types of smooth muscle have been identified:

1. Some smooth muscle generates action potentials spontaneously. Such cells exhibit slow-wave electrical activity. When these slow-wave potentials reach threshold, one or more spontaneous action potentials are generated, spread over neighboring cells and cause a contraction (Fig. 4.3).
2. Some smooth muscle generates action potentials in response to an appropriate stimulus.
3. Some smooth muscle responds to excitatory stimuli with a sustained, nonregenerating electrical response resembling a skeletal muscle end-plate potential.

Types 1 and 2 exhibit phasic contractions, whereas type 3 shows more sustained contractions. In those smooth muscle cells that normally generate an action potential (types 1 and 2 above), the current that causes the upstoke of the action potential is carried by Ca^{++}, not by Na^+.

Excitation-Activation-Contraction Coupling

The processes of excitation-activation-contraction coupling link electrical events of an action potential to the mechanical events of tension development and subsequent relaxation. Ionized calcium plays a crucial role in these processes, rising from a resting cytosolic concentration near 100 nM to a peak of about 1 μM during a normal contraction.

Striated Muscle

Sources of calcium

— ***Voltage-gated Ca^{++} currents*** $I_{Ca,L}$, the current that is carried through L-type calcium channels, carries most of the calcium that enters muscle cells from the extracellular space. Although it contributes no more than 10% of the total Ca^{++} needed for a maximal contraction, $I_{Ca,L}$ performs at least two vital functions:

- It supplies extracellular calcium for replenishing intracellular stores. In skeletal muscle this occurs mainly during sustained contractions. In cardiac muscle the duration of individual action potentials is sufficient to allow the Ca^{++}-transporting function of the channel to manifest itself even during a single twitch.
- It contributes toward maintenance of the action potential plateau in cardiac muscle and provides the trigger for releasing calcium from cardiac intracellular stores during systole.

— ***Ca^{++} leak*** The equilibrium potential for Ca^{++} is near +80 mV, and the resting sarcolemma is permeable to Ca^{++}. This results in some passive inward leakage of Ca^{++}.

— ***Na^{+}-Ca^{++} exchange*** This mechanism is of significance in cardiac muscle and smooth muscle. It is described in Chapter 6.

— ***Sarcoplasmic reticulum (SR)*** Most of the Ca^{++} held inside muscle cells is contained within the terminal cisternae of the SR and is associated there with the proteins *calsequestrin, calreticulin,* and *histidine-rich calcium-binding protein (HCP).*

Two different mechanisms govern communication between the T-tubule and the terminal cisternae of the SR. Both require coupling between the L type Ca^{++} channel, located in the plasma membrane, and Ca^{++}-release channels (ryanodine receptors), located in the SR membrane.

- In skeletal muscle activation of ryanodine receptors is directly proportional to the voltage difference across the plasma membrane.
- In cardiac muscle ryanodine-receptor activation is proportional to Ca^{++} current through the L type channels *(calcium-mediated calcium release).*

Uptake of calcium

- In skeletal muscle the main mechanism for Ca^{++} removal after an action potential is Ca^{++}-ATPase. It is present in both the sarcolemma and the membrane of the sarcoplasmic reticulum. It transfers 2 Ca^{++} per ATP and is modulated by agents such as the polypeptide phospholamban (in cardiac muscle and smooth muscle).

- In cardiac muscle and vascular smooth muscle the Na^{+}-Ca^{++} exchanger also contributes.

■ **Smooth Muscle** Smooth muscle contraction, like that in striated muscle, is initiated by elevations in cytosolic free [Ca^{++}]. The subsequent processes of tension development, though resembling those of striated muscle, differ from them in the details of Ca^{++} utilization.

● **Elevation of [Ca^{++}] in the cytosol** Increases in the concentration of intracellular calcium ions can occur after either a primary change in membrane potential (electromechanical coupling) or after chemical events that do not necessarily involve a change in membrane potential (chemomechanical coupling).

— ***Electromechanical excitation-activation-contraction coupling*** These processes are initiated by a sufficient depolarization of the cell membrane. They operate by activation of voltage-gated Ca^{++} channels that are coupled to ryanodine receptors in the sarcoplasmic reticulum.

— ***Chemomechanical excitation-activation-contraction coupling*** These processes account for drug-induced contractions of smooth muscle. Their two major components are (1) receptor-mediated intracellular Ca^{++} release and (2) modulation of the Ca^{++}-sensitivity of actin-myosin interactions.

Receptor-mediated Ca^{++} release. A major Ca^{++} source is release from intracellular stores by means of an inositol 1,4,5-trisphosphate (IP_3) receptor. It is found in the membrane of sarcoplasmic reticulum and in that of other Ca^{++}-containing intracellular structures.

Modulation of Ca^{++} sensitivity. Sensitization of the contractile apparatus to the effects of Ca^{++} expresses itself as an increase in developed force for a given [Ca^{++}]$_i$.

The Sliding Filament Theory of Muscle Function

Muscle has two functions: It changes length and develops tension. Both are achieved by periodic formation, rotation, and disengagement of cross-bridges between myosin and actin (Fig. 4.4). In addition to the structural and regulatory muscle proteins, these processes require Ca^{++} and ATP.

- Ca^{++} is essential for the removal of steric hindrance.
- ATP is required both for cross-bridge disengagement and for the energy-consuming processes of the power stroke.

■ **Striated Muscle**

● **Removal of steric hindrance**

- In the relaxed state there is no physical contact between myosin and actin because of the physical hindrance provided by the tropomyosin-troponin complex (see Fig. 4.1). However, when intracellular [Ca^{++}] exceeds 10^{-6} M, sufficient Ca^{++} ions are available to interact with troponin-C. As a result the conformation of the troponin-tropomyosin complex changes, and myosin heads are able to bind actin, forming a tightly locked myosin-actin cross bridge (Fig. 4.4*A*). In a working muscle this rigor state is short-lived and is terminated when a molecule of ATP is bound.

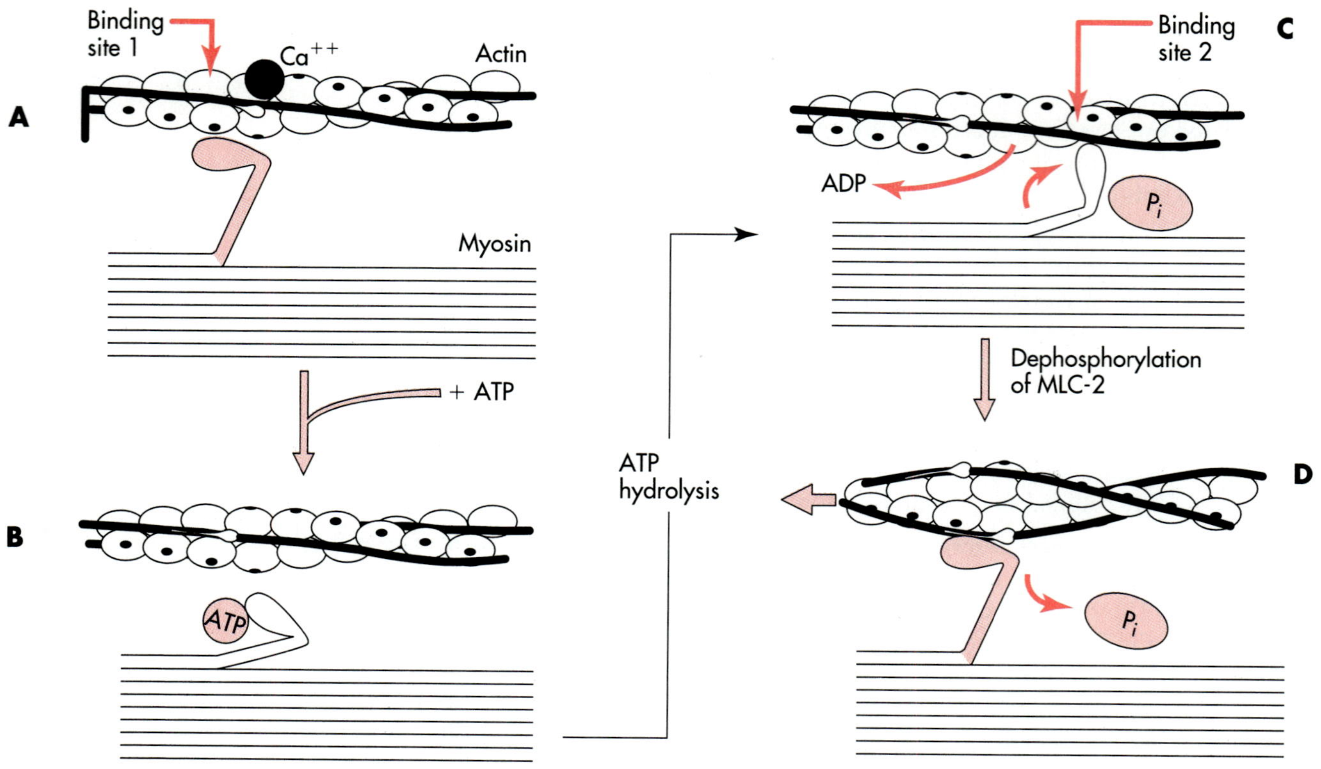

Fig. 4.4 The sequence of changes in the myosin-actin configuration during one cycle of actomyosin cross-bridge formation and detachment. *MLC-2,* Myosin light chain–2.

The power stroke

• Muscle contraction depends on changes in protein conformation and results from cyclic myosin-actin interaction during which the S1-S2 portion of the myosin molecule (see Fig. 4.1) tilts relative to the tail of the molecule. The sequence is described in Fig. 4.4 *B* to *D*.

- Figure 4.4 *B* shows that when a molecule of ATP binds to the myosin head then there will be a slight change in the region that constitutes the actin binding site. The change reduces binding affinity and loosens the myosin-actin bond so much that the myosin head may detach from actin. Once the myosin-actin bond has been loosened the myosin head is able to enfold the ATP molecule and ATP hydrolysis occurs.
- Figure 4.4 *C* shows that following ATP hydrolysis there is a large conformational change, allowing the myosin head to move along the actin molecule for a large distance (near 5 nm). Inorganic phosphate (P_i), a product of ATP hydrolysis, phosphorylates myosin light chain-2 (MLC-2) in the myosin head. In the absence of ATP the myosin head again binds tightly to actin, but at a new site.
- The proximity of actin to phosphorylated MLC-2 causes dephosphorylation of MLC-2 (Fig. 4.4 *D*) and this dephosphorylation

triggers the power stroke. The power stroke is a change in myosin shape that causes movement of the actin filament relative to myosin. It also returns the myosin head to its original conformation and leaves it tightly bound to actin in a rigor configuration. The cycle then repeats and stops before step *C* if there is not sufficient Ca^{++} present to remove steric hindrance.

- One cross-bridge rotation causes a displacement of only 5 to 10 nm. A sarcomere shortens by 100 to 300 nm during contraction. To achieve this degree of shortening, repeated release and reattachment of cross-bridges is necessary.
- The sequence repeats as long as sufficient Ca^{++} is present to remove the steric hindrance provided by troponin and tropomyosin and as long as sufficient ATP is present not only to supply the energy for the power stroke, but also to permit disengagement of the actomyosin cross-bridge.
- Lack of ATP leads to rigor because myosin cross-bridges require ATP to loosen their bond to actin.

Smooth Muscle The processes of cross-bridge formation, ATP hydrolysis, sliding filaments, and cross-bridge detachment are believed to be identical in smooth and striated muscle. However, there are many more biochemical steps involved in smooth muscle between Ca^{++} entry and eventual energy production by activation of myosin ATPase.

- In striated muscle the function of Ca^{++} is to remove the inhibitory effect that is exerted by troponin on actomyosin interactions and on ATPase activity. The action is initiated by the binding of Ca^{++} to troponin-C.
- Smooth muscle has no troponin. In its place the closely related, highly selective, cytosolic Ca^{++}-binding protein, *calmodulin,* is used. The steps that lead from elevated $[Ca^{++}]_i$ to cross-bridge formation in smooth muscle are summarized in Fig. 4.5:
    - The calmodulin molecule (*Cm*) (Fig. 4.5) has 4 high-affinity Ca^{++}-binding sites. When at least 3 of them are occupied, the molecule undergoes a conformational change allowing it to bind to myosin light-chain kinase (MLCK).
    - MLCK is activated when bound to activated calmodulin and is then able to catalyze ATP hydrolysis.
    - ATP hydrolysis leads to cross-bridge formation and conformational changes in the myosin head (see Fig. 4.4). The power stroke follows.
- Relaxation occurs either when Ca^{++} is removed from the cytosol or when myosin light-chain is dephosphorylated by the action of a myosin phosphatase.

Organization of Muscle and Muscle Types

Skeletal Muscle

Motor units The fibers in a whole skeletal muscle are arranged in functional groupings called *motor units.*

- A motor unit consists of one efferent spinal nerve (an α motor neuron) and all the muscle fibers innervated by its branches.
- Motor units differ in size (many fibers or few fibers) and in muscle fiber type (fast, slow, or intermediate). Size is anatomically determined. Fiber

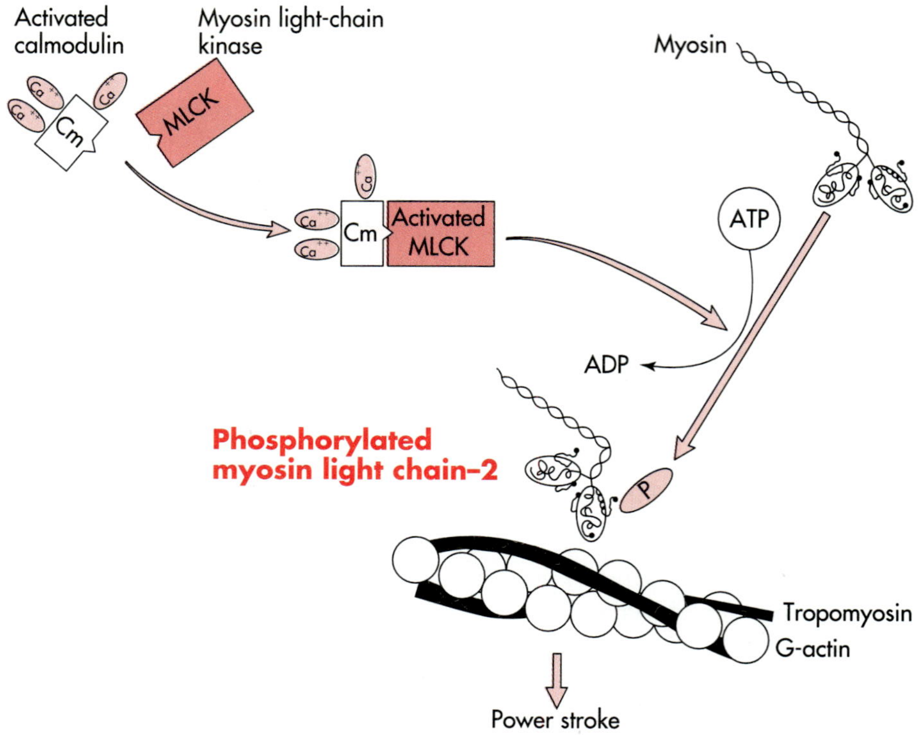

Fig. 4.5 Molecular events during smooth muscle activation. Calmodulin is activated by Ca^{++}, activates myosin light-chain kinase, and promotes phosphorylation of myosin light chain–2. *Cm,* Calmodulin; *MLCK,* myosin light-chain kinase; *P,* phosphorus.

type is driven by motor-unit function and expresses itself in morphologic, molecular, histochemical, and functional differences. The molecular differences include differences in isoforms of the contractile and regulatory proteins.

- Conversion of fibers to the other type does not occur under physiological conditions but is possible in experimental settings.

- **Types of skeletal muscle fibers** There are 3 types of muscle fibers, and all fibers in a motor unit belong to one of the 3 types.

— ***Fast fibers*** Two functional subtypes exist: *fast glycolytic* and *fast glycolytic-oxidative.*

Fast glycolytic

- The glycolytic pathway of ATP production (Fig. 4.6) dominates. It produces relatively little ATP but produces it quickly.
- There are few mitochondria.
- The fibers are:
  - large and have an extensive sarcoplasmic reticulum
  - white in appearance because they contain little myoglobin
  - capable of rapidly generating large forces but not of maintaining them

Fast glycolytic-oxidative. Such fibers are also called *intermediate fibers.* They are sparse in human muscles. They:

- are of small diameter

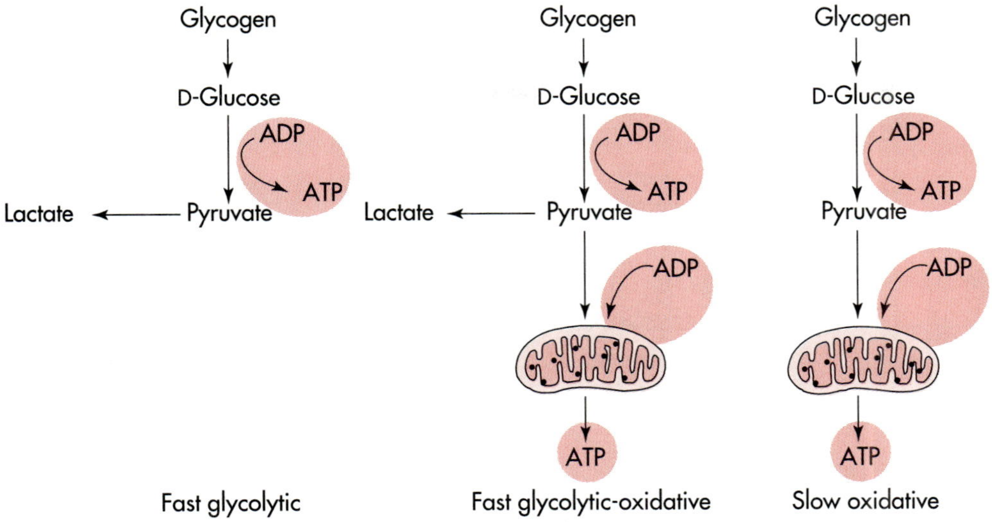

Fig. 4.6 The metabolic pathways of ATP production in skeletal muscle. Dominance of the glycolytic, oxidative, or mixed glycolytic-oxidative pathways in muscle fibers characterizes them as fast, slow, or intermediate fibers respectively. Intermediate fibers are rare in humans.

- contain a large number of mitochondria and have both oxidative and glycolytic activity (see Fig. 4.6)
- are red appearance because their myoglobin content is high

— ***Slow oxidative fibers***

- These are characterized by high oxidative activity (see Fig. 4.6), high capillary density, and high myoglobin content. The latter two features make them red in appearance.
- Their sarcoplasmic reticulum is less prominent than it is in fast glycolytic fibers.
- Every step of the excitation-activation-contraction process is slower in this muscle type than it is in the fast glycolytic type.

● **Types of whole skeletal muscle** Although all the fibers in a given motor unit are of the same type, any given region of muscle will show considerable anatomical intermixing of fibers from different motor units. As a result:

- Most human muscles contain both fast glycolytic and slow oxidative fibers but few intermediate fibers. The proportion of fibers is determined by the nature of the long-term muscle activity.
- Activity that is chronically in a phasic manner and at high frequencies (more than 40 per second) will lead to formation of fast fibers.
- Activity that is chronically tonic and at low frequencies favors formation of slow fibers.

— ***Type I, or red muscle*** Contains mostly slow oxidative fibers. Muscles controlling posture are red muscle.

— ***Type II, or white, muscle*** Contains mostly fast glycolytic fibers. Muscles controlling eye movement are white muscle.

Smooth Muscle

● **Types of smooth muscle**

— ***Unitary*** Cells are arranged in sheets with multiple gap junctions

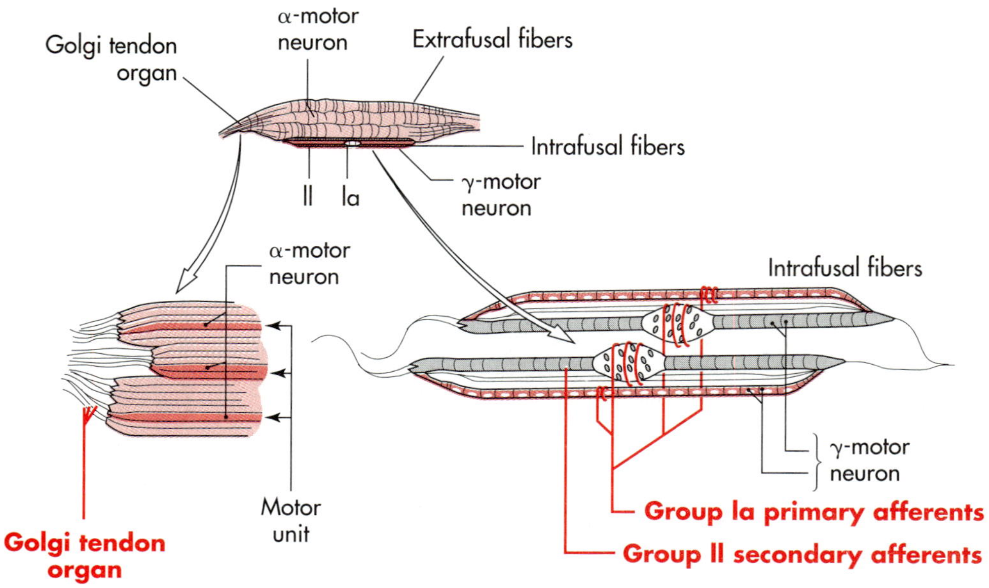

Fig. 4.7 Sensory structures and motor neurons in skeletal muscle. Extrafusal fibers are the working units of skeletal muscle, whereas intrafusal fibers are the sensory structures of *muscle spindles.* They monitor tension and length. Contraction of motor units is initiated by action potentials in α motor neurons. Contraction of intrafusal fibers is directed by γ motor neurons as part of central nervous programs that regulate motor control. Sensory information is conveyed to the central nervous system through afferents from Golgi tendon organs or intrafusal fibers.

for cell-to-cell ion transport. Spontaneous slow-wave fluctuations of membrane potential (see Fig. 4.3) are a feature of unitary smooth muscle.

— ***Multiunit*** Cells are arranged as discrete muscle fibers, with each fiber being innervated by an individual nerve. Electrical conductance between neighboring fibers is poor.

Regulation of Muscle Contraction

Skeletal Muscle

Neural regulation

— ***Effector structures*** Skeletal muscle contraction is initiated by action potentials in α motor neurons. When the motor nerve stimulus is adequate to elicit a muscle action potential, all the fibers in that motor unit contract synchronously. An extensive neuronal network ensures (1) integration of activity with neighboring motor units and (2) grading of activity relative to the desired force.

— ***Sensory structures*** Afferent information regarding muscle tension and length is sensed by receptors located in tendons *(Golgi tendon organs)* and by *muscle spindles,* which are special sense organs widely interspersed among the working fibers (Fig. 4.7).

Golgi tendon organ. Golgi tendon organs are the extensively branched termination of a large group Ib nerve fiber. They are woven into the tendons of a bundle of muscle fibers. Their firing patterns mirror muscle tension, averaged over several motor units.

Muscle spindle. Muscle spindles consist of specialized fibers, named *intrafusal fibers.* They are smaller than the working (extrafusal) fibers, which make up the bulk of the muscle.

- They are innervated by motor nerves (γ motor neurons) as well as sensory nerves.
- Spindles are transducers of muscle-shortening velocity and of muscle length.

- **Grading of contractile force** Force of contraction in a muscle is changed by two mechanisms: (1) one that changes the number of active motor units *(recruitment);* (2) one that changes the frequency of action potentials in motor nerves *(temporal summation).* In addition, in any one active fiber, maximal contractile force is affected by initial stretch (preload, or length-tension relationship) and by the biochemical environment (contractility, or degree of activation).

— ***Recruitment*** Small motor units are activated first because their neurons are small and therefore reach the critical number of total ion flux before it is reached in larger neurons. If the generated force is not sufficient for the task, activity is recruited in additional motor units.

— ***Temporal summation*** Skeletal muscle has a short cycle of electrical activity (10 to 80 ms) compared to its cycle of mechanical activity (160 to 250 ms). As a result:

- It is possible to stimulate a motor unit before the force generated by the preceding stimulus has returned to zero.
- If the rate of stimulation is so high that there is no time for any relaxation between neighboring action potentials, the muscle is said to be in a state of tetanization and is then developing the maximum force possible under the existing biochemical conditions.

Smooth Muscle

- **Modulation of smooth muscle contraction** At the tissue level, smooth muscle tension is influenced by neural and chemical factors as well as by physical stretch. At the cellular level these influences are translated into changes of cytosolic Ca^{++} concentration $[Ca^{++}]_i$ or sensitivity to a given $[Ca^{++}]_i$.

— ***Neural influences***

- Nerves modulate activity in unitary smooth muscle but initiate activity in multiunit smooth muscle. They exert their influence through neurotransmitters.
- The dominant smooth muscle neurotransmitters are acetylcholine and norepinephrine. When they activate their respective postsynaptic receptors:
  - they change membrane permeability to Ca^{++}
  - they also change membrane potential
  - they may cause muscle contraction without necessarily generating a smooth muscle action potential. (Whether a muscle action potential is generated depends only on whether the number of Ca^{++} ions needed for contraction is large enough to have moved the muscle cell membrane to its threshold for an action potential.)

• The final mechanical effect of acetylcholine or norepinephrine varies among different smooth muscles:

- In some, acetylcholine is excitatory, whereas norepinephrine is inhibitory.
- In others, their respective effects are each reversed.

— ***Chemical influences*** Local tissue factors such as Po_2, Pco_2, and $[H^+]$ as well as paracrine and endocrine agents influence smooth muscle activity.

— ***Physical stretch*** Stretching of unitary smooth muscle cells may activate mechanosensitive ion channels. It makes the average membrane potential less negative and thereby increases excitability. This causes:

- an increased number of spontaneous action potentials and leads to
- increased spontaneous muscle contraction

This automatic constrictor response of some smooth muscle to increased stretch is called the *myogenic reflex.*

— ***Cytosolic calcium concentration*** $[Ca^{++}]_i$ $[Ca^{++}]_i$ depends on membrane Ca^{++} currents and Ca^{++} release from intracellular stores, chiefly sarcoplasmic reticulum.

Transmembrane calcium current. Ca^{++} flux through membrane channels is modulated by agents that alter membrane potential (such as K^+) and by agents that affect channel conductance. Such agents include cAMP (which stimulates the L type channel) and cGMP (which inhibits the L type channel).

Ca^{++} flux across the sarcoplasmic reticulum (SR). SR release and uptake of calcium are influenced by many agents. The most important ones are:

- IP_3 (which promotes Ca^{++} release)
- cGMP (which inhibits Ca^{++} release)
- cAMP (which promotes both Ca^{++} entry through L type channels and Ca^{++} uptake by the SR)

— ***Calcium sensitivity*** Two mechanisms strongly influence the magnitude of developed tension at a given $[Ca^{++}]_i$:

- The degree of phosphorylation of myosin light-chain kinase (MLCK)
- The activity of myosin light-chain phosphatases (of which there are several)

Assessment of Striated Muscle Function

The purpose of muscle is to develop force and to change length over a period of time. Accordingly, techniques for the assessment of muscle function incorporate measurements of muscle tension and of shortening velocity (the first derivative of length with respect to time). Although it is easy to determine maximum muscle strength, it is difficult to assess the relationships among biochemical changes and their mechanical correlates. Thus special techniques are needed to determine indices of function such as magnitude and rate of force development or velocity of shortening.

In experiments on isolated muscle it is possible to separate the dynamics of

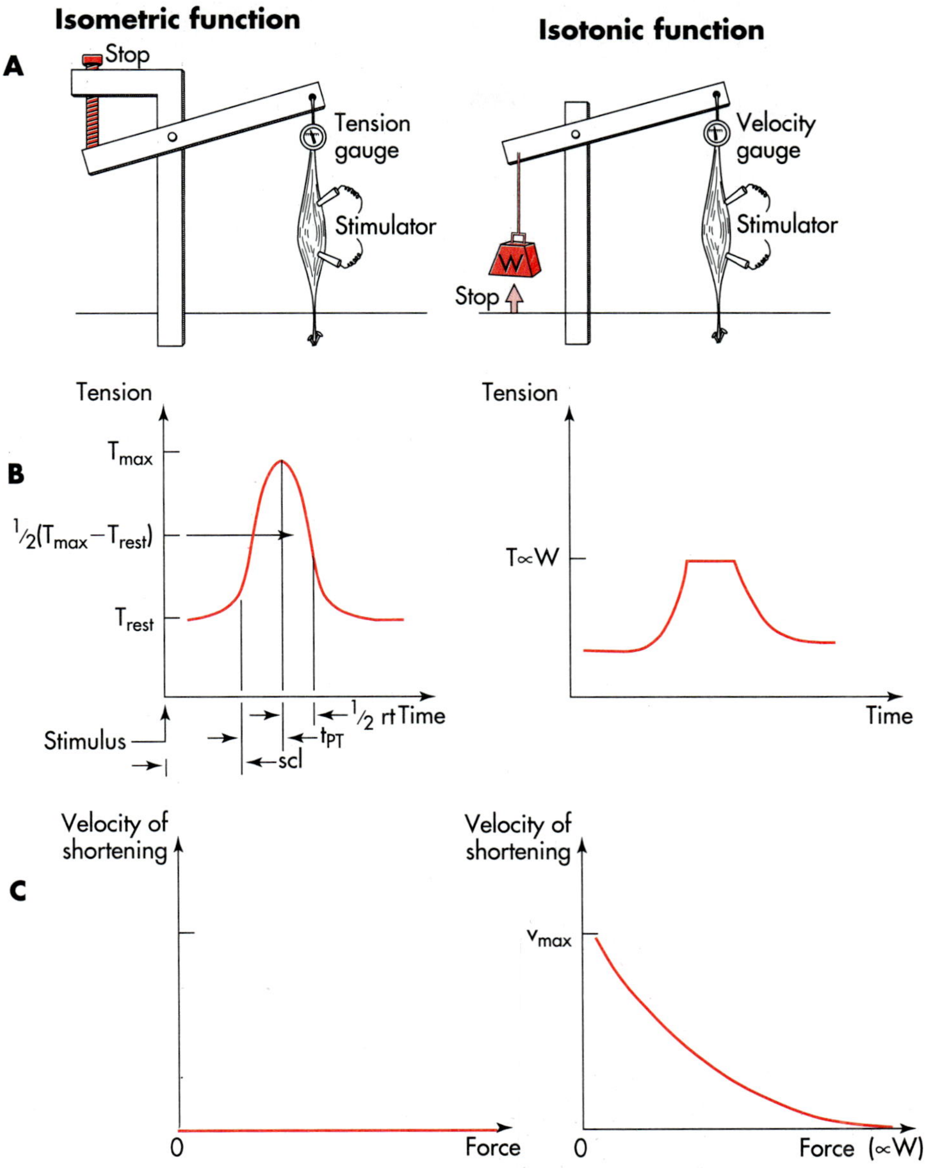

Fig. 4.8 Setups for the assessment of skeletal muscle. **A,** In an isometric test an adjustable stop permits initial stretch (preload) to be set. The stop also prevents shortening of the muscle when it develops tension in response to a stimulus. In an isotonic test an adjustable stop sets initial stretch and the muscle moves the weight, *W,* once it has developed sufficient tension. An isometric test gives information about the capability to develop force. An isotonic test gives information about the capability to shorten. **B,** Typical tension versus time graphs observed after application of a sufficient stimulus. ½ rt, One half relaxation time; scl = stimulus-contraction latency; T_{max}, maximum tension; t_{PT}, time to peak tension; T_{rest}, resting tension. **C,** Typical force-velocity graphs observed after application of a sufficient stimulus. An isometric test gives no information about the shortening capabilities of a muscle.

tension development from the dynamics of length changes. The essence of two such setups is shown in Fig. 4.8. The illustration also shows plots of tension versus time and velocity of shortening versus force, as they might be obtained in such experiments.

Isometric Function When muscle is fixed rigidly at both ends so that it cannot shorten, stimulation will be followed by the maximum tension development that is possible under the existing preload, contractility, and rate of stimulation. The pattern of change in tension with time after a single stimulus is called a *muscle twitch.* Its total duration is about 10 ms in a fast fiber, 100 ms in a slow fiber.

- **Isotonic Function** When muscle is suspended in such a way that during contraction it will move a weight (*W*) (see Fig. 4.8), it is possible to study shortening capabilities while contracting at a constant tension. In such a setting,

  - a plot of tension versus time gives limited information because the maximum tension that is developed will be determined by the weight that is being lifted.
  - Plots of force versus velocity of shortening give more information.
  - The performance of muscle can be characterized by v_{max}, the maximum velocity of shortening at zero load.

- **Whole Muscle Function**

  - **Heart** Techniques exist for assessing mechanical performance of the heart and for correlating biochemical changes with their mechanical correlates. These are outlined in Chapter 6.
  - **Skeletal muscle** Qualitative information about muscle function can be obtained when electrical activity is recorded with a pair of electrodes placed on the surface of the muscle. The record obtained consists of motor unit action potential trains and is called an *electromyogram* (EMG).

    - Changes in the amplitude of EMG deflections are directly related to changes in firing frequency of active motor units in the muscle.
    - Changes in the interval between neighboring pulses in the EMG are inversely related to changes in the number of active motor units in the muscle.

Effects of Exercise and Training on Skeletal Muscle

Most athletic endeavors involve three aspects: coordination, strength, and endurance. Training improves all three, or it can accentuate only certain aspects.

- **Coordination** With training, exercise-specific coordination improves in the central nervous system and leads to:

  - better coordination of action potentials in all involved muscles
  - greater inhibition of antagonist muscles

  The outcome is reduction of "wasteful" muscle activity and improved efficiency of exercise performance.

- **Strength** Pure strength training consists in brief, maximal efforts involving all motor units in the affected muscle.

  High tissue pressure associated with the intense effort tends to compress blood vessels in the muscle and thereby reduce oxygen supply. As a result:

  - Strength training favors fast, glycolytic motor units.
  - The prominent adaptive response of muscle to such training is hypertrophy. This is especially noticeable with isometric exercises.

- **Endurance** Endurance training consists in sustained, submaximal efforts. Such activity leads to some anatomical changes such as mild hypertrophy, increased capillarization, and moderate thickening of connective tissue sheaths. The major effect of endurance training on muscle is to increase the muscle's capacity for metabolic activity. This is seen as:

- an increase in the number of mitochondria as well as
- increased stores of oxidative enzymes.

Types of Skeletal Muscle Fibers Despite frequent use of motor units, conversion of slow units to fast units does not occur. The reason may lie in training-induced electrophysiologic changes:

- Average duration of action potentials increases gradually with training, but
- action potentials show a progressive diminution in their average frequency.

In laboratory experiments where conversion was accomplished, it occurred after maintained elevation of nerve action-potential frequency.

Multiple Choice Review Questions

1. Under the light microscope smooth muscle does not show the striated pattern that is characteristic of skeletal muscle. This difference is attributed to:
   a. Mechanisms of smooth muscle activation that are independent of Ca^{++}
   b. Absence of T-tubules in smooth muscle
   c. A less regular organization of thin and thick filaments in smooth muscle
   d. Fewer mitochondria in smooth muscle than in skeletal muscle

2. Energy released from the hydrolysis of ATP is utilized in one of the following aspects of muscle function. Which one?
   a. Release of Ca^{++} from the sarcoplasmic reticulum
   b. Generation of an end-plate potential in response to acetylcholine
   c. Detachment of actomyosin cross-bridges
   d. Rotation of myosin cross-bridges

3. Which one of the following statements referring to the ionic basis of tetanic muscle contraction is correct?

   Tetanic contractions occur in skeletal muscle during rapidly applied action potentials because:
   a. Action potentials in skeletal muscle are of such a brief duration that muscle cells are out of their electrically refractory periods before the twitch tension associated with an action potential has returned to zero. Action potentials fuse, allowing a longer time for calcium ions to be released from the sarcoplasmic reticulum. This increases the number of actomyosin cross-bridges.
   b. The shorter duration of action potentials permits a shorter time for calcium ions to be released from intravascular binding proteins.
   c. The shorter duration of action potentials permits a shorter time for potassium ions to be released from extracellular storage sites. This increases the number of actomyosin cross-bridges.
   d. Action potentials in skeletal muscle are of such a brief duration that muscle cells are out of their electrically refractory periods before the twitch tension associated with an action potential has returned to zero.

4. Activation of smooth muscle by Ca^{++} can best be ascribed to:
   a. Shortening of thick filaments in the presence of Ca^{++}
   b. Release of contractile proteins from sarcoplasmic reticulum
   c. Activation of calmodulin with subsequent activation of myosin light-chain kinase
   d. Binding of Ca^{++} to troponin-C and subsequent activation of myosin light-chain kinase

5. In a laboratory experiment functioning units of cardiac muscle, skeletal muscle, and smooth muscle are placed in the same bath. Each is provided with stimulating electrodes and with instruments to measure muscle tension and membrane potential. During the experiment each muscle is stimulated electrically. Then the Ca^{++} concentration in the bath is decreased and each muscle is stimulated again. These steps are repeated until the Ca^{++} concentration in the bath is near zero.

   Which of the following statements about muscle function at progressively lower extracellular Ca^{++} concentrations is correct?
   a. Of the three muscles, smooth muscle will be the first to stop responding to the electrical stimulus with a contraction and the last to stop displaying action potential activity in response to the stimulus.
   b. The plateau in the cardiac muscle action potential that follows each stimulus will become of shorter duration as extracellular Ca^{++} concentration decreases, and the cardiac muscle could be tetanized if the stimulation frequency were high enough.
   c. Skeletal muscle would stop responding to a stimulus with a contraction before smooth muscle stops responding.
   d. Cardiac muscle would be the last of the three muscle types to stop responding with a contraction because it has an active Ca^{++} pump in the cell membranes.

6. Which of the following statements is correct?
   a. Resting muscle cells produce most of their energy from the breakdown of glucose with a small contribution from the breakdown of fatty acids.
   b. The aerobic and anaerobic pathways of ATP production require equal amounts of fuel to produce the same amount of ATP.

c. During the first 30 minutes of very active work skeletal muscles rely mainly on fatty acids for energy substrate.
d. The strength of a muscle contraction is directly proportional to the total number of motor units that is activated simultaneously.

7. One of the genetic myopathies is characterized by a defect in glycogenolysis. Which one of the following symptoms would be expected to characterize this disorder?
   a. Presence of relatively short sarcomeres
   b. Absence of voltage-gated Ca^{++} channels
   c. Ability to perform low-level muscle activity but inability to sustain high levels of muscle activity
   d. Failure to propagate action potentials along motor neurons

8. Ca^{++}-channel blockers are sometimes used to control coronary artery spasticity in patients. Which one of the following statements best describes the basis for such a use of these blockers?
   a. Such drugs inhibit Ca^{++} uptake into the sarcoplasmic reticulum.
   b. Such drugs increase cardiac performance by reducing the duration of the cardiac action-potential plateau.
   c. Such drugs increase the afterload of the left ventricle by constricting smooth muscle in peripheral blood vessels.
   d. Such drugs inhibit excitation of vascular smooth muscle by inhibiting Ca^{++} entry from extracellular fluid into smooth muscle cells.

9. Each statement below concerning the contraction of myofibrils in skeletal muscle is true ***except*** (Fig. 4.9) that during a contraction:
   a. The size of the A band decreases.
   b. The size of the H band decreases.
   c. The size of the I band decreases.
   d. Z lines are drawn closer to the A band.

10. Each of the following statements referring to skeletal muscle is correct ***except:***
   a. A motor nerve fiber that exits from the spinal cord normally innervates several muscle fibers.
   b. All the muscle fibers innervated by one spinal efferent nerve fiber and its branches are collectively described as a motor unit.
   c. Small muscles that react rapidly and require fine control have a large number of muscle fibers per spinal efferent.
   d. Release of Ca^{++} from intracellular stores is not required for a contraction.

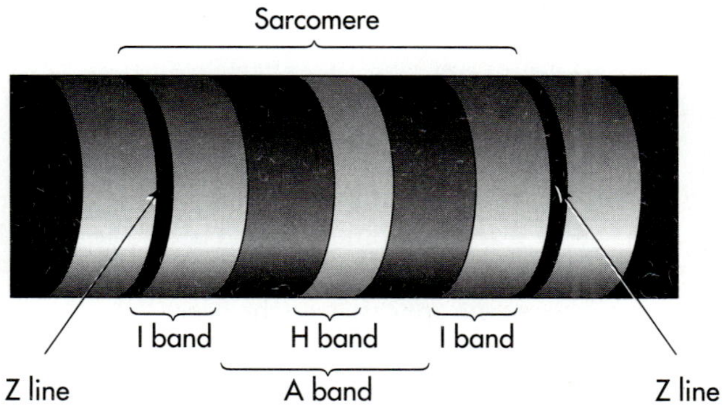

Fig. 4.9 Histologic appearance of a single myofibril.

Chapter 5

Respiratory Physiology

Overview

• Within the lungs a large number of terminal sacs, the *alveoli* and the epithelium of their enveloping capillary network provide an exchange interface between air and blood.

• The exchange is governed by concentration gradients.

• Gas concentration is expressed as *partial pressure.*

• There are 2 ways of transporting gases in blood. They are carried (1) in physical solution and (2) in chemical combination with specific carrier agents. (The amount of O_2 and CO_2 that can be physically dissolved in blood represents only 5% of the total amounts carried.)

• The partial pressure of a gas in the capillary blood surrounding an alveolus is set by a balance between the rate of its removal by blood flow and the rate of its replenishment by ventilation. As a result, gas exchange within any region of the lung depends on the prevailing local balance between ventilation and blood flow.

- **Inspiration** Mechanisms that decrease alveolar pressure below ambient atmospheric pressure cause dry air to be drawn into the upper respiratory passages where it is humidified and then drawn into the alveoli.
- **Expiration** The pressure gradient is reversed, and water-saturated air leaves the lung.
- **Control of Respiration** Rate and depth of respiration are adjusted in response to:
  - chemical stimuli such as concentration of O_2, CO_2, or $[H^+]$, or to
  - emotional stimuli such as fright or mirth.
- **Ventilation/Perfusion Ratio** As a result of gravitational effects, blood flow is unevenly distributed in the lungs and varies greatly within the respiratory cycle and with posture. In the normal upright lung the ventilation/perfusion ratio is higher at the apex than at the base.

Transport of Oxygen and Carbon Dioxide

Interactions Between Gases and Blood

Partial pressures of respiratory gases

• In a mixture of gases the partial pressure of any one of them is calculated as follows:

$$\text{Partial pressure of } x = \text{Total pressure (mm Hg)} \times \frac{\text{Percent contribution of } x}{100}$$

For example, the fraction of O_2 in dry air is 21%. Therefore, in dry air, at atmospheric pressure (760 mm Hg at sea level):

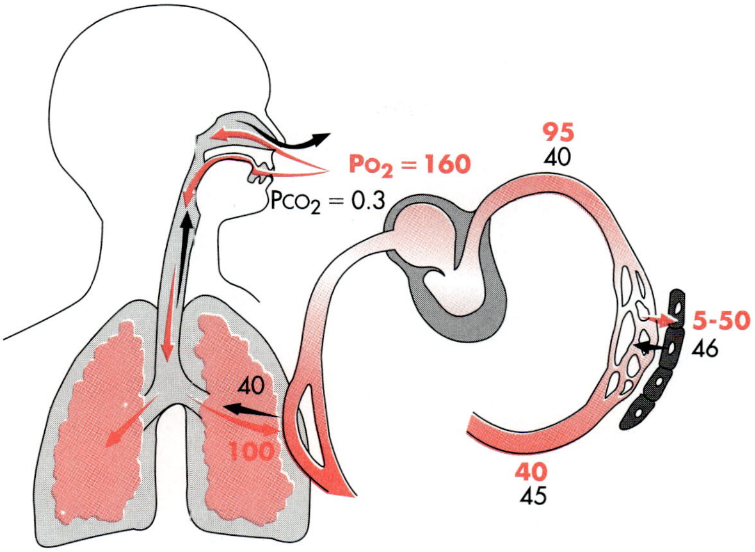

Fig. 5.1 Partial pressures (in mm Hg) of O_2 and CO_2 in the atmosphere, the alveoli, and at various sites in the vascular system. Values for Po_2 are shown in color.

$$Po_2 = 760 \times 0.21 = 160 \text{ mm Hg}$$

• By convention the composition of gas mixtures is usually quoted in terms of dry gas. Because air in the lungs is saturated with water vapor, water vapor pressure is subtracted from the total pressure in all calculations. Water vapor pressure varies with temperature and is equal to 47 mm Hg at normal body temperature.

For example, dry alveolar air is 14% O_2. As a result, at atmospheric pressure

$$\text{Alveolar } Po_2 = (760 - 47) \times 0.14 = 100 \text{ mm Hg}$$

Fig. 5.1 shows normal values for partial pressures of O_2 and CO_2 at different sites in the respiratory and cardiovascular systems.

- **Carbon dioxide exchange** Ninety percent of all CO_2 carried in blood is transported as HCO_3^- or H_2CO_3. About 5% is transported in chemical association with hemoglobin, and another 5% is carried as dissolved gas. Fig. 5.2 shows CO_2 exchange between tissue cells and blood, whereas Fig. 5.3 shows CO_2 exchange between blood and alveoli.

  — ***In the tissues*** CO_2 is produced by cellular metabolism and enters tissue capillaries down its partial pressure gradient. Carbonic anhydrase, found in red blood cells but not in plasma, catalyzes the conversion of CO_2 to HCO_3^-. CO_2 also reacts with the amine groups on hemoglobin.

  — ***In the lungs*** At the interface between capillary and alveolus the partial-pressure gradients cause CO_2 to diffuse out of the capillaries. This drives all binding reactions toward free CO_2.

- **Oxygen exchange** The bulk of oxygen transport occurs by reversible chemical binding with hemoglobin.

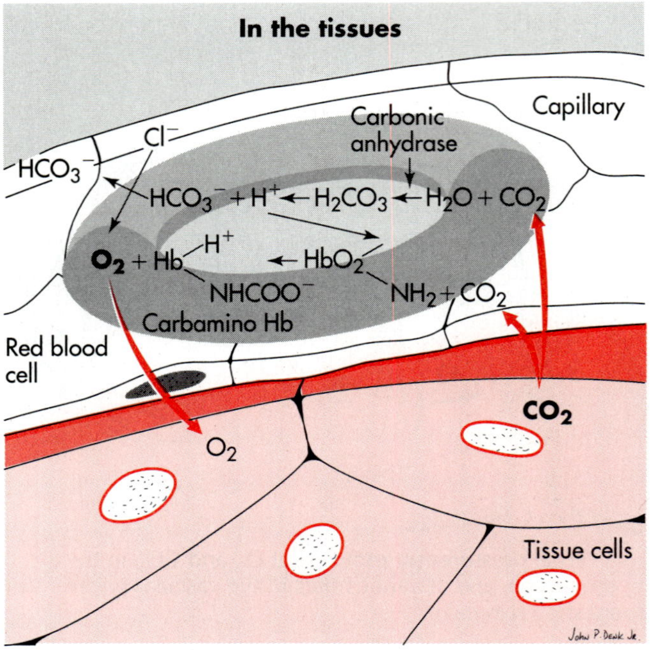

Fig. 5.2 Gas exchange between tissue cells and a red blood cell. O_2, carried reversibly by oxyhemoglobin, moves down its partial-pressure gradient, enters tissue cells, and leaves nonoxygenated hemoglobin behind. Nonoxygenated hemoglobin is then available to accept H^+ and CO_2. The combination of nonoxygenated hemoglobin, H^+, and CO_2 is called *carbaminohemoglobin.* CO_2 enters tissue capillaries down its partial pressure gradient. Carbonic anhydrase (found in red blood cells but not in plasma) catalyzes the formation of H_2CO_3. A fraction of H_2CO_3 dissociates to HCO_3^- and H^+. HCO_3^- diffuses out of the red blood cell and is replaced by Cl^- so that charge equality is maintained. This is called the *chloride shift.* The H^+ ions are buffered by nonoxygenated hemoglobin. CO_2 also reacts with the amine groups on hemoglobin.

— ***In the tissues*** Oxygen leaves oxyhemoglobin, moves down its partial pressure gradient, and enters tissue cells. It leaves nonoxygenated hemoglobin behind.

— ***In the lungs*** Oxygen moves down its partial pressure gradient, enters capillaries, and combines rapidly with hemoglobin in red blood cells to form oxyhemoglobin.

- **The hemoglobin dissociation curve** Depending on its ambient Po_2, each hemoglobin molecule can bind up to 4 molecules of O_2 in an easily reversible manner. In the lungs, at a Po_2 of 100 mm Hg, the binding is at 97% of its potential maximum, and hemoglobin is said to be 97% saturated. In the tissues, on the other hand, at a Po_2 near 20 mm Hg, hemoglobin saturation is only about 20%.

  The relationship between hemoglobin saturation and ambient Po_2 is expressed by the hemoglobin dissociation curve (Fig. 5.4). The figure also shows that hemoglobin saturation is subject to influences additional to those of Po_2:

  - An increase in temperature, Pco_2, or $[H^+]$, will shift the curve to the right because such changes reduce the affinity of hemoglobin for O_2.

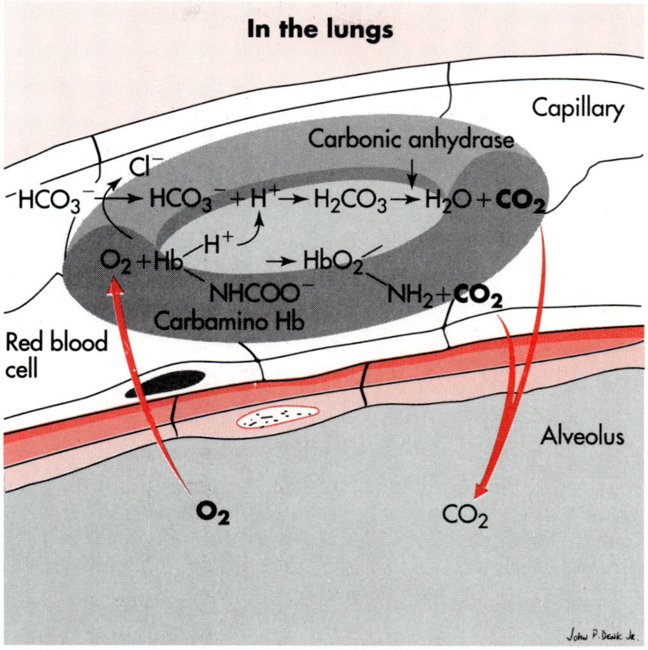

Fig. 5.3 Gas exchange between a red blood cell and an alveolus. O_2 moves down its partial-pressure gradient, enters capillaries, and combines rapidly with hemoglobin (Hb) in red blood cells to form oxyhemoglobin (HbO_2). In comparison with carbaminohemoglobin, oxyhemoglobin is a stronger acid and a weaker CO_2-binding agent. As a result, free H^+ and CO_2 are released from oxyhemoglobin. CO_2 diffuses down its partial-pressure gradient into the alveolus to be blown off in expired air. H^+, released from oxyhemoglobin, combines with HCO_3^- and quickly yields CO_2 because of the presence of carbonic anhydrase.

- Decreases in temperature, P_{CO_2}, or $[H^+]$, will shift the dissociation curve to the left because such changes increase hemoglobin's affinity for O_2, thereby inhibiting O_2 release.

Overall Gas Exchange and the Ventilation/Perfusion Ratio The crucial factor determining the partial pressure of any respiratory gas in pulmonary venous blood is the ventilation/perfusion ratio of the lungs. Lung units that are overperfused in relation to their ventilation will show high P_{CO_2} and low P_{O_2} in their capillary blood and contribute a disproportionately large amount of blood to the pulmonary outflow. Such units will cause a systemic elevation of P_{CO_2} and depression of P_{O_2}, just as if gas diffusion across the entire capillary/alveolar interface of the lungs were impaired.

Mechanics of Breathing

Throughout the respiratory cycle, the lung remains closely apposed to the inner wall of the thorax by a thin layer of fluid that separates the visceral pleura (on the lung surface) from the parietal pleura. This "attachment" permits mechanical forces of the lung or chest wall to be coupled to each other and to change lung volume during each respirator cycle.

Inspiration Two factors cause expansion of the chest cavity:

- Its vertical size is increased by active contraction of the diaphragm.

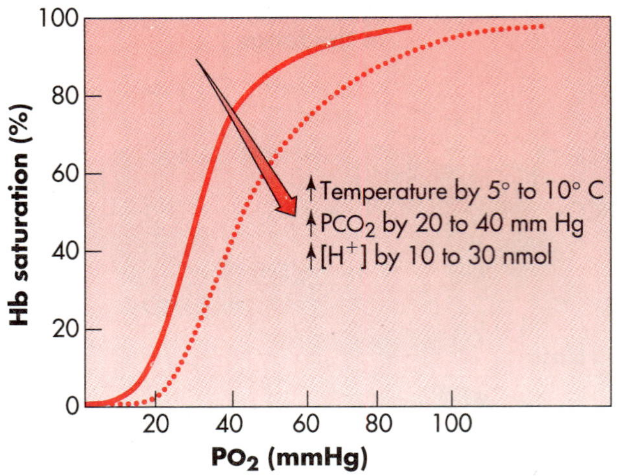

Fig. 5.4 The hemoglobin dissociation curve. The *plateau* at the upper end signifies a stable, near-maximal saturation despite wide variations in alveolar Po_2. The *steep portion* between 60 and 20 mm Hg results from the unloading of O_2 from hemoglobin as Po_2 decreases in tissues. The *dashed curve* indicates the effect of increasing temperature, Pco_2, or $[H^+]$ on the dissociation curve.

- Its cross section is increased by contraction of the external intercostal muscles, sometimes assisted by accessory muscles.

Expiration In quiet breathing expiration is driven by elastic recoil of the lungs and chest wall and is therefore a passive process. It may be assisted, if necessary, by active contraction of intercostal and abdominal muscles.

Inspiratory muscles continue to contract during part of expiration. Their gentle opposition to elastic recoil makes expiration time almost double the inspiration time.

Changes in Pressure and Volume During the Respiratory Cycle Mechanical aspects of respiratory function are usually described by curves showing lung volume over the range of respiratory pressures. Two sets of curves are used to demonstrate contributions of different mechanical properties of the component parts.

- *Static* pressure-volume curves, obtained under conditions of zero airflow at the moment of measurement, are used to demonstrate (1) the contributions of elastic properties alone and (2) the balance of forces that prevents lung volume from collapsing to zero at the end of expiration. The shape of such curves is influenced by *compliance* alone.
- *Dynamic* pressure-volume curves are used to describe mechanical relationships pertaining to normal breathing. The shape of these curves is influenced by both *compliance* and *resistance* in the components of the respiratory system.

Compliance Compliance is a measure of the readiness with which the components of the respiratory system will distend. It is defined as:

$$\text{Compliance} = \frac{\text{Change in volume}}{\text{Change in distending pressure}}$$

- Compliance is equal to the slope of a pressure-volume curve, and it changes with lung volume (Figs. 5.5 and 5.6) . Therefore, it is often calculated at a specific lung volume or normalized to lung volume. Such a measure is called *specific compliance.*

- Both *lung compliance* and *chest wall compliance* influence the pressure-volume behavior of the respiratory system.

— ***Lung compliance*** The compliance of the lungs is influenced by several factors:

- Tissue elements, with elastin and collagen being the major contributors
- Degree of tissue hydration
- Lung blood volume; distended interstitial fluid spaces or engorged blood vessels reduce lung compliance.
- Air-space configuration
- Surface forces at the air-fluid interface of the alveoli; surface forces influence lung compliance because the pressure required to maintain an air bubble at a given size within a fluid medium is directly related to the surface tension of the air-fluid interface (Laplace's law).

— ***Chest wall compliance*** The compliance of the chest wall is modulated by any factor that influences chest wall motion:

- Chest configuration
- Composition of the chest wall
- Mass of abdominal contents

- **Resistance** Three sources of resistance must be overcome during breathing:

1. Flow resistance in the airways is the most important resistive component and is most subject to increase with disease.
   - Airway resistance is calculated as:

$$\text{Airway resistance} = \frac{\text{Upstream pressure} - \text{Downstream pressure}}{\text{Air flow}}$$

   - Airway resistance increases greatly if air flow becomes turbulent and increases dramatically with small decreases in airway diameter.
2. Viscous resistance in the tissues of the chest wall contributes up to 20% of total resistance. This resistance component arises from friction between tissue elements of the chest wall.
3. Viscous resistance in the lungs contributes up to 15% of total resistance. The contribution of resistance to the shape of pressure-volume curves is most readily seen when the area under a dynamic pressure-volume curve is compared with that under a static pressure-volume curve covering the same range of volume.

- **Static pressure-volume characteristics of the lung** Static pressure-volume characteristics of the respiratory system are determined in experimental settings where pressures are maintained until air flow stops. Under this condition the lungs and the chest wall are held in their respective positions by the same pressure gradient because when there is no air flow and when the airway is open, alveolar pressure is equal to mouth pressure, which is equal to atmospheric pressure. It follows that the pressure gradient

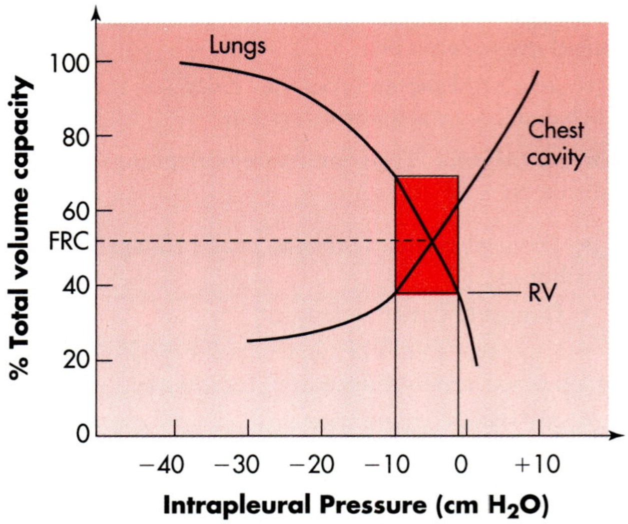

Fig. 5.5 Static pressure-volume curves of the respiratory system, obtained under conditions of zero air flow at each step of intrapleural pressure. Under those conditions lungs and chest wall are governed by the same pressure gradient. The area within the shaded rectangle indicates the pressure-volume range of normal breathing. *Both curves normally show hysteresis, less pressure being required to maintain a given volume during deflation than during inflation. Hysteresis has been omitted for the purpose of clarity.* *FRC,* Functional residual capacity; *RV,* residual volume.

expanding the lungs is the difference between intrapleural pressure and atmospheric pressure.

The pressure gradient that keeps the chest wall from expanding is also the difference between intrapleural pressure and atmospheric pressure.

Fig. 5.5 shows that the lungs have a natural tendency to collapse unless they are expanded because of a negative intrapleural pressure. It also shows that the chest wall has a natural tendency to expand unless it is restrained by a negative intrapleural pressure.

At steady state, the chest wall and lungs assume a volume that is called *functional residual capacity* (FRC) and is not zero but is at an intrapleural pressure that is near −5 cm H_2O. Fig. 5.5 also shows that:

- Expiration is never to the point where all air is removed from the lungs. It normally ends at a point where the net elastic force arising from the collapsing lung is exactly balanced by the net elastic force with which the chest cavity resists further volume reduction.
- Lung volume does not collapse to zero even when the distending pressure is zero. The reason is that small airways collapse before the alveoli do, leaving a *residual volume.*

Dynamic pressure-volume characteristics of the lung Dynamic pressure-volume curves are obtained under conditions when air flow is not zero. Under such conditions a portion of the pressure gradient is used to overcome airway resistance to flow and tissue resistance to movement. Accordingly:

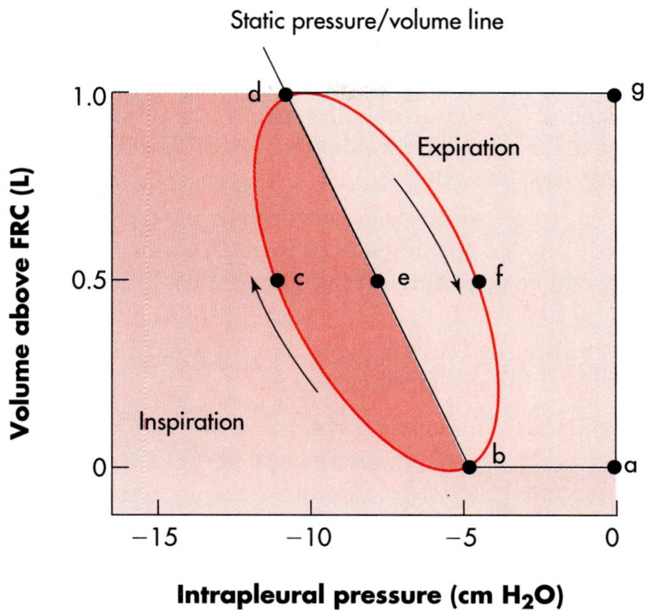

Fig. 5.6 Dynamic pressure-volume characteristics of the lungs and the work of breathing. The area *abcdga* represents the work of inspiration. It represents work against elastic forces *(abedga)* and against resistance *(bcdeb)*. The area *dfbed* represents the work of expiration, that is, work required to overcome resistance. Work done against resistance *(bcdfb)* is mostly lost as heat.

- During inspiration, pressure in excess of static pressure is required at each volume to overcome resistance (Fig. 5.6).
- During expiration, less pressure than predicted by static conditions is required at any given volume because some energy is supplied by elastic recoil (see Fig. 5.6).

The work of breathing

- The area under a dynamic pressure-volume curve represents the *work of breathing,* and several work components can be identified (see Fig. 5.6).
- Inspiratory work is work done to overcome elastic forces and work done to overcome resistance.
- Expiratory work is only the work required to overcome resistance. This work is normally accomplished by the energy stored in the stretched elastic structures and is dissipated during passive exhalation.

Surface tension, pressure, and the shape of alveoli Breathing amounts to periodic inflation and partial deflation of a large number of small bubbles. If alveoli simply obeyed the laws governing surface forces in air bubbles surrounded by water, it should not be possible to keep them inflated, and air breathing could not occur in humans.

- In an air bubble lined with water, surface tension creates an inwardly directed force that tends to collapse the bubble. The size of the bubble depends on the opposing hydrostatic pressure of the air within the bubble. At equilibrium:

$$\text{Distending pressure} \propto \frac{\text{Surface tension}}{\text{Bubble radius}}$$

• Since surface tension per unit of surface area is determined only by the nature of the liquid, it is the same for all bubbles in water. As a result, small air bubbles require a higher distending pressure than larger bubbles do. Analogously to the air-bubble example, the pressure inside small alveoli should be higher than that inside large alveoli, and smaller alveoli should empty into larger alveoli.

• Furthermore, since alveoli are open at one end and are connected to other alveoli at that opening air can escape from them when surface tension shrinks them, no opposing internal pressure can build up, and they should collapse.

• In the healthy lung the *surfactant system* and *anatomic interdependence* prevent alveoli from collapsing.

— ***Surfactant*** Pulmonary surfactant is mostly phospholipids, synthesized in the endoplasmic reticulum of type II cells that line the alveoli and secreted from there into the alveoli.

• The surface tension of pulmonary surfactant is only one fifth that of extracellular fluid. As a result the pressure gradients required to expand the lung are lower than they would be without surfactant.

• The pulmonary surfactant system has the unusual property of reducing surface tension more in small alveoli than in large alveoli. As a result small alveoli can be maintained by the same hydrostatic pressure as large alveoli, and there is no tendency for small alveoli to empty into large alveoli.

— ***Anatomical interdependence*** Alveoli are anatomically interdependent in two respects:

1. Alveolar outflow tracts are interconnected.
2. Pulmonary interstitial pressure influences the pressure gradient across the alveolar walls.

• Interconnected outflow tracts are a liability because they create potential conduits to adjacent alveoli. However, these conduits are not normally functional because in the presence of surfactant there are no gradients in pressure among neighboring alveoli.

• Alveoli do not collapse at the end of expiration when P_{alv} is zero because pulmonary interstitial fluid pressure is −5 cm H_2O. That is, the pressure inside alveoli is greater than that outside alveoli.

• The ultimate factor preventing alveolar collapse is that the lung as a whole cannot collapse because of close mechanical coupling between its outer lining (the visceral pleural membrane) and the inner lining of the thorax (the parietal pleural membrane).

- **Lung volumes** The volume of air held by the maximally filled lungs can be divided into 4 nonoverlapping *volumes.* Measures of lung air content that include more than one volume are called *capacities.* Table 5.1 lists their definitions and magnitudes.

  The volume of air that is moved into or out of the lungs during a breath (tidal volume) can be increased or decreased by calling on inspiratory or expiratory reserve volumes. However, not all of the tidal volume is available to ventilate the alveoli because some of it is used to fill the dead space.

- **Dead space** The dead-space volume (about 150 ml) normally consists of the anatomical volume of those respiratory structures that are not involved

Table 5.1 ***Respiratory Volumes and Capacities***

Parameter	Definition	Typical Resting Value (mL)
Volumes		
Tidal volume (V_T)	The volume of gas inspired or expired in a single respiratory cycle	500
Inspiratory reserve volume (IRV)	The maximum volume of gas that can be inhaled starting at the end of a normal inspiration	3000
Expiratory reserve volume (ERV)	The maximum volume of gas that can be exhaled starting from the end of a normal expiration	1100
Residual volume (RV)	The volume of gas that remains in the lungs after a maximum expiration	1200
Forced expiratory volume (FEV_1)	The maximum amount of gas that can be expired *in the first second* of a forced exhalation from total lung capacity to residual volume	4000
Anatomic dead space (V_D)	The volume of the conducting, nonexchanging portion of the respiratory system	150
Capacities		
Total lung capacity (TLC)	The total amount of gas in the lungs at the end of a maximum inspiration (the sum of RV, ERV, V_T, and IRV)	5800
Vital capacity (VC)	The maximum volume of gas that can be inspired after a maximum expiration (the sum of the ERV, V_T, and IRV)	4600
Inspiratory capacity (IC)	The maximum amount of gas that can be inspired starting from the FRC (the sum of V_T and IRV)	3500
Functional residual capacity (FRC)	The amount of gas in the lungs at the end of a normal expiration (the sum of the ERV and RV)	2300
Forced vital capacity (FVC)	The amount of gas that can be expelled from the lungs by expiring as forcibly as possible after a maximum inspiration	5000

in respiratory exchange processes. The physiological dead space (measured by a gas-equilibration technique) may be slightly different from the anatomical dead space.

Pulmonary Blood Flow

Effects of Gravity Mean pulmonary arterial hydrostatic pressure is only about 15 mm Hg (20 cm H_2O) above atmospheric pressure. This value is so low that the effects of gravity on local hydrostatic pressure are significant:

- Pulmonary vessels that are located in the apex, more than 10 cm above heart level, will have an intravascular hydrostatic pressure near 0 mm Hg.
- Vessels near the base of the lungs are, in the upright posture, situated below heart level and have higher pressure than those at the apex.

As a result of this vertical intravascular hydrostatic pressure gradient blood flow and blood volume in certain regions of the lungs can be affected by respiratory pressure fluctuations and by postural changes (Fig. 5.7).

- In apical regions where alveolar pressure exceeds pulmonary venous pressure, blood flow will be determined by the difference between pulmonary arterial pressure and alveolar pressure, not by the difference between arterial and venous pressures (see Fig. 5.7).

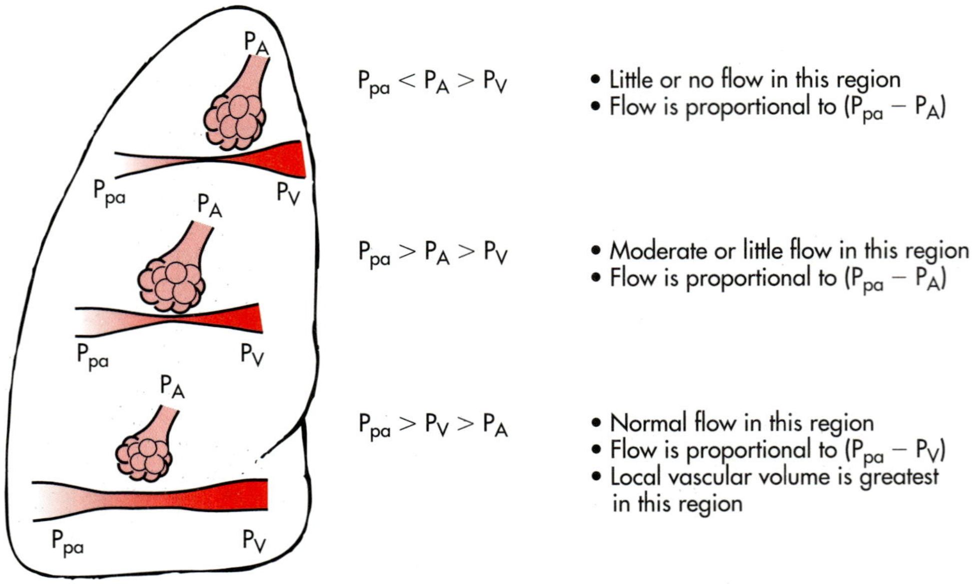

Fig. 5.7 Gravity affects intravascular pressures but not alveolar pressure. As a result, blood flow is not necessarily determined by arteriovenous pressure differences in every region of the lung. P_A, Alveolar pressure; P_{pa}, pulmonary arterial pressure; P_v, pulmonary venous pressure.

- Near the base of the lungs elevated hydrostatic pressure tends to expand blood vessels and thereby tends to increase the local blood volume.

Effects of Chemical Agents In most vascular beds hypoxia causes pronounced vasodilatation.

- However, low Po_2 in the blood perfusing pulmonary arterioles does not cause vasodilatation.
- Furthermore, pulmonary arterioles constrict when *alveolar* Po_2 is low. Hypoxic pulmonary vasoconstriction has protective significance, in that perfusion is directed toward well-oxygenated regions of the lungs.

Ventilation and Perfusion in the Lung Adequate gas exchange in the lungs requires that *ventilation* and blood flow *(perfusion)* be matched to each other. Both ventilation and perfusion decrease from the base toward the apex in the upright lung, but the change in ventilation with height is much less than that in perfusion. As a result,

- the ventilation/perfusion ratio is lower near the base of the lungs than at the apex
- the extent of capillary-alveolar gas diffusion is lower near the base of the lungs
- pulmonary venous Po_2 is lower and pulmonary venous Pco_2 is higher near the base of the lungs

These unequal vertical gradients in ventilation and perfusion are responsible for the observation that Po_2 in pulmonary venous blood is generally lower than alveolar Po_2.

Control of Respiration

The basic aim of respiratory regulation is to maintain arterial Po_2 and $[H^+]$ within their normal limits. Under resting conditions this can be accomplished by central nervous mechanisms that set basal respiratory rate and depth. Basal ventilation is

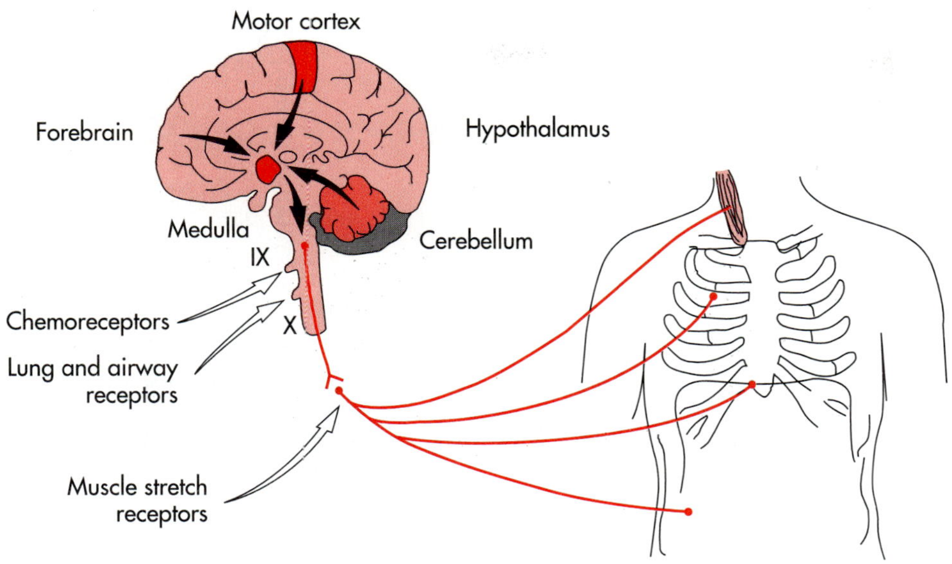

Fig. 5.8 Nervous regulation of ventilation. Central and peripheral nervous components are shown.

modified by emotional, mechanical, and chemical factors in such a way that rate and depth of respiration are homeostatically linked to arterial $[H^+]$.

■ **Neuronal Connections** The rate and depth of the basic respiratory rhythm are set by medullary neurons *(nucleus tractus solitarii* and *nucleus retroambiguus)* that generate spontaneously a program of motor instructions to the muscles of respiration. As summarized in Fig. 5.8, the basic program is modulated by other central nervous areas and by peripheral needs.

● **Central nervous modulation of ventilation**

- The *hypothalamus* accomplishes whole-body autonomic integration and adjusts breathing for moment-to-moment needs.
- The *forebrain* initiates breathing patterns that have emotional connotations (such as surprise gasps or languorous sighs).
- The *motor cortex* issues breathing-program modifications for the purpose of generating speech and for volitional control over breathing.
- The *cerebellum* participates in breathing modulations associated with postural changes.

● **Peripheral modulation of ventilation** Respiratory adaptations to peripheral needs are accomplished through 3 classes of receptor afferents (see Fig. 5.8): *chemoreceptors, irritant receptors,* and *stretch receptors.*

— ***Chemoreceptors***

• Chemosensitive neurons are located centrally (in the medulla) as well as peripherally (in the carotid and aortic bodies).

• They respond, with different sensitivities, to changes in P_{O_2}, P_{CO_2}, or $[H^+]$ of tissue fluid or blood and elicit homeostatically appropriate reflex changes in ventilation.

— ***Irritant receptors***

• Such receptors are located throughout the airways and lungs.

• They respond to mechanical or chemical irritation and elicit complex reflexes like sneezes and coughs.

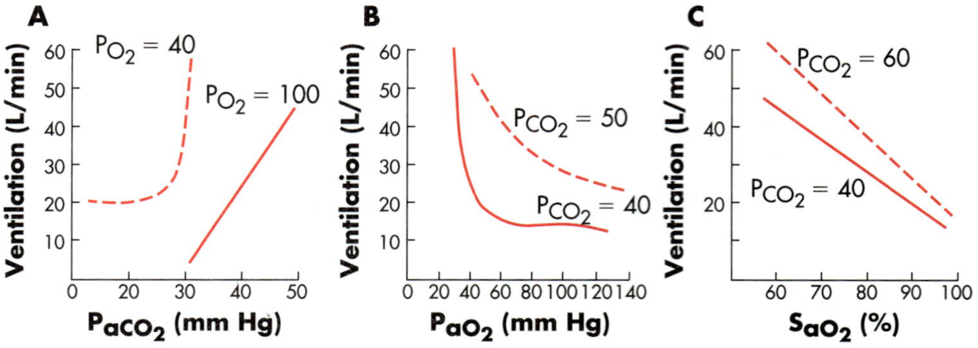

Fig. 5.9 Effects of arterial blood gases on ventilation. **A,** The effect of arterial Pco_2 on ventilation depends on the existing arterial Po_2. **B,** The effect of arterial Po_2 on ventilation depends on arterial Pco_2. **C,** Ventilation is linearly and inversely related to arterial O_2 saturation (Sao_2).

— *Stretch receptors*

- Neurons receptive to mechanical deformation are found in the lungs and in the respiratory muscles. They monitor lung volume and in general cause both inhibition of inspiration and promotion of expiration upon lung inflation (the *Hering-Breuer reflex*).
- Their role is easily demonstrated in animals where severing of afferent stretch-receptor paths leads to a characteristic change in breathing pattern (slow, deep breathing that is believed to be regulated by chemical factors). Such a change does not occur in humans, an indication that mechanical factors may be a less important regulatory influence in us.

Chemical Regulators of Ventilation

Effects of CO_2, H^+, and O_2 on chemoreceptors These three agents behave in an interactive, not in an algebraically additive manner.

- Medullary chemoreceptors do not respond significantly to changes in Po_2. They do, however, provide the primary afferent signal in responses to hypercapnia. The stimulus is not CO_2 *per se* but the extracellular concentration of H^+.
- Responses to Po_2 are initiated primarily by carotid body chemoreceptors.

Effects of arterial blood gases on ventilation

- At normal arterial Po_2, ventilation is directly related to arterial Pco_2, and the relationship is governed mostly by medullary chemoreceptors (Fig. 5.9, *A*). About 75% of the increased ventilatory drive in conscious hypercapnic humans originates in central chemoreceptors; the remainder derives from peripheral chemoreceptors.
- At normal arterial Pco_2, ventilation is affected little by changes in arterial Po_2 until Pao_2 falls to near 50 mm Hg (Fig. 5.9, *B*). At that point ventilation increases steeply with small further decreases in Pao_2. Thus peripheral chemoreceptors provide a last resort defense mechanism during hypoxemia. The curvilinear relationship between ventilation and Pao_2 can be made linear if ventilation is plotted against arterial oxygen saturation instead of partial pressure (Fig. 5.9, *C*). Keep in mind, however, that despite this linear relationship the chemoreceptors must sense Pao_2 rather than So_2.

Multiple Choice Review Questions

1. Which one of the following will cause a substantial increase in the total carrying capacity for O_2 in arterial blood (mL of O_2/100 mL of blood)?
   a. Breathing pure oxygen
   b. Increasing blood temperature
   c. Increasing hydrogen-ion concentration
   d. Increasing alveolar ventilation
   e. Increasing the hematocrit

2. Which one of the following statements regarding intrapleural pressure is true?
   a. It is defined as the pressure inside the lungs.
   b. It is defined as the difference between alveolar and atmospheric pressures.
   c. It becomes more negative when the residual volume increases.
   d. It prevents the lungs from collapsing at the end of expiration.
   e. It is generally more positive than alveolar pressure.

3. Which one of the following statements regarding Cl^- concentration in red cells of venous blood compared to Cl^- concentration in red cells of arterial blood is true in a healthy person?
   a. It is lower because Cl^- is lost during capillary transit of red cells.
   b. It is lower because red cells increase in volume at the venular end of capillaries.
   c. It is higher because Cl^- is gained during capillary transit.
   d. They are identical.

Questions 4 to 7 refer to the following scenario:

A patient has a shallow stab wound in the right side of the chest wall that allows air to reach the intrapleural space (right-sided pneumothorax).

4. Which one of the following statements regarding residual volume (RV) in the affected right half of the lung in this patient is true?
   a. It is greater than RV in the normal half of the lung.
   b. It is greater than the functional residual capacity in the affected right half of the lung.
   c. It is smaller than RV in the normal half of the lung.
   d. It is the same as RV in the normal half of the lung.

5. If the anatomic dead space in this patient is 100 mL, the rate of respiration is 20/min and tidal volume is 300 mL, what is the alveolar ventilation in liters per minute in this patient?
   a. 3
   b. 4
   c. 5
   d. 6

6. If the alveolar ventilation in this patient were normal, which one of the following statements would describe the patient's overall ventilation-to-perfusion ratio?
   a. It would be increased above normal.
   b. It would be decreased because the right half of the lung is smaller.
   c. It would be increased above normal because air would enter the right lung directly through the chest wall.
   d. It would be normal.

7. Which one of the following combinations of partial pressures (in mm Hg) would most likely apply to the blood leaving the affected right half of the lung?
   a. $Po_2 = 100$; $Pco_2 = 40$
   b. $Po_2 = 115$; $Pco_2 = 65$
   c. $Po_2 = 115$; $Pco_2 = 25$
   d. $Po_2 = 50$; $Pco_2 = 45$
   e. $Po_2 = 65$; $Pco_2 = 25$

8. A healthy person is flying in an airplane that has been pressurized to 10,000 feet above sea level (atmospheric pressure = 500 mm Hg). Which one of the following statements is true in this person while flying?
   a. Alveolar Po_2 will not be different from normal because the fractional composition of the air remains unchanged ($O_2 = 21\%$).
   b. Alveolar Po_2 will be increased because water vapor pressure will be decreased.
   c. Alveolar Po_2 will be decreased but not enough to have a significant effect on hemoglobin saturation in the passenger's arterial blood.
   d. Hemoglobin saturation in the passenger's arterial blood will be less than 80%.

9. Asthma attacks are characterized by widespread narrowing of the airway. Results of pulmonary function tests during an acute asthma attack will demonstrate all of the following *except:*

a. decreased forced expiratory volume in 1 second (FEV_1)
b. increased forced vital capacity (FVC)
c. decreased FEV_1/FVC
d. normal or increased total lung capacity (TLC)

10. If all type II alveolar cells in a person suddenly produced only surfactant with lower than normal surface tension, each of the following would happen *except:*

a. The static pressure-volume curve for the lungs would be shifted to the right.
b. The residual volume would increase.
c. The alveolar pressure would decrease.
d. The static pressure-volume curve for the chest wall would be shifted to the left.

11. Arterial P_{CO_2} is normally more tightly controlled than arterial P_{O_2}. Which one of the following statements correctly identifies the major reason for this observation?

a. Peripheral chemoreceptors do not respond to changes in P_{O_2}.
b. The ventilatory response to hypercapnia is linear over the full range of physiological levels of P_{CO_2}.
c. The respiratory control centers stop functioning when arterial P_{O_2} levels fall below 60 mm Hg.
d. Changes in P_{O_2} have to be converted into hematocrit changes by erythropoietin before they can become a respiratory stimulus.

Chapter 6

Cardiovascular Physiology

Overview

The cardiovascular system consists of the heart, the blood vessels, and the regulatory mechanisms that govern their functions.

- **Structure** The system circulates blood in a continuous loop, and the left ventricle of the heart is generally taken as the starting point of the continuous loop. The function of the left ventricle is to pump blood into the root of the aorta.

    - From the aorta the blood is distributed to individual vascular beds through a system of successively branching arteries, small arteries, and arterioles.
    - From the arterioles blood flows into capillaries, the small blood vessels that provide an interface between blood and the cells of tissues.
    - Blood returns from the capillaries to the right atrium of the heart through a confluent network of venules, small veins, veins, and the inferior and superior venae cavae.
    - The right side of the heart pumps blood from the right ventricle to the lungs, and from there it returns to the left side of the heart to begin the loop again.

- **Function**

    - **Overall purpose** The purposes of the cardiovascular system are

        - To provide to all tissues under all survivable conditions the supply of nutrients that is required to support their metabolic activities and
        - To remove from all tissues the products of their metabolic activity.

    - **Heart function**

        - Heart action is initiated by spontaneous, rhythmic electrical activity in specialized cells of the heart, called *pacemaker cells.*
        - A specialized conduction system allows pacemaker activity that originates from a single cell to be conveyed to all regions of the heart.
        - In cardiac muscle cells the processes of excitation-activation-contraction coupling convert electrical activity into mechanical action (muscle contraction).
        - Contraction in the muscular wall of the ventricles creates pressure and propels blood into the large vessel that is connected to each ventricle.

    - **Blood vessel function** The vessel network connected to each ventricle has three main functions:

        - To distribute ventricular outflow.
        - To smooth out any pulsations so that the flow reaching the tissues is constant despite the pulsatile nature of heart action.

Table 6.1 *Summary of the Contribution of Different Portions of the Cardiovascular System to Overall Cardiovascular Function*

Anatomic Region	Contribution to System Function
Chambers of the heart	Pumping action
Arterial system	"Reservoir" for pressure
Arterioles	Smoothing of flow pulsations; resistance adjustment
Capillaries	Exchange between blood and tissue
Venous system	"Reservoir" for volume
Vascular endothelial cells	Secretion of vasoactive chemicals

- To secrete from its endothelial cells a variety of substances whose biological effects are required for the normal function of the organism.

- **System function** Normal mechanical function of the cardiovascular system depends on the contributions that are made by different components as a result of their structure or mechanical behavior (Table 6.1).

Cardiac Electrophysiology

The electrical activity of cardiac cells can be measured as periodic changes in the potential difference between their cytoplasm and the extracellular region.

- The polarity, amplitude, and pattern of change of cardiac cell electrical activity is determined only by transport of ions across the cell membrane.
- Passive-transport mechanisms through ion-selective channels exert a dominant influence over short-term electrical behavior of cardiac cells, measured on a time scale of a few milliseconds.
- Active-transport mechanisms are important for long-term electrical behavior.

Ion channels Voltage-gated, *ion-selective channels* in the surface membrane are of special importance for the contraction-relaxation cycle of cardiac muscle cells. Their periodic conductance changes are responsible for the intracellular fluctuations in ionic and metabolic constituents that form the basis of heart action.

- Conductance changes in cardiac muscle cells occur on a time scale of milliseconds as a result of channel activation and inactivation.
- Channel conductance is also influenced over longer periods by the action of neurotransmitters or hormones.
- The dominant ion currents involved in cardiac myocyte action are Na^+, Ca^{++}, and K^+.

- **Sodium channels** The Na^+ channel in cardiac muscle is comparable in structure and function to that in other excitable tissues.
- **Calcium channels**
  - The L type Ca^{++} channel is prominent in cardiac muscle.
  - S-A nodal pacemaker cells and Purkinje cells also show a significant number of the T type Ca^{++} channels.
  - Table 6.2 shows the important properties of the two channel types.

Table 6.2 *Significant Properties of L and T Types of Ca^{++} Channels in Cardiac Cells*

Channel Type	Current	Threshold	"Open" Time	Duration of Inactivation	Response to β-Adrenergic Receptor Stimulation
L	$I_{Ca,L}$	Near −40 mV	Long	Very long	Increased current
T	$I_{Ca,T}$	−50 to −65 mV	Short	Short	Not significant

- **Potassium channels** Several functional subtypes of K^+ channels have been identified. Four of them are of importance in cardiac muscle cells:
  - Channels that carry the current $I_{K\text{-}Ach}$.
    - They respond to extracellular acetylcholine with a burst of openings.
  - Channels carrying the current I_{K1}.
    - They are the predominant carrier of the basal potassium current in human myocytes.
    - They are insensitive to acetylcholine.
  - Channels carrying the "delayed-rectifier" current, I_k.
    - I_K differs from I_{K1} primarily in that its threshold membrane potential is much more positive (between −50 and +50 mV).
    - This channel conducts the outwardly directed current that causes repolarization at the end of an action potential.
    - The I_K channel is not affected by acetylcholine, but β-adrenergic agonists move its threshold 5 to 10 mV more negative, increase the magnitude of the maximally activated current, and retard channel deactivation.
  - Channels that carry the current $I_{K\text{-}ATP}$.
    - They normally make a small contribution to basal potassium current because they are inhibited by ATP. However, they may gain in importance at times when ATP levels are low, as during coronary ischemia.

Membrane Potentials

- **Resting membrane potential** In the resting state most cardiac cells maintain a stable resting membrane potential between −80 to −95mV. It arises from passive ion fluxes and electrogenic transport.

— ***Passive diffusion mechanisms*** Four currents contribute significantly:

- An inwardly directed sodium current (I_{Na}).
- An outwardly directed potassium current (I_{K1}). I_{K1} is the major current providing stability of membrane potential at rest.
- A small chloride current that is directed outwardly, against the concentration gradient, because the chloride equilibrium potential is positive with respect to the resting membrane potential.
- A small inward calcium leakage current ($I_{Ca,leak}$).

— ***Active transport mechanisms*** The Na^+, K^+ pump transports $2K^+$ into the cell while extruding $3Na^+$ ions for each molecule of ATP hy-

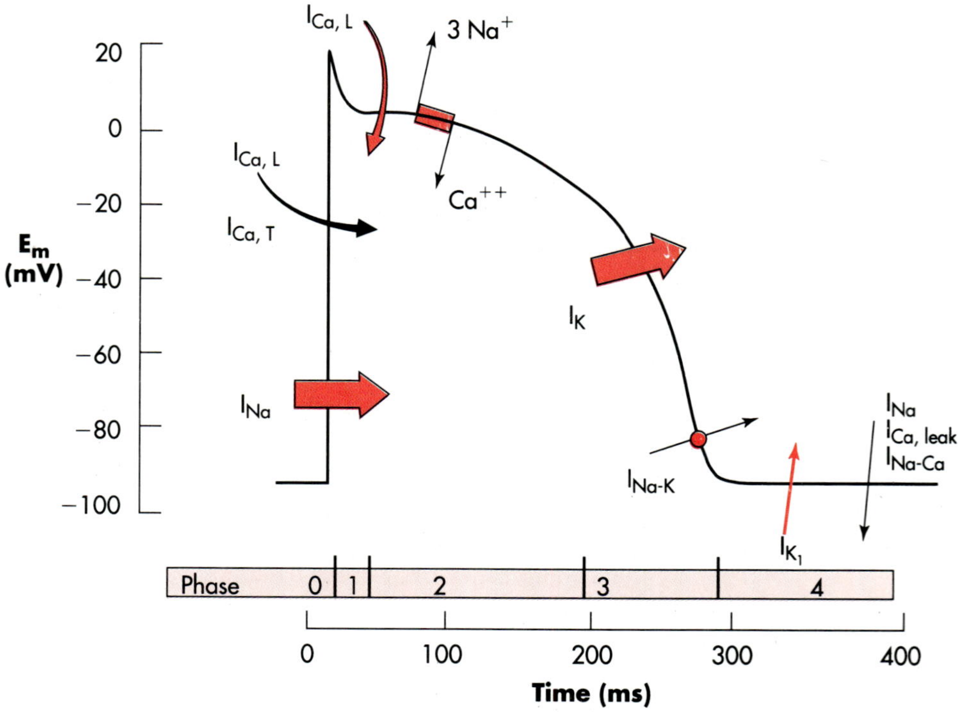

Fig. 6.1 Ion currents dominating the 5 phases of the cardiac action potential. Arrows pointing into the area under the trace signify currents that enter the cell. Those pointing outward signify currents leaving the cell. E_m, Membrane potential; Current, flow of positive charge. Key currents are shown in color.

drolyzed. The coupling ratio of 3:2 prevails over a wide range of conditions. There also is an ATP-consuming Ca^{++} pump that moves 2 Ca^{++} out for each ATP.

— ***Coupled transport*** The Na^+, Ca^{++} exchanger operates through a sarcolemmal protein and is driven by the Na^+ electrochemical gradient. It helps to maintain resting membrane potential by transporting $3Na^+$ inwardly for each Ca^{++} out. Reversal to a Na^+-out/Ca^{++}-in state occurs at a membrane potential just slightly positive to the normal resting membrane potential.

Cardiac action potential

— ***General features***

- Cardiac muscle cells, being excitable cells, are capable of responding to an appropriate stimulus with quick activation of a sufficient number of membrane channels that permit entry of Na^+.
- Na^+ entry causes rapid depolarization as shown by a steep action potential upstroke, identified as *phase 0* (Fig. 6.1).
- A characteristic feature of cardiac muscle action potentials is the *plateau (phase 2)*, a phase of maintained depolarization (or hardly perceptible repolarization) that may last 200 to 300 ms. While the plateau is maintained, the cell cannot be reexcited. This prolonged refractoriness prevents cardiac muscle tetanization because it ensures relaxation of muscle-twitch tension before a subsequent action potential.
- The plateau terminates with the onset of late rapid repolarization *(phase 3)*.

— ***Ion currents*** The action potential is produced by transmembrane flow of ionic currents. These are regulated by a variety of ion-selective membrane channels. Fig. 6.1 shows the ion currents dominating each phase of the action potential.

Upstroke (phase 0)

- When a suitable stimulus depolarizes the membrane to the threshold potential, Na^+ channels are activated, and the membrane potential rapidly moves toward the Na^+ equilibrium potential.
- Channels carrying $I_{Ca,T,}$ the transient Ca^{++} current, are activated at E_m more positive than −50 to −65mV.
- At E_m more positive than −40mV, channels carrying $I_{Ca,L,}$ the long-lasting Ca^{++} current, are activated and remain activated in phases 1 and 2.

Early rapid repolarization (phase 1)

- At the peak of the upstroke the membrane potential reaches between +20 and +40 mV and then undergoes rapid partial repolarization by a short-lived outward current, I_{to}, carried mostly by K^+ moving out and to some extent by Cl^- moving in.
- $I_{Ca,T}$ is inactivated in phase 1.

The plateau (phase 2)

- The plateau arises from what is nearly a balance between depolarizing and repolarizing currents during this phase.
- The major depolarizing influences:
  - An inward Ca^{++} current, $I_{Ca,L,}$ carried through channels that were opened during phase 0 provides the major inflow of positive charge.
  - The $3Na^+$, Ca^{++} exchanger, operating in the Ca^{++} inward mode at this $E_{m,}$ provides the major outflow of positive charge, $I_{Na\text{-}Ca.}$

Late rapid repolarization (phase 3)

- The plateau terminates because $I_{Ca,L}$ decays when the L type channels carrying it are inactivated by processes dependent on time, voltage, and intracellular $[Ca^{++}]$.
- Repolarization then occurs rapidly because of the dominant influence of outward K^+ currents, mainly by means of I_K, the "delayed-rectifyer" current. (All K^+ channels in cardiac cells "rectify" in that they show a tendency to lower their conductance with cell depolarization.)
- As the membrane potential approaches its resting value:
  - I_K diminishes as the electrochemical gradient for K^+ decreases.
  - The Na^+, Ca^{++} exchanger returns to its Ca^{++}-out mode.
  - The Na^+, K^+ pump current gains in relative importance.

Diastole (phase 4)

- In diastole nonpacemaker cells maintain a stable, resting membrane potential because of an exact balance between inward and outward currents.

- The significant outward currents are:
  - $I_{Na\text{-}K}$, the current resulting from active $3Na^+$, $2K^+$ pumping
  - I_{K1}, a K^+ current through the inwardly rectifying K_1 channel*
- The significant inward currents are:
  - $I_{Na,}$ a Na^+ leakage current in the direction of the steep electrochemical gradient for Na^+
  - $I_{Ca,leak}$, a Ca^{++} leakage current
  - $I_{Na\text{-}Ca}$ the Na^+, Ca^{++} exchanger current

- **Pacemaker potential** Pacemaker cells do not have a stable membrane potential in diastole.

  - Their diastolic potential gradually approaches threshold, and they depolarize when the threshold for an action potential is reached.
  - Pacemaker cells are concentrated in the sinoatrial (SA) node and in Purkinje fibers, including the atrioventricular (AV) node. Sinus rhythm is normally driven by SA-node cells because they are the earliest to depolarize.
  - SA-nodal pacemaker cells differ from other myocytes in that:
    - Their maximum diastolic potential (MDP) is more positive than normal myocyte resting membrane potential.
    - They have virtually no I_{K1} channels, the channels that normally carry the major diastolic stabilizing current.
    - The action potential upstroke is caused by L type Ca^{++} current.
    - They contain $I_{f,}$ a nonselective, voltage-gated channel for inward transport of monovalent cations.

— ***Modulation of SA-node pacemaker rate***

- Physiologic modulation of pacemaker activity is directed mostly by the autonomic nervous system and is achieved by extremely subtle changes in transmembrane ion currents.
- Sympathetic stimulation increases intracellular cAMP. The consequent activation of protein kinase A increases pacemaker frequency partly by reducing action potential duration but mostly by increasing the slope of the phase 4 rise in membrane potential.
- Parasympathetic stimulation slows pacemaker frequency by muscarinic mechanisms that include membrane hyperpolarization, a less rapid action potential upstroke, and a decreased slope of the phase 4 rise in pacemaker membrane potential.

Spread of Cardiac Action Potentials and Genesis of the ECG

Two aspects of cardiac anatomy help to create, just before each heartbeat, a large, organized electrical field in and around the heart:

*The channel is termed *inwardly rectifying* because it is capable of carrying K^+ into the cell whenever E_m is more negative than E_K. In that setting its conductance is greater than it is while carrying outward flow during the periods when E_m is more positive than E_K.

• The specialized conduction systems of the heart, that is, the bundle of His–Purkinje system and cross connections between neighboring muscle fibers, ensure that large numbers of cells depolarize in synchrony.

• The spread of depolarization along relatively fixed, predetermined paths ensures that the orientation of the electric field with respect to the body surface changes little from beat to beat.

• The temporal changes in magnitude and direction of the electrical fields that accompany cardiac depolarization and repolarization can be detected as potential differences between any two points on the body surface.

• Several surface points have become conventional for the attachment of recording electrodes. Recording potential differences between them results in bipolar lead ECGs such as leads I, II, and III.

• It is also conventional to record ECGs at several surface points, not with reference to one other surface point but with reference to a point that is derived electronically by the recording apparatus from voltages measured at two or three other surface points. This results in *unipolar ECGs.* (Leads aV_R, aV_L, aV_F and precordial leads V_1 to V_6 are typical unipolar leads.)

ECG traces show deflections that are typically identified as "waves" labeled *P, Q, R, S,* and *T.*

The relationship between instantaneous cardiac electrophysiologic events and ECG traces is best understood in terms of *cardiac vectors* and their projections onto a geometric line that connects the end points of individual leads.

Depolarization Vector Electrical depolarization of cardiac cells involves each cell undergoing a change in its membrane potential from its normally negative value to a slightly positive value. Hence the process of cardiac depolarization can be imagined as a wave of positivity sweeping over the tissue and can, with the help of the following conventions, be represented by a vector:

• Depolarization is represented by an arrow that is identified with a + sign at its head.

- The direction of the arrow is the same as the direction in which the depolarization wave moves through the tissue.
- The length of the arrow is directly proportional to the number of cardiac cells that are depolarizing at that instant.

• A depolarization vector can be drawn to represent at any one instant the spatially averaged cardiac electrical activity at that instant.

• Similarly, all the depolarization vectors occurring over a period of time (for example, over the duration of atrial depolarization) can be averaged to yield a time- and space-averaged depolarization vector (for example, a *P* vector).

Repolarization Vector Repolarization events, represented by the T-wave of the ECG, can be similarly derived by application of the concept of a repolarization vector. Repolarization is represented by an arrow that is identified with a − sign at its head. The direction of the arrow is the same as the direction in which the repolarization wave moves through the tissue, returning the membrane potential of cells to their negative, resting value.

Table 6.3 shows the electrophysiologic events that constitute a single cardiac cycle. Vector representations of each event and the corresponding deflection in the frontal-plane, scalar ECG are also shown.

Table 6.3 *Electrical Events of the Cardiac Cycle*

Time (ms)	Cardiac Electrophysiologic Event	Vector Representation and ECG Lead Projection	Corresponding ECG Deflection
0-70	S-A node 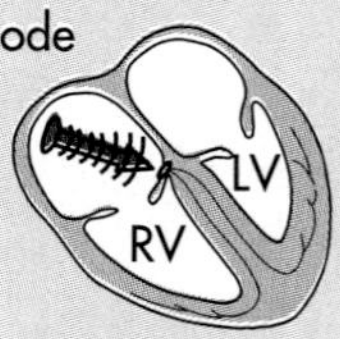 Action potentials that were spontaneously generated by the dominant pacemaker (the S-A node) spread through atrial tissue	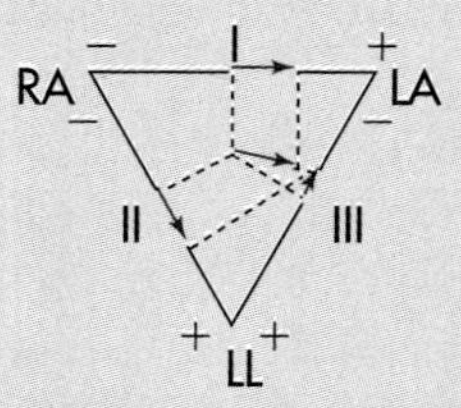	P-wave • Upward deflection in lead I • Smaller upward deflection in lead II • Upward, biphasic or downward deflection in lead III
70-150	A-V node delay		P-R segment • Normally isoelectric • Measured from the end of the P-wave to the onset of the QRS complex. (Note that the P-R segment is not the same as the P-R interval. The P-R interval is defined as the interval from the onset of the P-wave to the beginning of the QRS complex.)
150	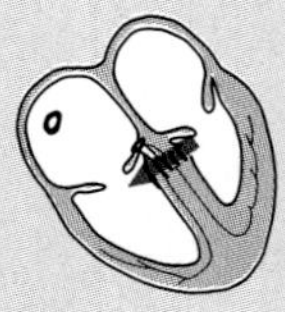 Beginning of ventricular depolarization, which normally occurs at a point high in the interventricular septum	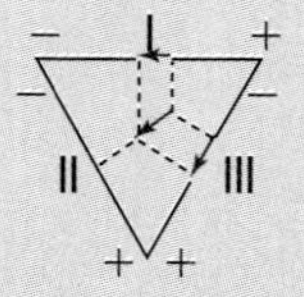	Q-wave • Defined as the first downward deflection preceding the R-wave. • Present in lead I • Small or absent in lead II • May be absent in lead III because the Q vector would produce an upward deflection in lead III
152-165	152 ms 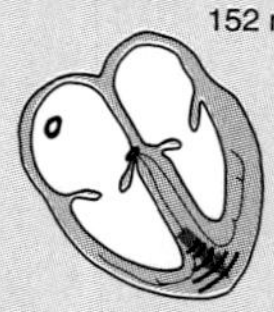 155-165 ms 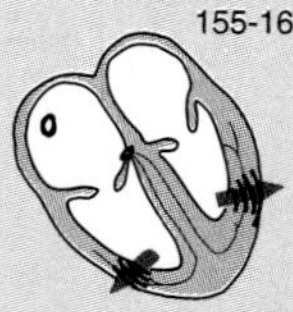 Depolarization of both ventricles	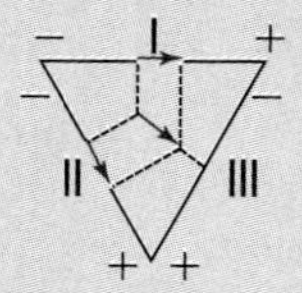 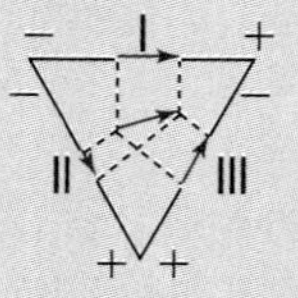	R-wave • Large upward deflection in each of leads I, II, and III
180-200	180 ms 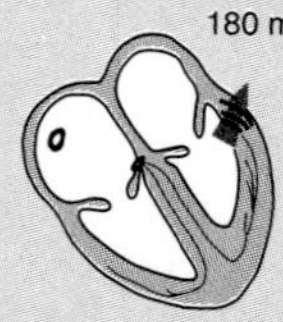 Depolarization of last portions of the ventricles and of the septum tip	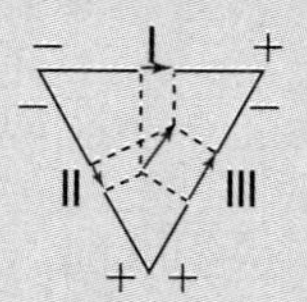	S-wave • Defined as the first downward deflection following R • Downward deflection in leads I, II, and III
250-600	Repolarization of the ventricles		T-wave • Upward deflection in leads I, II, and III

Cardiac Muscle Functions

■ **Cardiac Metabolism, Oxygen Consumption, and Work** Although cardiac myocytes can use a variety of metabolic substrates,

• fatty acids are the key source of energy required for the performance of cardiac work. They supply, under normal conditions, two thirds of myocardial ATP.

• Sustained increase in cardiac work requires an increase in the rate of ATP production and utilization.

• The cell's need for ATP is regulated, in part, by control of the Krebs cycle and in part by the rate of ADP phosphorylation (Fig. 6.2).

■ **Excitation-Activation-Contraction Coupling**

• Excitation-activation-contraction coupling links electrical events of the myocyte to the mechanical events of tension development.

• A vital aspect of such coupling is the elevation of intracellular ionized calcium from its resting value near 100 nM to a peak of about 1 μM during a normal contraction of ventricular cells.

● **Sources of calcium** Four significant sources of Ca^{++} are involved in excitation-activation-contraction events. They are voltage-gated current, $I_{Ca,L}$, Ca^{++}-leakage current, Na^{++}, Ca^{++} exchange, and Ca^{++} release from sarcoplasmic reticulum.

— ***Voltage-gated Ca^{++} currents*** Most of the calcium that enters the myocyte at the start of an action potential is carried by $I_{Ca,L}$. As far as the mechanical events of the heart are concerned, $I_{Ca,L}$ supplies no more than 10% of the total Ca^{++} needed for a maximal contraction, but it performs one vital function in excitation-activation: It provides the trigger for releasing calcium from intracellular stores during systole.

— ***Calcium leak***

- The large electrochemical gradient for Ca^{++} causes some continuous, passive inward leak of Ca^{++}.
- Conductance of the leakage channel is not dependent on the membrane potential.

— ***Sodium-calcium exchange***

• The Na^{+}, Ca^{++}-exchange mechanism uses the sodium electrochemical gradient as the source of energy for moving calcium across the cell membrane.

• Its stoichiometry is to exchange $3Na^{+}$ ions for each Ca^{++}, making the pump an electrogenic pump.

• Its direction and magnitude depend on the difference between the existing membrane potential (E_m) and the reversal potential of the exchanger ($E_{Na\text{-}Ca}$).

- $E_{Na\text{-}Ca}$ is determined by the ion equilibrium potential for Na^{++} (E_{Na}) and Ca^{++} (E_{Ca}) according to the formula,

$$E_{Na\text{-}Ca} = 3E_{Na} - 2E_{Ca}.$$

- When $E_{Na\text{-}Ca}$ is positive to E_m, net positive charge must flow into the cell and Ca^{++} will be extruded; the exchanger must operate in the Ca^{++}-outward mode.

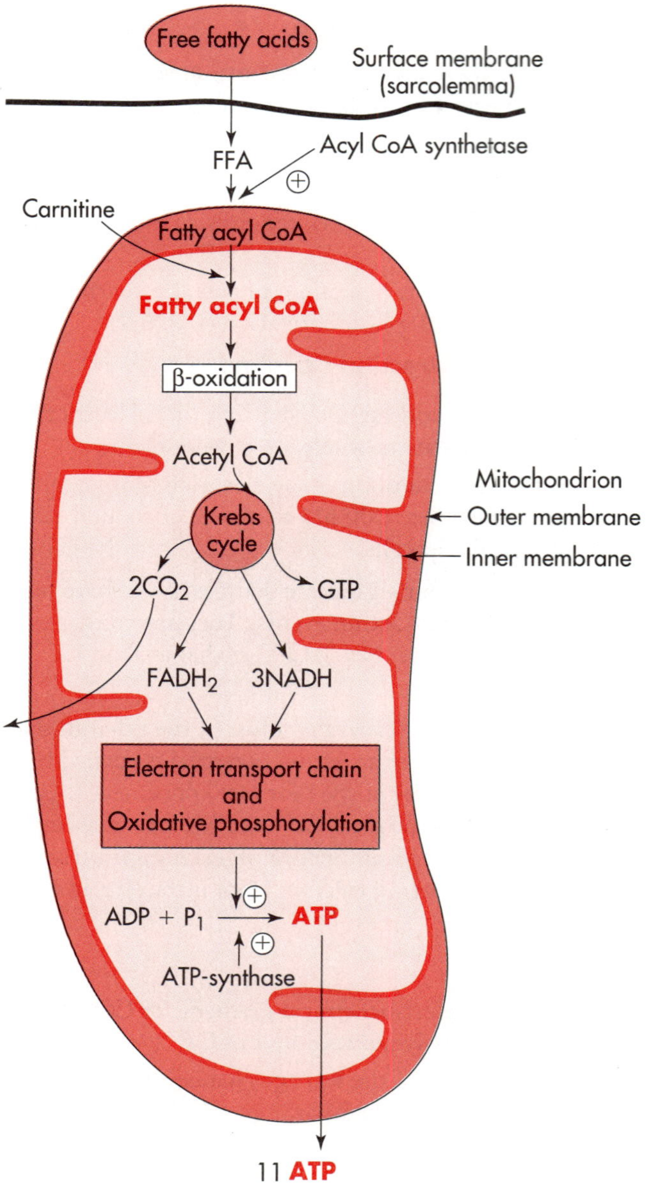

Fig. 6.2 Metabolism of fatty acids in cardiac myocytes. Five stages can be identified: 1. Free fatty acid transport to the outer mitochondrial membrane and their activation to fatty acyl CoA 2. Carnitine-based transport into mitochondria and formation of acetyl coenzyme A 3. The Krebs cycle 4. The electron-transport chain and oxidative phosphorylation 5. Phosphorylation of ADP. *ADP,* Adenosine diphosphate; *ATP,* adenosine triphosphate; *FADH$_2$,* reduced flavin adenine dinucleotide; *FFA,* free fatty acids; *NADH,* reduced nicotinamide adenine dinucleotide; *P_i,* inorganic phosphate.

- When $E_{Na\text{-}Ca}$ is negative to E_m, net positive charge must flow out of the cell; the exchanger must operate in the Ca^{++}-inward mode.

- The normal value for $E_{Na\text{-}Ca}$ in human cardiac muscle is −40mV. E_m fluctuates between its resting value of −90 mV and up to +35 mV at the peak of the action potential. This allows the Na^+, Ca^{++}-exchange mechanism to bring calcium into the cell during one part of the cardiac action potential and extrude it in another.

- The Na^+, Ca^{++} exchanger supplies Ca^{++} too slowly to be significantly involved either in supplying trigger calcium for the sarcoplasmic reticulum or in supplying Ca^{++} for actin-myosin interactions. Its major role may be to modulate the supply of Ca^{++} for subsequent storage in the sarcoplasmic reticulum.

— *Sarcoplasmic reticulum (SR)*

- This membrane-lined structure functions as the major intracellular source and store for ionized calcium.
- The primary release mechanism is calcium-mediated calcium release, involving two types of Ca^{++} channels.
    - The SR membrane near the T-tubule contains Ca^{++}-sensitive receptors that are capable of being blocked by *dihydropyridine.* When these receptors are activated by "trigger" Ca^{++} that has crossed the T-tubule membrane through voltage-gated L and T type channels, there follows a fulminant release of Ca^{++} from the SR stores through a different Ca^{++} channel.
    - The Ca^{++} channels through which this release occurs can be blocked by *ryanodine.*
- The actual amount of Ca^{++} that is relased during a particular contraction is a function of both the amount of extracellular Ca^{++} supplied (mostly by $I_{Ca,L}$) and the amount of Ca^{++} stored in the sarcoplasmic reticulum.
    - In a single contraction the sarcoplasmic reticulum is of major importance as a source of Ca^{++}. However,
    - Ca^{++} transport mechanisms across the cell surface membrane are of great importance over any interval that contains more than one contraction-relaxation cycle because these membrane-transport mechanisms build up the intracellular stores.

● **Uptake of calcium** When the action potential has passed and the myocyte has repolarized, calcium is removed from the cytosol by two main routes: (1) the Na^+, Ca^{++} exchanger and (2) ATP-dependent Ca^{++} pumping. ATP-dependent Ca^{++} pumps are found in both the sarcolemma and the membrane that lines the sarcoplasmic reticulum.

● **Tension development**

- Interaction of free intracellular Ca^{++} with protein constituents of cardiac muscle cells initiates the processes that lead to the development of tension.
- These protein constituents are actin, myosin, tropomyosin, and troponin. In the resting state there is no physical contact between myosin and actin because the tropomyosin-troponin complex is configured in a way that prevents such contact (Fig. 4.4). However, when Ca^{++} interacts with troponin-C, the conformation changes in a way that exposes the binding site (Fig. 4.4), permitting a momentary cross bridge to form between myosin and actin.
- Muscle contraction is brought about by a cyclic sequence of con-

formational changes within the S1 and S2 regions of the myosin molecule. This sequence is described as the *sliding-filament hypothesis* in Chapter 4.

Cardiac Performance Each contraction of cardiac muscle is ultimately translated into the cyclic emptying of each ventricle and the ejection of a bolus of blood called the *stroke volume.*

- The vigor and effectiveness of ventricular emptying are termed *cardiac performance.*
- Four factors determine cardiac performance:
  - The degree of sarcomere stretch before actomyosin activation *(preload)*
  - The resistance to ventricular contraction *(afterload)*
  - The interval between contraction-relaxation cycles *(heart rate)*
  - The efficacy of actomyosin interaction *(contractility)*

Preload Changes in diastolic ventricular stretch change cardiac performance according to the *Starling-Frank law* of the heart. It states that until an optimal sarcomere length is reached, increasing sarcomere length increases ventricular performance in the next several heart beats.

— ***Cellular basis of the preload mechanism*** Within the range of sarcomere lengths where increased stretch increases performance, the enhancement results from three cellular changes:

- Increased stretch causes a more optimal alignment of myosin heads with their binding sites on the actin filament and thereby allows more actomyosin cross bridges to form.
- Increased sarcomere stretch increases cytosolic [Ca^{++}]. This arises partly from an increase in sarcomere surface area and partly from a slightly greater asynchrony between Ca^{++} release and uptake mechanisms.
- An increase in sarcomere length changes Ca^{++} sensitivity in such a way that a greater force is developed at any one $[Ca^{++}]_i$.

Afterload The healthy heart is able to increase performance in the face of increased afterload. This is sometimes called the *Anrep effect.* The Anrep effect is believed to be caused by metabolic factors, secondary to reactive hyperemia* in the subendocardial layers.

Heart rate Peak ventricular tension increases as heart rate increases over a broad range (the *Bowditch effect,* or *Treppe* ['staircase'] *phenomenon*).

— ***Cellular basis of the Bowditch effect*** The Bowditch effect arises from the dependence of systolic $[Ca^{++}]_i$ on residual Ca^{++} from the previous beat.

- Shorter intervals between successive beats reduce the time for uptake of Ca^{++} and thereby increase end-diastolic $[Ca^{++}]_i$.
- When the interval between successive beats is extremely short, the duration of the action potential is also short. This reduces the time

*The term *reactive hyperemia* refers to an increase in tissue blood flow after a period of reduction or cessation of flow. In the case of increased ventricular afterload, the reduction of coronary flow is believed to result from increased mechanical compression of the subendocardial layers.

for systolic influx of Ca^{++} and sets an upper limit to the positively inotropic effect of heart rate.

- **Contractility** Contractility refers to any factor influencing myocardial performance when preload, afterload, and heart rate are not changed. An increase in contractility, most easily brought about by β_1-adrenoceptor stimulation, is associated with the following:

  - Increased velocity of wall shortening
  - Reduced time to peak tension
  - More rapid rate of relaxation
  - Shorter duration of systole
  - Greater extent of fiber shortening
  - Greater stroke volume
  - Higher ejection fraction
  - Decreased end-systolic volume

— ***Cellular basis of contractility*** The underlying cellular mechanisms by which changes in contractility affect myocardial performance are rooted predominantly in Ca^{++} dynamics (instantaneous effects) and the properties of the contractile proteins themselves (long-term effects).

Calcium dynamics

- $[Ca^{++}]_i$ influences both the total number of actomyosin cross bridges that are formed during the excitation-activation phase and the rates of protein phosphorylation that drive the activation-contraction phase.
- Stimulation of β_1-adrenoceptors leads to increased $[Ca^{++}]_i$.
- Stimulation of β_1-adrenoceptors also has a small negative influence on calcium sensitivity (a measure of the degree of actomyosin activation at a given $[Ca^{++}]_i$). However, this effect is more than compensated by the larger number of cross bridges that results from the increased $[Ca^{++}]_i$.
- Keep in mind, however, that decreased Ca^{++} sensitivity means weaker binding of Ca^{++} to troponin-C and hence a more rapid release of Ca^{++}. This is expressed most strongly during the relaxation phase when Ca^{++} is released from troponin and is responsible for more rapid ventricular relaxation after β_1-adrenoceptor stimulation.

Properties of the contractile proteins

- The contractile properties of myosin can be modulated at the level of the heavy chains as well as at the level of the light chains.
- Modulation at the heavy-chain level expresses itself as differences in ATPase activity among the 3 myosin isozymes, V1, V2, and V3.
  - V1 has the highest ATPase activity, and V3 has the lowest.
  - On the other hand, V1 requires more energy to develop a given force than the other two need.
  - In some species the relative abundance of the three isozymes is changed by age, plasma titer of thyroid hormone, or degree of heart failure.

• Modulation at the myosin light-chain level occurs by mechanisms that affect the balance between light-chain phosphorylation and dephosphorylation.

Blood Vessel Function

■ **Vascular Smooth Muscle** Smooth muscle cells are present in blood vessel walls, encircling the vessel circumference in a helical fashion. Their contraction and relaxation allow blood vessels to change their diameter and elastic properties.

● **Electrophysiology**

— ***Ion channels*** Vascular smooth muscle (VSM) has a variety of selective and nonselective ion channels. Two functional types can be identified: *voltage-gated channels* and *ligand-gated channels.*

Voltage-gated channels

• These consist mostly of Ca^{++} channels (both L and T types) and K^{+} channels.

• Voltage-gated Na^{+} channels have also been identified in some smooth muscle. When they are present, they contribute to the change in membrane potential during the upstroke of the action potential.

Ligand-gated channels

• Such channels are influenced by chemical rather than electrical changes.

• VSM membranes contain a great variety of ligand-gated receptors.

— ***Resting membrane potential*** VSM cells have a more positive resting membrane potential than cardiac muscle cells have. It is between −40 and −60 mV and arises from greater resting Na^{+} current in VSM.

— ***Action potentials***

- Action potential upstrokes in VSM are caused by Ca^{++} currents, not by Na^{+} currents.
- Three electrically distinct types of vascular smooth muscle have been identified:

  1. One that generates action potentials spontaneously.
  2. One that generates action potentials in response to an adequate stimulus.
  3. One that responds to excitatory stimuli with a sustained, nonregenerating electrical response that resembles a skeletal muscle end-plate potential.

- Types 1 and 2 exhibit phasic contractions whereas type 3 shows more sustained contractions.

● **Excitation-activation-contraction coupling** Smooth muscle contraction is initiated by elevations in cytosolic free $[Ca^{++}]$. These can occur either

- after a primary change in membrane potential *(electromechanical coupling)* or
- after chemical events that do not necessarily involve a change in membrane potential *(chemomechanical coupling).*

— ***Electromechanical excitation-activation-contraction coupling*** Electromechanical coupling is initiated by a sufficient depolarization of the cell membrane. This can be caused by an action potential or by another, nonregenerative electrical event such as the type 3 sustained depolarizations described above.

— ***Chemomechanical excitation-activation-contraction coupling***

• Chemomechanical coupling accounts for VSM contractions that are not initiated by changes in membrane potential.

• It operates by mechanisms promoting intracellular Ca^{++} release (predominantly by inositol trisphosphate, IP_3) or modulating Ca^{++} sensitivity of the actin-myosin interaction.

- Control of developed tension

— ***Short-term mechanisms*** Two mechanisms are used to modulate the tone of VSM:

• Changes in the concentration of free, ionized calcium in the cytosol ($[Ca^{++}]_i$) or

• changes in the Ca^{++} sensitivity of the contractile components.

Mechanisms affecting cytosolic calcium concentration. $[Ca^{++}]_i$ depends on the magnitude of the calcium current through membrane channels ($I_{Ca,leak}$ and $I_{Ca,L}$) as well as on calcium release from the sarcoplasmic reticulum (SR).

• Membrane transport of Ca^{++} is enhanced by cAMP and inhibited by cGMP.

• SR release of Ca^{++} is stimulated by IP_3 and inhibited by cGMP.

Mechanisms affecting calcium sensitivity. The effectiveness of the intracellular mechanisms for tension development at a given $[Ca^{++}]_i$ is governed by the activities of several intracellular enzymes controlling reactions of phosphorylation and dephosphorylation of myosin light-chain kinase (MLCK).

— ***Long-term mechanisms*** A characteristic feature of smooth muscle is its ability for prolonged maintenance of contractile force with low energy consumption. No homeostatic mechanisms that are specifically directed at regulating long-term mechanisms in VSM have been verified though many have been investigated in the search for causes of hypertension.

Vascular Endothelium

• The endothelial lining of blood vessels forms substances that maintain an antithrombic surface and modulate the tone of the underlying smooth muscle.

• Both relaxing and constricting factors are synthesized by the endothelium and participate in the local regulation of tissue blood flow:

- Endothelium-dependent relaxation

• Three dilating factors are derived from the endothelium: nitric oxide (NO), prostacyclin (PGI_2), and endothelium-derived hyperpolarizing factor (EDHF).

• Synthesis of NO within endothelial cells is promoted by elevated cytosolic $[Ca^{++}]_i$, which follows increases in shear stress from flowing blood or activation of a variety of membrane receptors by

Table 6.4 *Relaxing Factors Derived from the Endothelium*

Trigger	Endothelial Factor	Mediator for VSM Relaxation
Shear stress	↑ NO	↑ cGMP
	↑ PGI_2	↓ cAMP
Various receptor agonists	↑ NO	↑ cGMP
	↑ PGI_2	↓ cAMP
	↑ EDHF	Hyperpolarization

EDHF, Endothelium-derived hyperpolarizing factor; *NO,* nitric oxide; *PGI_2,* prostacyclin; *VSM,* vascular smooth muscle.

their respective ligands. NO causes vascular smooth muscle relaxation by an increase in cGMP levels within VSM.

- Prostacyclin (PGI_2) synthesis is also enhanced by shear stress and several receptor agonists. It relaxes VSM by decreasing cAMP within VSM.
- Endothelium-derived hyperpolarizing factor (EDHF) is a short-lived product of M_1 muscarine receptor activation. It may be epoxyeicosatrienoic acid (EET), an intermediate product of arachidonic acid metabolism. It relaxes smooth muscle by hyperpolarizing its membrane potential.

• Nitric oxide makes by far the greatest contribution. Its role appears to be the continuous regulation of resistance vessels.

Table 6.4 summarizes the triggers for release of endothelium-derived relaxing factors and the second messengers responsible for vascular smooth muscle relaxation.

Endothelium-dependent constriction

• The vascular endothelium also produces smooth muscle–constricting factors under certain circumstances.

• Such production varies greatly among species and also among different vascular beds within a given species.

• The major contracting factors are prostaglandin H_2 (PGH_2), thromboxane (TXA_2), the endothelins, locally produced angiotensin II, and an as-yet-unidentified contracting factor, EDCF, that is produced and released in response to hypoxia.

- Prostaglandin H_2 (PGH_2) and thromboxane (TXA_2) both bind the TXA_2 receptor. Both are produced when cyclo-oxygenase is activated by mechanical or chemical stimuli. Preference for production of one or the other is determined by local concentrations of promoters and inhibitors. PGH_2 has an extremely short half-life.
- Endothelin-1 is cleaved from its precursor, preproendothelin, and this process is modulated by several receptor-coupled agents.
- Angiotensin II is produced in endothelial cells because such cells contain converting enzyme.
- Endothelium-derived contracting factor (EDCF) causes vasoconstriction during hypoxia.

• Table 6.5 summarizes the control of endothelial constrictor release.

Table 6.5 ***Constricting Factors Derived from the Endothelium***

Trigger	Endothelial Response	Mediator of VSM Contraction
Shear stress	↑ PGH_2/TXA_2	Receptor-mediated mechanisms
Various receptor agonists	↑ PGH_2/TXA_2	
	↑ Endothelin I	
Angiotensin I	↑ Angiotensin II	
Hypoxia	↑ EDCF	

EDCF, Endothelium-derived contracting factor; *PGH_2,* prostaglandin H_2; *TXA_2,* thromboxane; *VSM,* vascular smooth muscle.

Hemodynamics of the Cardiac Cycle

Cyclic contraction and relaxation of cardiac muscle cells leads to cyclic changes in pressure* within the chambers of the heart. These pressure changes lead to changes in chamber volume and to expulsion of blood into adjacent structures. The direction of flow is set by the cardiac *valves,* which are interposed between atrium and ventricle (= the mitral and tricuspid valves) or between ventricle and outflow vessel (= the aortic and pulmonic valves).

The cardiac cycle can be considered in three sequential phases (Fig. 6.3):

- Atrial contraction
- Ventricular contraction
- Ventricular filling

Left Atrial Contraction

- The atrial pressure rises ("a" wave) when atrial muscle contracts.
- Blood is added to the left ventricle. As a result,
- left ventricular volume increases to the level of left ventricular end-diastolic volume (LVEDV) and left ventricular pressure rises with a.

Left Ventricular Contraction

- As the ventricle begins to contract, ventricular pressure rises above atrial pressure and the mitral valve closes, producing the first heart sound (S_1).
- At the beginning of ventricular contraction the aortic valve is still closed because aortic pressure is still higher than left ventricular pressure.
- As the ventricle continues to contract, ventricular pressure continues to increase during isovolumetric (constant volume) contraction.
    - The increasing ventricular pressure pushes against the closed mitral valve leaflets, bulges them, and causes the c wave in the atrial pressure trace.
    - At the same time the contracting ventricular muscle fibers pull the fibrous valve skeleton toward the apex of the heart. This enlarges the atrial chamber and causes the x descent in the atrial pressure trace.
- When ventricular pressure just exceeds aortic pressure, the aortic valve opens, at which time, ventricular volume decreases rapidly and aortic flow increases as the stroke volume is ejected.

*Pressures in the cardiovascular system are always given with reference to atmospheric pressure. For example, a vascular pressure of 25 mm Hg means 25 mm Hg above atmospheric pressure.

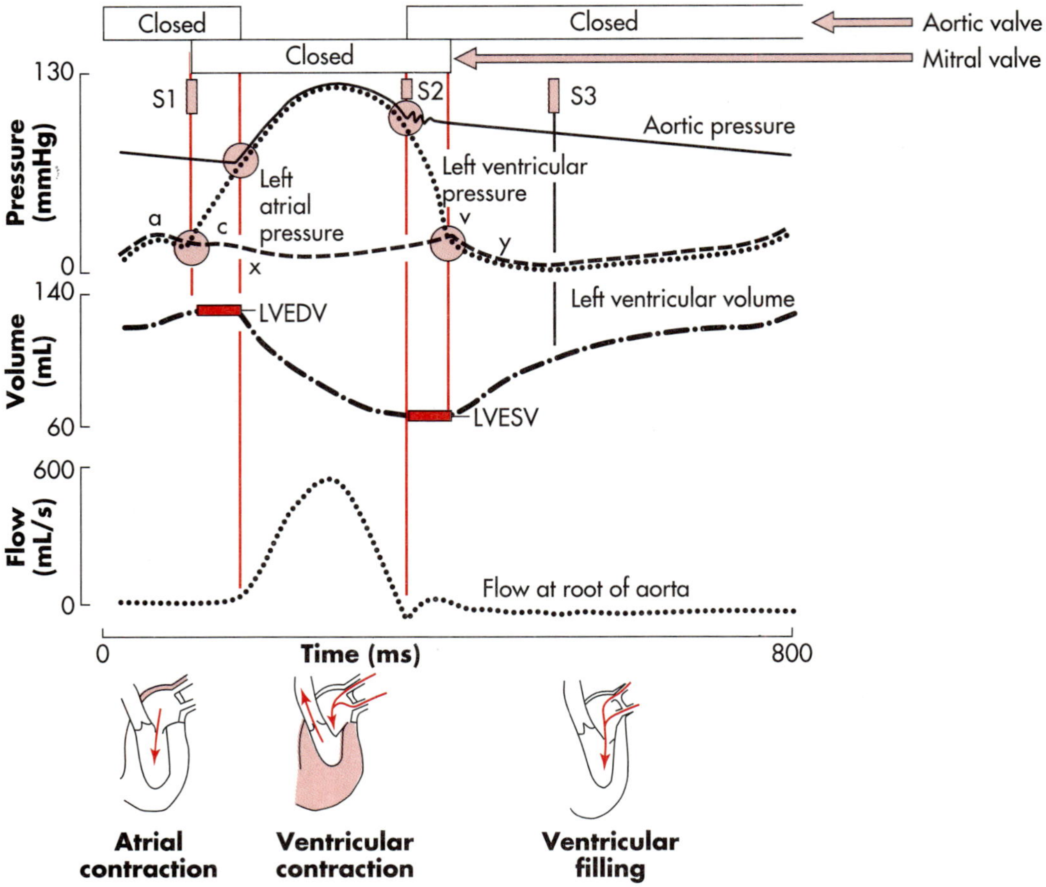

Fig. 6.3 Changes in hemodynamic variables on the left side of the heart during one cardiac cycle. The upper bars show the status of relevant valves at different times during the cycle. The course of the following variables is shown: aortic pressure, left ventricular pressure, left atrial pressure, left ventricular volume, and flow into the root of the aorta (that part of the aorta lying immediately distal to the aortic valve). Events on the right side of the heart follow similar patterns, but the systolic right ventricular pressure is near 40 mm Hg. *LVEDV,* Left ventricular end diastolic volume; *LVESV,* left ventricular end systolic volume; S_1, first heart sound; S_2, second heart sound; S_3, third heart sound. a, c, x, v, and y refer to features of the atrial pressure tracing.

- Ventricular pressure, aortic pressure, and aortic flow continue to rise until the ventricle begins to relax.
- Throughout the duration of ventricular contraction and initial ventricular relaxation the atrium fills by inflow from the pulmonary veins. This causes atrial pressure to rise after the x descent.

Left Ventricular Filling

Initial ventricular relaxation

- As the ventricle begins to relax, ventricular pressure falls and the rate of ventricular ejection slows greatly.
- Soon ventricular pressure falls below aortic pressure because the ventricles relax rapidly while blood inertance and blood vessel compliance retard pressure decline in the aorta.
- During this phase there continues to be flow from each ventricle in the direction of the energy gradient (not the pressure gradient!) until the aortic valve closes.

- After the aortic valve has closed, there is a clear separation between aortic pressure and ventricular pressure. This separation begins in the period of isovolumetric relaxation.

- **Significant ventricular filling**

  - The mitral valve opens when the falling ventricular pressure intersects the rising atrial pressure ("v" wave).
  - Ventricular volume increases rapidly as the ventricle fills in response to a gradient created between the full atrium and the rapidly relaxing ventricle.
  - **At normal resting heart rate most ventricular filling occurs in this phase.**
  - Rapid ventricular relaxation is responsible for the descent in ventricular pressure (y descent in atrial pressure) even while ventricular volume increases rapidly.

Heart Sounds

- S_1 (see Fig. 6.3) identifies the first heart sound.
  - Its timing corresponds with closure of the atrioventricular valves.
  - Separate mitral and tricuspid components of the sound (= splitting of the sound) may be revealed under some circumstances.
- S_2 identifies the second heart sound.
  - Its timing corresponds to closure of the aortic and pulmonic valves.
  - Separate aortic and pulmonic components of the sound (= splitting of the sound) may be revealed under some circumstances, especially during inspiration.
- S_3, when it is present, occurs at the end of the period of rapid ventricular filling.

Peripheral Vascular Mechanics

Mechanisms to Ensure Steady Flow Elasticity and viscoelasticity of the blood vessel walls are the mechanical factors that produce smooth, uninterrupted tissue blood flow even though aortic blood flow is greatly pulsatile.

- **Arterial elasticity** Blood vessel elasticity provides elastic recoil to propel flow in diastole. It derives mostly from collagen within blood vessel walls.
- **Viscoelasticity** Viscoelasticity of blood vessels acts like a shock absorber in smoothing out cardiac cycle fluctuations in flow and pressure. It derives mostly from relaxed vascular smooth muscle.

Mechanism to Permit Preferential Flow Distribution Blood flow depends on driving pressure* according to the formula:

$$\text{Blood flow} = \frac{\text{Driving pressure}}{\text{Resistance}}$$

*Driving pressure is the difference between pressure at the inflow end and pressure at the outflow end.

- **Resistance in blood vessels**

  • Resistance to flow in a single vessel is influenced by blood viscosity, vessel length, and, most significantly, vessel diameter:

$$\text{Resistance} \propto \frac{\text{Blood viscosity} \times \text{Vessel length}}{\text{Diameter}^4}$$

  • Since vessel diameter enters the calculation as a fourth power, changes in blood vessel diameter have a profound influence on flow resistance.

- **Resistance in vascular beds** Within vascular beds blood vessels may be connected *"in series"* or *"in parallel."* The pattern of their connection influences the degree of precision required of the mechanisms that exist for changing flow resistance.

  — ***Blood vessels in series***

  • Flow occurs sequentially from one vessel to the next

$$R_{\text{Total}} = R_1 + R_2 + R_3 + \ldots + R_n$$

  • Increasing the resistance in any one vessel (or portion of a vessel) decreases the flow through all the vessels. Vascular beds that show a "series" arrangement of blood vessels can change their resistance only by precise diameter adjustments of individual vessels.

  — ***Blood vessels in parallel***

  • In this arrangement of blood vessels, flow is divided and occurs simultaneously in several vessels.

$$1/R_{\text{Total}} = 1/R_1 + 1/R_2 + 1/R_3 + \ldots + 1/R_n$$

  • Increasing the resistance in any one vessel decreases the flow in that vessel and decreases the flow through all vessels combined unless the driving pressure increases. However, the vessels with unchanged resistance will receive a larger proportion of the total flow.

  • Vascular beds that show a "parallel" arrangement of blood vessels can change their resistance by *vasomotion* (total closing or total opening of selected blood vessels).

  • Opening of previously closed blood vessels is called *recruitment.*

  • Most vascular beds are arranged in parallel. As a result,

    • increased flow resistance in one vascular bed will cause blood flow to be redistributed to other vascular beds;
    • decreased flow resistance in one vascular bed will preferentially direct flow toward that bed.

Laplace's Law

• Hollow elastic structures that are being stretched by an internal distending pressure develop circumferential tension within their walls.

• Other hollow structures, such as the cardiac chambers, develop active tension in their walls in order to create a pressure within the chamber.

• In both settings the law of Laplace relates circumferential wall tension to internal pressure, radius, and wall thickness. For a cylindrical structure like a blood vessel the law takes the form shown in Fig. 6.4.

• The law of Laplace may be viewed from a slightly different perspective when the region of interest is a cardiac chamber in which pressure is being gen-

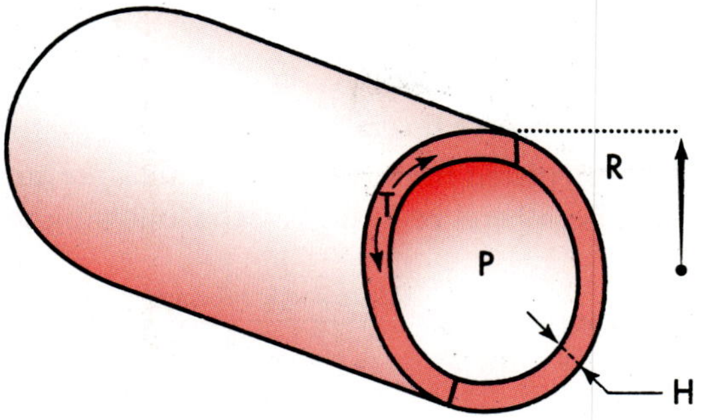

$$T = \frac{P \times R}{2H}$$

Fig. 6.4 The law of Laplace as it applies to a cylindrical structure like a blood vessel. *H*, wall thickness; *P*, distending pressure; *R*, radius; *T*, circumferential tension.

erated as a result of wall tension. In that case the formula may be used to calculate the pressure *(P)* that is generated within the chamber as a result of wall tension *(T)*:*

$$P = (T \times 2H)/R$$

Systemic Pressures and Flows

- **While the heart is stopped** The total volume of blood within the cardiovascular system is slightly larger than the system can hold without stretch. As a result, there exists blood pressure even when the heart has been stopped. The pressure distending the cardiovascular system when the heart is stopped is called *mean circulatory filling pressure* (MCFP). In humans MCFP is about 7 mm Hg at every point in the cardiovascular system. It is influenced mainly by blood volume and blood vessel stiffness (compliance).
- **While the heart is beating**
  - The beating heart takes blood from the venous side and adds it to the arterial side. Therefore,
  - the pumping action of the heart must lower the pressure below 7 mm Hg on the venous side and raise it above 7 mm Hg on the arterial side.
  - At steady state the magnitude of the arteriovenous pressure difference is just large enough to maintain a steady average flow through the system, against the flow resistance that is offered by the blood vessels.
    - In humans this steady state is achieved at a mean pressure near 100 mm Hg above atmospheric pressure at the root of the aorta.
    - As the blood flows toward the periphery, the pressure is gradually dissipated and reaches about 0 mm Hg at the inflow to the right atrium.

*This version of the law of Laplace is incorrectly oversimplified to stress the point that, for example, left ventricular pressure is dependent on ventricular wall thickness, radius, and wall tension. This view of the law of Laplace helps in understanding the effects of ventricular dilatation (increased R), hypertrophy (increased H), or failure to develop sufficient wall tension (decreased T) on the ability of the ventricle to generate sufficient pressure (P) to open the aortic valve during systole.

— *Cardiac output*

- Cardiac output is the average flow in the system. It is averaged over a period of several heart beats.
- It is calculated from the arteriovenous pressure gradient and the total peripheral resistance by the following relationship:

$$\text{Cardiac output} = \frac{\text{Mean aortic pressure} - \text{Mean right atrial pressure}}{\text{Total peripheral resistance}}$$

— *Venous return*

- Venous return is the average flow returning to the right atrium from the venae cavae.
- It can differ from cardiac output for only a few heart beats because the heart and the lungs have restricted abilities to sequester or supply blood.
- Venous return is profoundly altered by external pressures (such as those arising from the respiratory pump or the muscle pump) and by gravity (resulting in orthostatic, or postural, hypotension).

The respiratory pump. During inspiration there is augmentation of the pressure gradient from extrathoracic veins to intrathoracic veins. The change in pressure gradient causes increased venous return during inspiration.

The muscle pump. Muscle activity compresses veins from the outside and "massages" blood toward the heart in veins that are equipped with valves.

Orthostatic hypotension

- Changes in posture can alter the placement of some portions of the cardiovascular system relative to the absolute center of gravitational attraction. As explained below, this can change hydrostatic pressure within blood vessels.
- Blood vessels are continuous columns of fluid. In contrast, the space surrounding the outside of blood vessels is not a continuous column of fluid and therefore does not undergo the same pressure changes as the inside of blood vessels.

As a result the pressure difference across blood vessel walls (the transmural pressure) can change during postural changes.

- Arteries are strong vessels, little affected by postural changes.
- Veins are readily distended (or compressed) by changes in transmural pressure. For that reason,

- venous return is momentarily decreased as blood is pooled in the venous system on standing up suddenly, and it is increased momentarily as pooled blood is released from the venous system on lying down.

Arterial Blood Pressure Arterial blood pressure changes with distance downstream from the heart, and, within each cardiac cycle, the arterial pressure fluctuates between *systolic* and *diastolic* pressure.

- The average pressure during a cycle is the *mean arterial blood pressure.*
- *Pulse pressure* is the difference between systolic and diastolic pressure.

Systolic blood pressure

- Systolic blood pressure is the highest pressure reached during a cardiac cycle.
- Given a diastolic arterial blood pressure upon which the effect of cardiac action is superimposed, systolic arterial blood pressure is influenced by:
  - Cardiac performance
  - Aortic compliance
  - Total peripheral resistance (TPR)
    - TPR determines the rate of outflow from the arterial system to the capillary network (called the *peripheral outflow)* and thereby the net volume added to the arterial reservoir during systole.
    - The contribution of TPR effects to systolic pressure is about 20% of that attributable to cardiac performance. As a result,
- cardiac performance and aortic compliance are the major factors that determine how high systolic pressure will rise above diastolic pressure in a cardiac cycle.

Diastolic blood pressure

- The diastolic blood pressure is the lowest pressure reached during a cardiac cycle.
- Given a systolic level from which the pressure begins to decline, diastolic arterial blood pressure is determined by these factors:
  - Total peripheral resistance (TPR), which determines the rate of peripheral outflow from the arterial system during diastole and by that the rate of decline of ABP during ventricular diastole.
  - Heart rate, which determines the duration of the diastolic interval and by that the point at which the steady decline of pressure in diastole is halted.

Pulse pressure

- The pulse pressure is the difference between systolic and diastolic pressure.
- Under most conditions it is determined mainly by cardiac performance and aortic compliance.

Mean arterial blood pressure

- The average pressure during a cardiac cycle.
- Approximations to its calculation vary with the shape of the arterial pressure pulse.
- For pulse shapes observed in the aorta,

$$\text{Mean arterial blood pressure} = \text{Diastolic pressure} + \tfrac{1}{3}\ \text{Pulse pressure}$$

Effects of gravity

— ***Effects of gravity on blood pressure*** Gravity affects the pressure existing within blood vessels just as it affects the pressure that exists at various depths under the ocean surface.

- Within a continuous column of fluid and with respect to a measurement reference point within the column, pressure changes with distance from the reference point:

- Pressure increases for points closer to the absolute center of gravitational attraction (center of the earth).
- Pressure decreases for points farther away from the absolute center of gravitational attraction.

• The pressure measured at any point in the circulation is the sum of the pressure created by the pumping action of the heart (P_{Heart}) and gravitational forces arising from the weight of the column of blood that stands between the measuring point and the reference point for cardiovascular measurements.

P_{Heart} at any point is the pressure of heart action remaining after the losses from flow resistance are subtracted from the pressure created at the root of the aorta.

• The gravitational forces are given by the product ρGh (ρ, density of blood; G, the constant of gravitational acceleration; h, the height of the column of blood between the measuring point and the reference point for cardiovascular measurements.

• The tricuspid valve is taken to be the relative gravitational reference point for all cardiovascular pressures. This is the point to which all measuring devices are zeroed, no matter whether the body position is upright or supine.

• As a result of the above fundamental laws, the pressures measured at any point in the cardiovascular system are affected by gravity as shown in Fig. 6.5.

— ***Effects of gravity on blood flow***

• At any point in the circulation the total energy of a unit volume of blood of density, ρ, flowing at velocity, v,* consists of the sum of hydrostatic energy ($P_{Heart} + \rho Gh$) and gravitational potential energy (ρGh) and kinetic energy ($\frac{1}{2}\rho v^2$). That is, per unit volume of blood,

$$\text{Total energy} = \text{Hydrostatic energy} + \text{Gravitational E} + \text{Kinetic E}$$
$$= (P_{Heart} + \rho Gh) + \rho G(-h)^{\dagger} + \tfrac{1}{2}\rho v^2$$

• Gravitational changes in hydrostatic energy are always equal and opposite to changes in gravitational energy and the net effect of gravity on total energy at any point is zero.

• Since blood flows from regions of higher energy to regions of lower energy (*not necessarily* from regions of higher pressure to regions of lower pressure), the position of the unit volume with respect to the measurement reference point has no steady-state effects on blood flow. (There may, however, be transient orthostatic effects after changes in position.)

• The effects of gravity on the total energy of blood at various points in the cardiovascular system are shown in Fig. 6.6.

---

*Flow velocity, v, is assumed to be averaged across the lumen and over time.

†The negative sign arises from the fact that moving, for example, in the direction of decreasing hydrostatic energy (points above the tricuspid level) also means moving the same distance in the gravitational reference frame but in the direction of increasing gravitational potential energy.

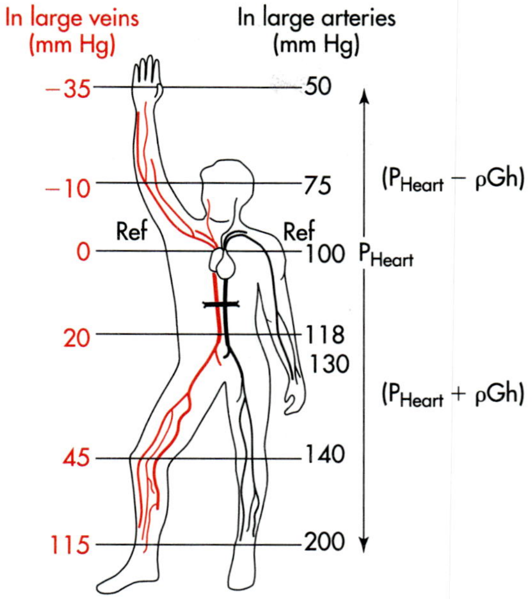

Fig. 6.5 The pressure at any point is the sum of the pressure created by the heart (P_{Heart}) and the pressure created by the weight of blood in the gravitational field *(ρGh)*. For example, at the elevated wrist:

On the arterial side:

$P_{Heart} = (100 - 3)$ mm Hg (if one assumes a frictional loss of 3 mmHg on the arterial side between heart and wrist.)

On the venous side:

$P_{Heart} = (100 - 88)$ mm Hg (if one assumes a frictional loss of 88 mm Hg for the whole circuit between heart, arteries, and veins to the wrist.)

On both the arterial and the venous sides:

$\rho Gh = 1.06 \times 980 \times (-60) = -62{,}328$ dyne/cm^2
$= -47$ mm Hg

As a result, at the elevated wrist:

Arterial blood pressure = 50 mm Hg
Venous blood pressure = −35 mm Hg

It follows that (1) elevation of a limb will reduce blood loss from a severed artery and (2) opening a vein in an elevated limb may create an air embolus.

Cardiovascular Regulation

Regulation occurs at two levels in the cardiovascular system:

- ***Individual tissues*** respond to their own metabolic needs and attempt to adjust their overall vascular resistance in such a way that they receive an adequate blood flow to satisfy their metabolic demands.
- ***Whole-body regulation*** involves the entire system simultaneously and appears to operate in two basically different modes, depending on whether conditions are normal or critical.
    - When conditions are normal, the objective of whole-body cardiovascular regulation is to ensure that *all* tissues receive sufficient blood flow to meet their metabolic needs.
    - When conditions are critical, the objective of whole-body cardiovascular

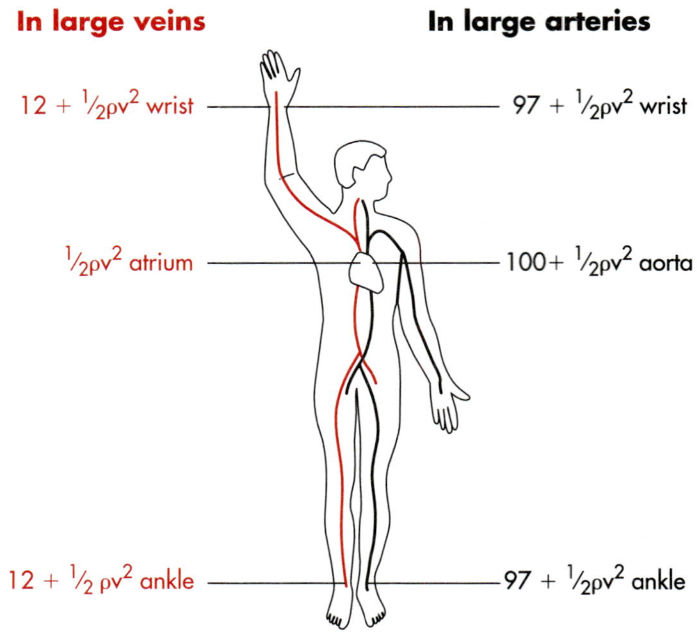

Fig. 6.6 Total energy at any point is the sum of hydrostatic energy ($P_{Heart} + \rho Gh$), gravitational potential energy (ρGh), and kinetic energy ($\frac{1}{2}\rho v^2$). The hydrostatic energy of a volume of blood at the elevated wrist is decreased by 47 mm Hg (Fig. 6.5). However, the volume of blood at the elevated wrist has acquired additional potential energy of 47 mm Hg on both the arterial and the venous sides by virtue of being placed farther from the absolute gravitational reference point (the center of the earth). It follows that the steady-state effect of gravity on the total energy of a volume of blood is zero. It follows also that changes in gravitational factors would have only short-term effects on blood flow if the cardiovascular system were made of rigid tubes.

regulation is to ensure survival of the organism by diverting all available flow to the brain and to the coronary vessels.

- Cardiovascular regulation involves both the control of cardiac function and the control of vascular function.

Control of Cardiac Function

- With each beat, each ventricle ejects a stroke volume of about 70 mL (in adults).
- The total volume of blood ejected per minute from each ventricle (i.e., the cardiac output, CO) is the product of stroke volume (SV) and heart rate (HR):

$$\text{HR} \times \text{SV} = \text{CO}$$

- Cardiac output can be adjusted by adjustment of the heart rate or stroke volume.
- Table 6.6 shows the normal ranges of HR and SV in healthy humans.
- It follows from the values shown in Table 6.6 that cardiac output can range from 3.5 L/min to near 35 L/min in humans.*

*Values up to 55 L/min have been recorded in some world-class swimmers and cyclists. One such athlete, Miguel Indurain, has a resting heart rate of 28 min^{-1}!

Table 6.6 *Normal Ranges of Heart Rate and Stroke Volume in Humans*

		Exercise	
	Rest	Fit Person	Athlete
Heart rate (min^{-1})	50	180	220
Stroke volume (mL)	70-80	110	150
End-diastolic volume (mL)	130	150	170
End-systolic volume (mL)	50-60	40	20

- **Control of heart rate**

• Each heart beat is initiated by electrical activity in the dominant pacemaker cell, which is normally located in the region of the SA node.

• The dominant pacemaker generates a spontaneous action potential whenever its membrane potential reaches threshold. Therefore,

- heart rate is determined by the interval between successive points of intersection of the phase 4 pacemaker potential with the threshold potential.
- The duration of that interval depends on only four factors, and a change in any of them will change heart rate:
  - duration of the pacemaker action potential
  - slope of the phase 4 pacemaker potential
  - level of threshold potential
  - level of maximum diastolic potential

— ***Short-term mechanisms*** Minute-to-minute changes in HR are achieved mainly by changing the slope of the phase 4 membrane potential in the dominant pacemaker cell.

— ***Long-term mechanisms*** These mechanisms are related to changes in extracellular electrolyte concentrations:

• Changes in extracellular [K^+] have pronounced effects on the maximum diastolic potential, the level from which E_m has to rise toward the threshold.

• Changes in extracellular [Ca^{++}] influence the threshold level, with the threshold membrane potential being made more positive by increased extracellular [Ca^{++}].

- **Control of stroke volume** Stroke volume is the difference between end-diastolic volume and end-systolic volume.

— ***Short-term mechanisms for regulating stroke volume*** In the short term (minutes to hours) stroke volume is determined by the completeness of systolic ventricular emptying, which, in turn, varies directly with myocardial performance. Four factors are important:

- ventricular preload
- ventricular afterload

- heart rate
- ventricular contractility

The cellular mechanisms by which these factors influence cardiac performance are described previously in the discussion of cardiac performance.

Preload

- Ventricular preload is determined by the degree of stretch of ventricular muscle fibers before they contract.
- At rest, preload effects help adapt cardiac performance to postural changes and match left-sided heart output to right-sided heart output during respiratory changes in venous return.
- During exercise, preload-initiated responses contribute to the mechanisms that increase cardiac output, though reflex mechanisms that operate by means of catecholamines can overshadow preload effects.

As the effectiveness of autonomic modulation of cardiac performance decreases with advancing age, the preload mechanism becomes of increasing importance among the factors that match cardiac output to the oxygen demands of the body.

DETERMINANTS OF PRELOAD. The degree of ventricular filling in diastole is determined by 3 factors: filling pressure, ventricular compliance, and time available for ventricular filling.

- Filling pressure is the difference between atrial pressure and ventricular pressure.
- Ventricular compliance is a measure of the ease with which the ventricle expands while it accepts diastolic inflow. It is affected by the properties of cardiac muscle itself or by factors that alter conditions in the pericardial space. Such factors take a long time to change and do not therefore contribute to short-term regulation.
- Time available for ventricular filling is inversely related to heart rate. Significant reductions in this interval occur at heart rates in excess of 150 min^{-1}, and such rates can be associated with compromised ventricular diastolic filling.

Afterload. Afterload is determined by the force that opposes ventricular emptying.

The major component of this force is the aortic input impedance.

- A decrease in afterload increases stroke volume and allows ventricular ejection at a lower wall tension.
- An increase in afterload causes the opposite effects. Their negative effect is blunted in the healthy heart by its ability to increase performance in the face of increased afterload (Anrep effect).

Heart rate. The heart rate affects contractility by changing the duration of the diastolic interval and, with that, the period of time during which Ca^{++} can be sequestered in intracellular stores.

Contractility

- This parameter, which cannot be measured directly, is influenced most profoundly by changes in the biochemical milieu within the sarcomere.

- In physiologic settings contractility is influenced mostly by sympathetic discharge, circulating epinephrine, and coronary blood flow.

— ***Long-term mechanisms for regulating stroke volume (cardiac hypertrophy)***

- Hypertrophy is the normal adaptive process by which the heart adjusts to a chronic stimulus.
- A chronic pressure overload leads predominantly to cell thickening by parallel deposition of new sarcomeres.
- A chronic volume overload leads predominantly to cell lengthening by serial deposition of new sarcomeres.
- Physiologic hypertrophy is characterized by normal or even increased contractility.
- Pathologic hypertrophy is characterized by decreased contractility.

Control of Vascular Function

Vascular impedance

- The afterload of the left ventricle contains elements of:
  - compliance
  - inertance
  - resistance
- Their combined effects determine vascular impedance and determine at each instant the relationship between pressure and flow at any site in the cardiovascular system:

$$\text{Pressure} = \text{Impedance} \times \text{Flow}$$

- Vascular impedance varies from site to site, and its measurement is technically difficult. However,
- in most circumstances the resistance component of vascular impedance undergoes the greatest change. Hence, linking driving pressure and tissue blood flow by only hydraulic resistance gives a sufficient approximation.

Total peripheral resistance The average perfusion pressure of all vascular beds is mean arterial blood pressure (ABP). It is determined by flow (cardiac output, CO) and resistance to flow (total peripheral resistance, TPR):

$$\text{ABP} = \text{TPR} \times \text{CO*}$$

— ***Determinants of resistance*** Vascular resistance in any tissue is changed when the degree of constriction of vascular smooth muscle is changed.

In a single blood vessel

- Hydraulic resistance is inversely related to vessel diameter to the fourth power.

*The product TPR × CO is not really equal to mean ABP but is equal to the difference between mean ABP and mean right atrial pressure. The error is small and justified by the simplicity of the resulting mnemonic.

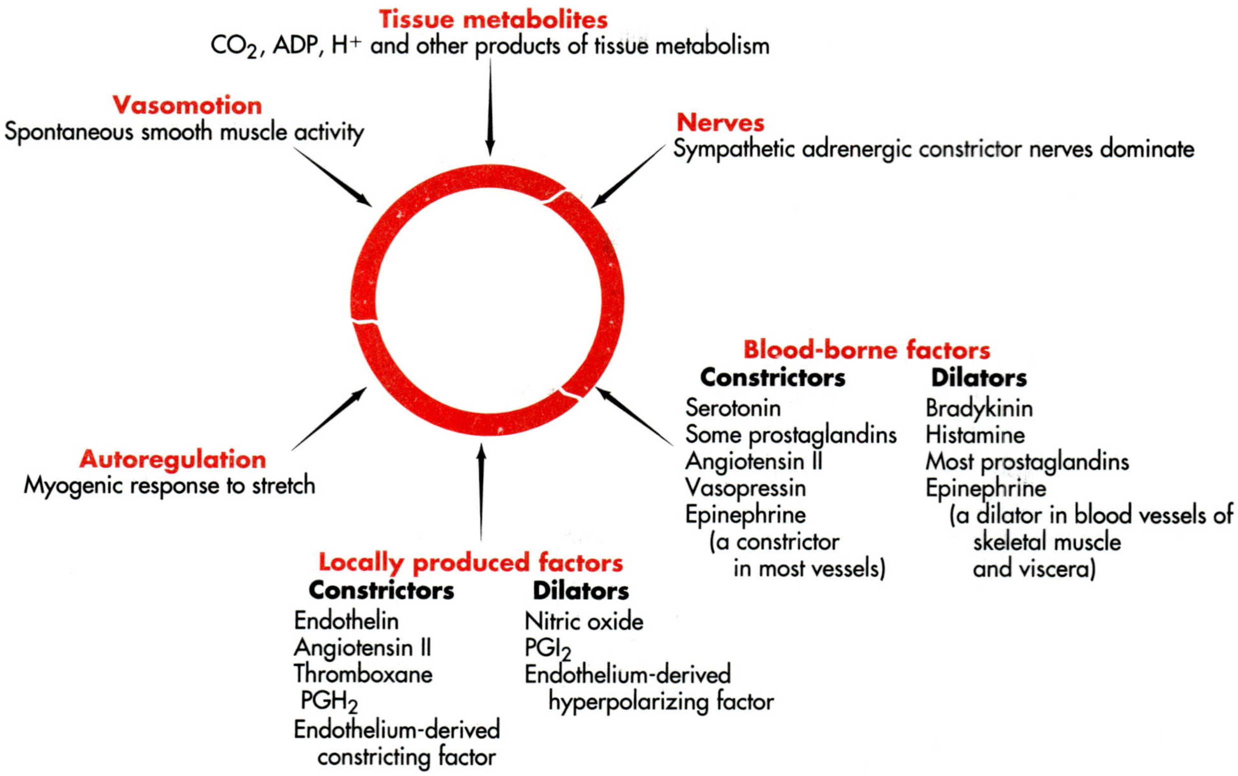

Fig. 6.7 Determinants of vascular smooth muscle tone.

- In the short term, changes in vessel diameter are brought about by changes in the tone of vascular smooth muscle, located primarily in arterioles. The determining factors are summarized in Fig. 6.7.
- In the long term, vessel diameter may be changed by any process that leads to narrowing of the lumen. Plaque formation is an example of such a process.

In vascular beds. The vascular bed of individual tissues consists of a large number of blood vessels that are connected in parallel. As a result, the effective tissue vascular diameter is determined both by

- the diameter of individual blood vessels (smooth muscle tone) and
- the total number of vessels being perfused. Spontaneous activity in precapillary sphincters determines whether a given capillary is open or closed and thereby determines the number of vessels being perfused. In a resting person such random activity accounts for 80% of total peripheral resistance.

— ***Adjustment of total peripheral resistance (TPR)***

- Total peripheral resistance, as opposed to the vascular resistance of an individual tissue, is adjusted in concert with cardiac output for the purpose of regulating mean arterial blood pressure. Such regulation is necessary because mean arterial blood pressure is an important determinant of tissue blood flow.
- The adjustment of peripheral resistance involves a conflict between local tissue demands and the central need for regulating mean arterial blood pressure.

• Each tissue decreases its *local resistance* in response to metabolic products, endothelium-derived vasodilator substances, and locally important neurotransmitter agents.

• The adjustment of TPR for the purpose of regulating mean arterial blood pressure is feedback-regulated by neuronal receptors that are located in certain areas of the vascular system.

• As a result of the competing interaction of local and central needs,

- tissues that are metabolically active have low resistance to blood flow because in them vasodilator actions of metabolites counteract centrally directed sympathetic vasoconstrictor influences. On the other hand,
- blood vessels in tissues that are not active are constricted by sympathetic nerves.

• In the optimally regulated system every tissue receives the blood flow it requires for its full range of metabolic activities.

Integrated Control of Arterial Blood Pressure

Overall scheme

• In most tissues, blood flow is increased as local tissue metabolism increases because locally produced metabolic factors relax vascular smooth muscle and decrease vascular resistance.

• If a local change in flow or volume is large enough to affect systemic arterial blood pressure, there will be a change in the activity of stretch receptors that are located most prominently in the cardiopulmonary region.

• The cardiovascular regulatory system responds to patterns of stretch receptor discharge with appropriate adjustments in heart rate, stroke volume, and total peripheral resistance. These adjustments are designed to accomplish five specific aims.

— *Specific aims of cardiovascular regulation*

1. To adjust cardiac output so that it is sufficient to meet whole-body metabolic needs.
2. To adjust total peripheral resistance in such a way that it and the existing cardiac output result in the appropriate arterial blood pressure (which is deduced from wall stretch in the areas where stretch receptors are located).
3. To adjust regional blood flows to meet local needs while protecting cerebral and coronary blood flow.
4. To adjust total and regional venous capacitance for blood volume sequestration or mobilization, as needed.
5. To adjust the rate of net fluid transfer across the walls of the capillary bed by altering the ratio of precapillary resistance to postcapillary resistance.

Cardiovascular regulation is summarized in the following mnemonic:

$$\begin{array}{r} HR \times SV = CO \\ \times \\ TPR \\ \| \\ ABP \end{array}$$

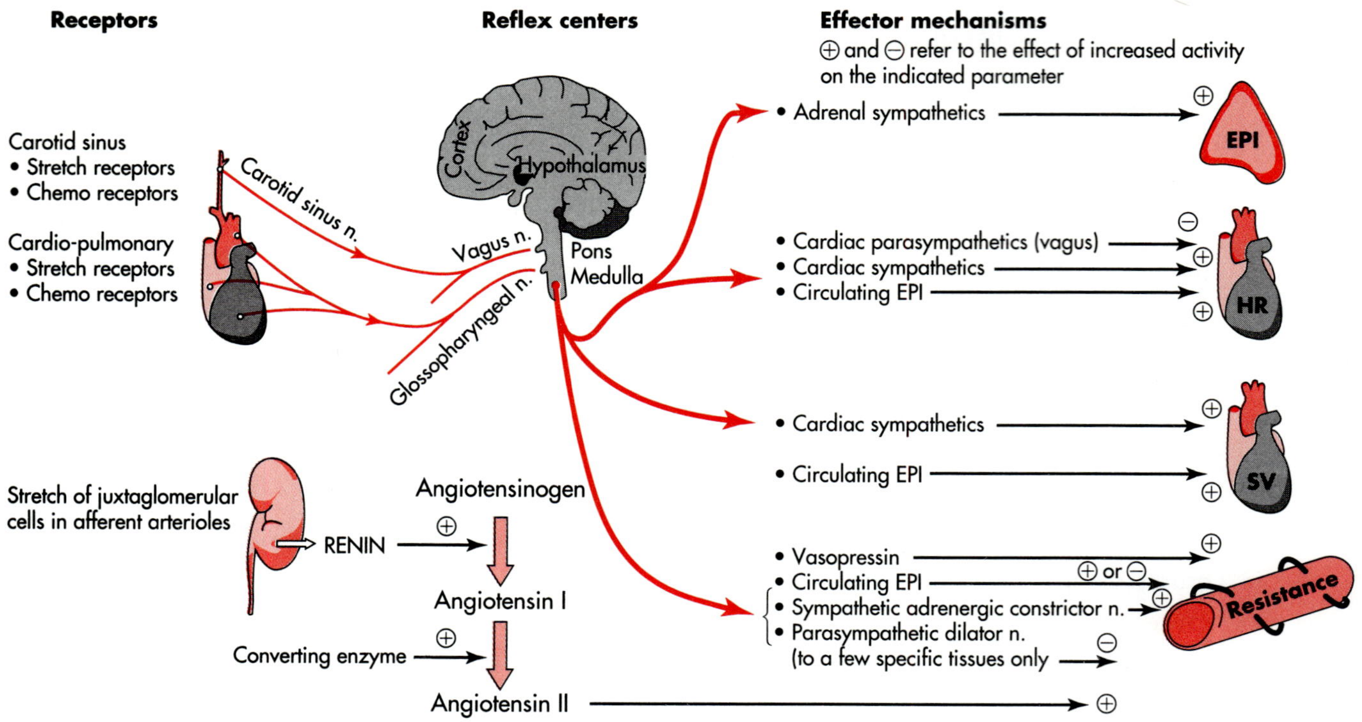

Fig. 6.8 Summary of cardiovascular reflex regulation. The vagus and glossopharyngeal nerves along with renal juxtaglomerular cells form the major afferent mechanisms. Reflex centers are located at levels as low as the spinal cord and as high as the frontal cortex. The adrenal medulla, cardiac pacemaker cells, cardiac muscle cells, and vascular smooth muscle constitute the effector mechanisms.

Reflex regulation Some cardiovascular responses are of an intrinsic nature, requiring no nerves or hormones. Examples are responses to change in preload or afterload. However, the bulk of cardiovascular regulation is governed by a reflex arc consisting of:

- receptors
- afferent paths
- reflex centers
- efferent paths
- effector mechanisms

Fig. 6.8 shows a summary of the components of cardiovascular reflex regulation. The system has a centrally directed neurohormonal component involving mostly the autonomic nervous system as well as a hormonal component that is based mostly in the kidneys. This latter component gives rise to the renin-angiotensin system.

— *Afferent mechanisms*

- Afferent fibers arise from receptors in the *carotid sinus* and *cardiopulmonary* region. They enter the midbrain via the *vagus nerves* and *glossopharyngeal nerves.*
- Input from stretch receptors normally dominates in importance over that from chemoreceptors.
- Input from carotid chemoreceptors has its greatest effect on ventilation. This will secondarily affect cardiovascular function by means of the respiratory pump mechanism.

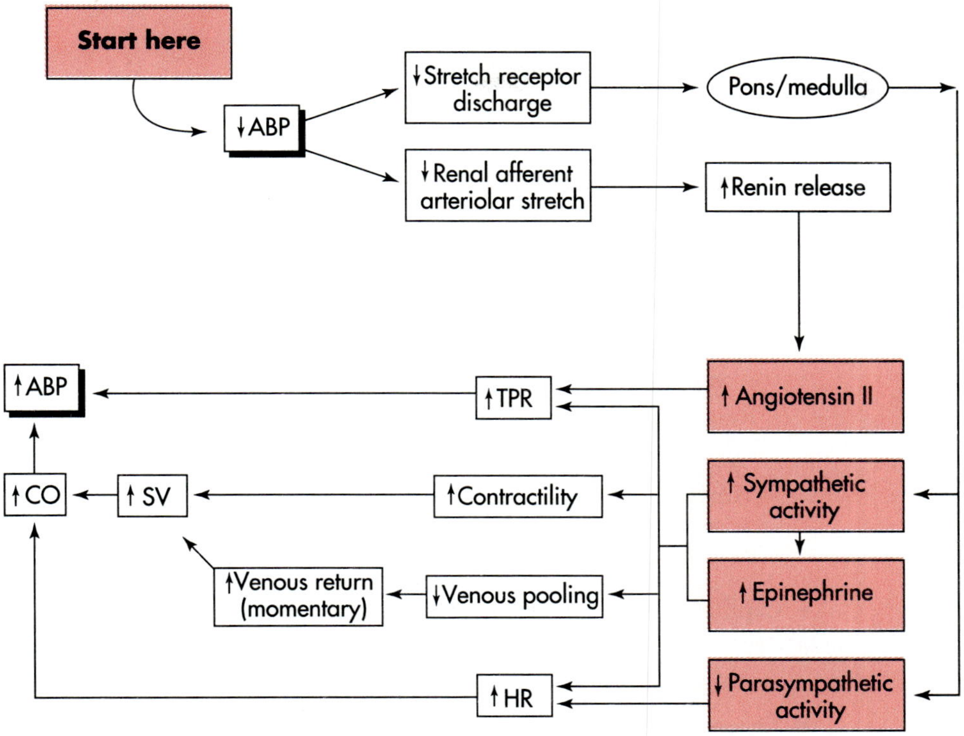

Fig. 6.9 Response of the regulatory components to a fall in arterial blood pressure. *ABP,* Mean arterial blood pressure; *CO,* cardiac output; *HR,* heart rate; *TPR,* total peripheral resistance; *SV,* stroke volume.

- Juxtaglomerular cells in renal afferent arterioles respond to decreased wall stretch with increased renin secretion and subsequently increased formation of angiotensin II.

— ***Reflex centers***

- The cardiovascular reflex centers are located in the ***midbrain*** (pons and medulla).
- Integration of cardiovascular function with other physiological systems is achieved through tracts from the ***hypothalamus*** to the midbrain.
- Emotional correlates derive from limbic and cortical tracts (see Chapter 3).

— ***Efferent mechanisms*** The autonomic nervous system and blood form efferent paths for cardiovascular regulation. Fig. 6.9 illustrates the operation of the regulatory components when their aim is to restore mean arterial blood pressure toward normal during an episode of imposed hypotension.

Microcirculation

■ **Structure** The essential elements of the microcirculation are shown in Fig. 6.10.

● Arterioles (diameter, 20 to 200 µm)

- These vessels contain smooth muscle units and therefore serve as elements capable of modulating both local flow and tissue vascular resistance.

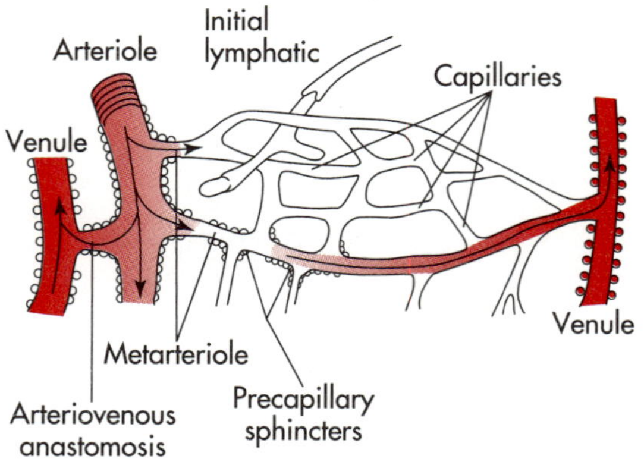

Fig. 6.10 A typical microcirculatory unit. The unit consists of an arteriole, perhaps a small number of metarterioles, several capillaries, a venule, perhaps an arteriovenous anastomosis (AV shunt), and several initial lymphatics.

- Capillaries often do not directly join arterioles but branch from metarterioles.

Metarterioles

- These vessels are of smaller diameter than arterioles.
- They can form a direct flow channel between an arteriole and a nearby venule.

Capillaries (diameter, 5 to 10 μm)

- These vessels lie distally to the smooth muscle fibers that form the *precapillary sphincter* at the junction between capillary and metarteriole.
- Capillaries contain only endothelial cells, no smooth muscle.
- Capillaries are distinguished from one another by the size and number of junctions between neighboring endothelial cells.
    - Nonfenestrated (continuous) capillaries have few endothelial junctions, and their clefts are small. Such capillaries are "tight." They are found in brain and muscle.
    - Fenestrated capillaries have more endothelial junctions than nonfenestrated capillaries have. Fenestrated capillaries are found in renal glomeruli and in the splanchnic bed.
    - Discontinuous capillaries have many and large endothelial junctions. The liver, spleen, and bone marrow contain such capillaries.

Venules Venules form the collecting conduits for blood flow from the capillary bed.

- Venules contain smooth muscle units. This feature permits them to be elements that control tissue blood flow.
- Venules have permeable walls. As a result they participate in the exchange functions of the microcirculation.

Initial lymphatics

- Lymphatic microvessels are anchored to surrounding tissue by microfilaments.

- Larger lymphatic vessels exhibit spontaneous contractile activity. They also have valves. This permits unidirectional pumping of lymph.

Arteriovenous anastomoses (A-V shunts)

- A-V shunts are muscular bypass channels.
- They are found in only some tissues, most prominently in those requiring, on occasion, a blood flow far in excess of metabolic needs (e.g., the skin).

Function The three major functions of the microcirculation are local hemodynamic regulation, homeostasis of the endothelial micromilieu, and exchange of substances across the capillary wall.

Local hemodynamic regulation Hemodynamic regulation at this level is governed by smooth muscle *tone.*

- The tone of smooth muscle in the arteriole, metarterioles, and venule is its degree of constriction. It determines total flow through the microvascular unit.
- The tone of precapillary sphincters determines what fraction of metarteriolar flow will pass through true capillaries.
- The tone of precapillary (i.e., arteriolar and metarteriolar) smooth muscle relative to postcapillary (i.e., venular) smooth muscle determines how much of the pressure generated by the heart will be conveyed to the capillary lumen.
    - Dilatation of precapillary smooth muscle increases hydrostatic pressure within the capillary.
    - Dilatation of postcapillary smooth muscle decreases capillary hydrostatic pressure.

Homeostasis of the endothelial micromilieu Capillary endothelial cells synthesize many factors. As described elsewhere, their functions include modulation of flow (see Vascular Endothelium), and modulation of hemostasis (see Chapter 2).

— ***Modulation of flow***

- Locally produced endothelial factors can act on upstream or downstream smooth muscle to alter its tone.
- The rate of synthesis for many of these factors is influenced by local flow rates or the local chemical milieu.

— ***Hemostasis*** Capillary endothelial cells synthesize factors that are essential for maintaining normal conditions at the blood-endothelium interface.

Exchange of substances across the capillary wall Substances move across capillary endothelium at rates that are determined by concentration gradients, pressure gradients, and endothelial permeability.

- Lipid-soluble substances move through endothelial cytoplasm.
- Other substances move through cell junctions or through pinocytotic vesicles.

— ***Exchange of fluid*** Fluid is lost to the interstitium because capillaries are leaky and because there is a hydrostatic pressure within capillaries. If such lost fluid were not recaptured, the blood volume would quickly be depleted.

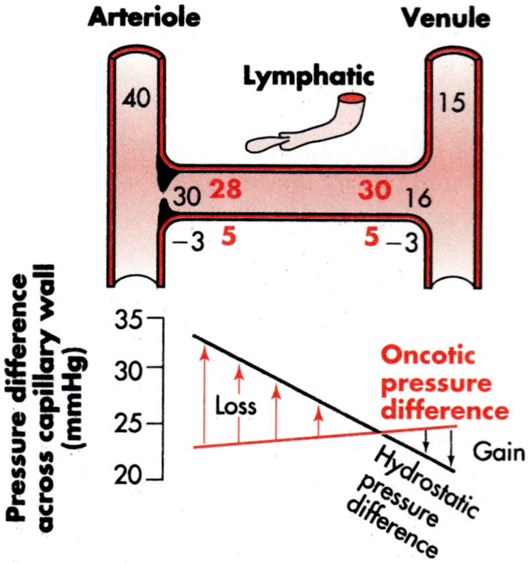

Fig. 6.11 Transmural pressures when the precapillary sphincter is open (with typical values). Hydrostatic pressures are shown in black, and oncotic pressures are shown in color. At the arteriolar end hydrostatic pressure exceeds oncotic pressure. As a result there is net fluid loss from the capillary to the interstitium. At the venular end hydrostatic pressure may be lower than oncotic pressure. As a result there may be net fluid gain to the capillary from the interstitium at this end. As a result of the existing pressures and their changes along the length of the capillary, fluid leaves the capillary along most of its length when there is blood flow in the capillary.

Balance of fluid exchange across capillary endothelium is maintained by two mechanisms:

- A hydrostatic pressure gradient tends to push fluid out of the capillary and an opposing protein osmotic (= oncotic) pressure gradient tends to draw fluid into the capillary.
- Excess interstitial fluid is removed by means of lymphatic uptake.

Exchange of fluid when the precapillary sphincter is open. Fig. 6.11 shows typical values for hydrostatic pressures and oncotic pressures in the capillary environment when the precapillary sphincter is open.

Exchange of fluid when the precapillary sphincter is closed. Fig. 6.12 shows typical values for hydrostatic pressures and oncotic pressures in the capillary environment when the precapillary sphincter is closed.

The above illustrations show that periodic closure of capillaries (vasomotion) ensures a balance between capillary fluid loss and fluid gain.

Features of Regional Circulations

■ **Cerebral Circulation** The cerebral circulation is characterized by autoregulation and by the blood-brain barrier.

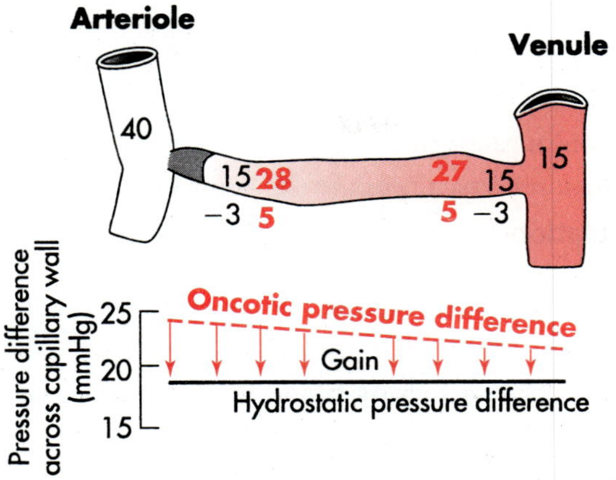

Fig. 6.12 Transmural pressures when the precapillary sphincter is closed (with typical values). Hydrostatic pressures are shown in black, and oncotic pressures are shown in color. When the precapillary sphincter is closed, there is no blood flow in the capillary. As a result, hydrostatic pressure is constant along the length of the capillary. As a further consequence, capillary hydrostatic pressure is set by the prevailing venular pressure because the capillary and venular regions are continuous at this point. The existing pressures along the length of the capillary cause fluid to enter the capillary along most of its length when there is no blood flow in the capillary.

Autoregulation (constancy of total blood flow)

- Total cerebral blood flow is kept nearly constant in the range of 60 to 160 mm Hg arterial blood pressure. The dominant mechanism is a myogenic response to vascular smooth muscle stretch.
- Regional cerebral flow varies with regional metabolic activity.
- Although blood vessels are innervated, nerves play a minor role in the control of vascular resistance; control by metabolic factors dominates:
  - Cerebral blood vessels are very sensitive to plasma P_{CO_2}, because CO_2 readily diffuses into vascular smooth muscle cells. There it forms H_2CO_3 initially and then H^+ ions. Intracellular H^+ causes vasodilatation.
  - Cerebral blood vessels are less sensitive to plasma $[H^+]$ because the blood-brain barrier prevents plasma H^+ from diffusing into smooth muscle cells.

Blood-brain barrier

- Most cerebral capillaries are of the nonfenestrated type, and their endothelial cells form tight intercellular junctions that prevent diffusion of many substances from the blood to the brain cells (the blood-brain barrier).
- This blood-brain barrier is breached in only a few areas of the brain. There blood-borne substances have access to the brain and substances produced in the brain can reach the bloodstream.
- The following areas lack a blood-brain barrier:
  - Pituitary gland
  - Pineal gland

- Median eminence
- Area postrema
- Choroid plexus
- Subfornical organ
- Organum vasculosum of the lamina terminalis

Coronary Circulation The three significant features of the coronary circulation are the presence of mechanical compression of blood vessels during systole, the dominance of metabolic factors rather than neuronal factors in the control of vascular resistance, and a very high oxygen-extraction ratio.

Mechanical factors affecting blood flow Tissue pressure provides a strong mechanical impediment to flow during systole, particularly in subendocardial vessels. Therefore coronary blood flow is maximal during ventricular diastole.

Control of coronary vascular resistance

- Control of flow is driven by local, metabolic factors including adenosine concentration and tissue O_2 demand.
- Tissue O_2 demand and consumption are determined mostly by the wall tension that needs to be developed for ejection of stroke volume.
- Since O_2 extraction in cardiac myocytes is normally high,* increased O_2 demand is met mostly by increased coronary flow rather than by increased O_2 extraction.

Cutaneous Circulation The cutaneous circulation shows a large number of arteriovenous anastomoses. Their purpose is to permit regional flows that are far in excess of the metabolic needs of the tissue.

Anatomical features The cutaneous circulation is characterized by

- a capillary plexus subserving tissue nourishment and
- a venous plexus subserving body temperature regulation.

Control of blood flow

- Sympathetically controlled arteriovenous anastomoses permit large blood flow through the venous plexus if required for thermoregulation.
- Cutaneous blood vessels are dilated in circumstances in which there is increased parasympathetic nervous activity to structures within the skin. The coincidence of such parasympathetic activity and the accompanying effect on local blood flow has led to the postulate that "active vasodilatation" is among the components that regulate cutaneous blood flow.
- It is likely that this "active" vasodilator influence is derived from chemicals that are liberated secondarily to cholinergic activation of sweat glands. Bradykinin has been implicated as the chemical mediator causing the vasodilatation.

Splanchnic Circulation (Gastrointestinal Tract, Liver, Spleen, and Pancreas) This large vascular bed has two major circulatory functions:

- It serves as a transport system for digestive functions of the splanchnic area.

*Coronary venous blood typically contains 4 to 5 mL of O_2 per 100 mL of blood, whereas systemic venous blood typically contains 14 to 15 mL of O_2 per 100 mL of blood.

- It serves as a reservoir in which blood volume can be sequestered or from which blood volume can be instantly mobilized.

- Anatomical features
  - Capillaries in this bed are generally of the fenestrated type and therefore highly permeable. This is especially true of the liver.
  - Most liver blood flow is provided by the portal vein; some is provided by the hepatic artery.
- Control of vascular resistance
  - Vasodilator influence derives in part from digestive enzymes, e.g.: gastrin and cholecystokinin, in part from
  - the purinergic nervous system whose neurotransmitter agent is adenosine and in part from
  - the β-adrenergic receptor agonist action of circulating epinephrine.
  - Vasoconstrictor influence derives mostly from sympathetic adrenergic influence.

Skeletal Muscle Circulation

- When skeletal muscle is resting, its blood vessels are strongly influenced by centrally directed, autonomic stimuli. As a result, resting skeletal muscle is a major locus of peripheral resistance adjustments. On the other hand,
- blood vessels in exercising skeletal muscle are controlled by local metabolic factors and serve the local metabolic needs of the tissue.

Pulmonary Circulation Pulmonary vascular pressures at any level from artery to veins are lower than comparable systemic pressures.

- **Hydrostatic implications of low vascular pressures** Blood flow distribution within the lung is greatly affected by the orientation relative to gravity (upright versus supine) and by alveolar pressure (inspiration versus expiration). These aspects are more fully explained in Chapter 5, Respiratory Physiology.
- **Fluid exchange across pulmonary capillaries** The dynamics of fluid exchange across pulmonary capillaries are governed by the same factors as in peripheral capillaries, but there are quantitative differences:
  - Pulmonary capillary hydrostatic pressure is only about 7 mm Hg, and pulmonary interstitial hydrostatic pressure is about −8 mm Hg.
  - Pulmonary capillaries are more permeable to protein than systemic capillaries are. Therefore the oncotic pressure of pulmonary interstitial fluid is near 14 mm Hg. (Pulmonary plasma oncotic pressure is 28 mm Hg, identical to systemic plasma.) As a result of these pressure values,
  - the net pressure forcing fluid out of patent pulmonary capillaries is only about 1 mm Hg.
  - Similarly, when pulmonary capillaries are closed, there is a large net force for the absorption of pulmonary interstitial fluid.
- **Pulmonary vascular resistance**
  - *Anatomical features*
    - The pulmonary capillary bed is very large. This extent makes the effective diameter also very large in comparison to the effective diameter of the systemic circulation, and it results in a low hydraulic resistance.

- Arterioles and venules have little smooth muscle. This inadequacy has implications for the control of resistance.

— ***Control of pulmonary vascular resistance***

- Sympathetic and parasympathetic effects are present, but they are weak.
- Pulmonary vascular resistance is controlled mostly by passive influences such as:
  - the effect of gravity on intravascular distending pressure
  - the compressive effect of alveoli on blood vessels
- Hypoxia and hypercapnia cause vasoconstriction in the lung.
- This unusual response of pulmonary vascular smooth muscle has the effect of shifting blood flow away from poorly ventilated alveoli.

Multiple Choice Review Questions

1. Loss of blood can be expected to cause poor blood flow in many areas of the body and can lead to cell death and subsequent K^+ spillage in those areas. Each of the following could be a result of increased extracellular K^+ concentration, *except:*
   a. The upstroke of the cardiac action potential would be steeper.
   b. Repolarization of cardiac cells would be caused by K^+ ions.
   c. Spontaneous electrical (pacemaker) activity would be generated in S-A nodal (pacemaker) cells.
   d. The duration of the cardiac muscle action potential might be decreased a little.
   e. Ca^{++} would enter cardiac muscle cells during the plateau of the action potential.

2. Please identify the one *correct* statement.
   Cells in the sinoatrial node are normally the dominant pacemaker because they:
   a. Are most richly supplied with nerve endings.
   b. Are incapable of developing contractile tension.
   c. Are the first region in the embryonic heart to show contractile activity.
   d. Have the steepest rise of diastolic membrane potential.
   e. Are the origin of the conduction system of the heart.

3. If the *ductus arteriosus* did not close after birth, then, with a patent *ductus,* blood would flow:
   a. From pulmonary artery to aorta during diastole.
   b. From aorta to pulmonary artery only during systole.
   c. Only during diastole.
   d. Continuously except for a shift in direction between systole and diastole.
   e. Continuously in one direction.

4. Which one of the following statements regarding steady-state venous pressure at the top of the head is *true* in a person who is being suspended, head down?
   a. It is increased mainly because of increased cardiac output.
   b. It is decreased mainly because of increased cardiac output.
   c. It is decreased mainly because blood volume is reduced by the translocation of plasma fluid to the pulmonary interstitium.
   d. It is increased mainly because of gravitational factors.

5. Which one of the following changes in the QRS complex of the ECG would you expect to see in a person who is being suspended, head down?
   a. Decreased amplitude in lead I; increased amplitude in lead II.
   b. Increased amplitude in lead I; increased amplitude in lead II.
   c. Increased amplitude in lead III; decreased amplitude in lead II.
   d. Decreased amplitude in lead III; decreased amplitude in lead II.

6. Identify the one *correct* statement.
   Systemic vasodilatation and bradycardia may be caused by:
   a. Decreased blood pressure in the carotid sinus.
   b. Cutting the carotid sinus nerves.
   c. Cutting the vagus nerves.
   d. Stimulating the carotid sinus nerve at a high physiological frequency.
   e. Occluding the common carotid artery in the neck.

7. Each of the following mechanisms influences the rate of fluid exchange between blood and interstitial fluid across the capillary wall, *except:*
   a. Fluid removal through lymphatic vessels.
   b. Protein metabolism by the liver.
   c. Pressure generation by the left ventricle.
   d. Transport of leukocytes through the capillary membrane.

8. Each of the following statements about the microcirculation is true, *except:*
   a. Recruitment (opening of previously closed capillaries) increases the area of endothelium that is available for blood/tissue exchange.
   b. Recruitment is a result of relaxation of precapillary sphincters.
   c. Recruitment results in decreased resistance to blood flow.
   d. Relaxation of vascular smooth muscle surrounding arterioles results in decreased resistance to blood flow.
   e. Dilatation of arterioles without simultaneous recruitment of capillaries has no effect on capillary hydrostatic pressure.

Chapter 7

Body Fluids and Electrolytes

Body Fluid Compartments

Body water exists within anatomically defined compartments.

- At the largest level of subdivision, body water exists as *intracellular fluid* and *extracellular fluid.* They are separated from each other by the cell surface membrane.
- Extracellular fluid is further subdivided into *interstitial fluid* and *intravascular fluid (plasma).* They are separated from each other by the capillary endothelium.

Fluid Intake and Fluid Output

Water and electrolytes normally enter by mouth.

- The daily intake in a Western society is 50 to 350 mmol of Na^+ (generally in the form of NaCl), 50 to 100 mmol of K^+ (generally as the phosphate salt), and 1 to 2 L of water (in a temperate climate).
- Some of the intake travels through the gastrointestinal system and leaves with stool. Most of the intake is absorbed into blood vessels of the intestinal wall and becomes part of the intravascular fluid space.

Internal Distribution of Fluid Intake

- From the plasma of the intravascular space:
  - Some water leaves through the lungs.
  - Some water and electrolytes leave through the sweat glands and the kidney.
  - The remaining water and electrolytes exchange with interstitial fluid across the capillary endothelium.
- Interstitial fluid exchanges with intracellular fluid across cell surface membranes.

Electrolyte Composition of Body Fluids

- **Intracellular fluid** K^+ and protein^{n-} are the predominant ions. Many proteins exist as negatively charged ions. Their valence can be greater than 1.
- **Extracellular fluid** Na^+, Cl^-, and HCO_3^- are the predominant ions, and their normal extracellular concentrations (in mmol) in humans are 140, 100, and 25 respectively.
- **Intravascular fluid (plasma)** Plasma has a higher protein^{n-} concentration than interstitial fluid has. This has two consequences:
  - Because the protein molecules are charged, charge equality demands that all diffusible ions (including Na^+, Cl^-, and HCO_3^-) be distributed at slightly unequal concentrations between interstitial fluid and plasma. This is known as the *Gibbs-Donnan phenomenon.*
  - Plasma osmolality exceeds interstitial osmolality by about 1 mOsm/

kg. Although this difference is too small to be measured, its water-attracting effect *(oncotic pressure)* is equivalent to a hydrostatic pressure of 25 mm Hg.

Physicochemical Properties of Body Fluids

- **Osmolality** The osmolality of a solution is defined as the number of osmotically active particles per kilogram of water. It is usually measured by determining the freezing point of the solution; this is a fussy, time-consuming process.

  - The concept of osmotic activity in a solution implies that the solution is separated from a differently composed solution by a *selectively permeable membrane.* (Such membranes are often called "semipermeable membranes.") Biological membranes are selective in that they permit the free passage of water but offer varying degrees of resistance to the passage of other constituents of the solution.
  - The ability of a solution to attract water from a neighboring fluid compartment and through the selectively permeable membrane that separates them is directly related to and expressed as its osmolality.

- **Osmotic pressure and oncotic pressure**

  - The osmotic pressure of a solution is equal to the hydrostatic pressure (measured in millimeters of mercury [mmHg] or equivalent units) that would have to be exerted on it to prevent water from an adjacent compartment to be drawn through the selectively permeable membrane by the osmotic gradient.
  - Oncotic pressure is a term that is used to refer to the osmotic pressure arising from plasma proteins.

- **Tonicity** The ultimate measure of whether water transport has occurred in a biological system is whether there has been a change in cell volume.

  - An *isotonic* solution is one that will not cause a change in the volume of cells that have been placed into it.
  - Iso-osmotic solutions are not necessarily isotonic. For example, a solution of urea in water, iso-osmotic with respect to intracellular fluid, will nevertheless cause cell swelling as urea equilibrates down its concentration gradient across the cell membrane (most mammalian cell membranes are quite permeable to urea). Therefore, urea solutions that are isosmotic with respect to intracellular fluid are not isotonic.

- **Specific gravity** The specific gravity of a solution is its weight compared to the weight of an equal volume of distilled water. Specific gravity can be used to estimate solute concentration and is often used for that purpose because it can be so much more easily determined than osmolality.

  Caution must be used in correlating specific gravity and osmolality when there exists the possibility that high-molecular-weight substances such as glucose or certain radiolabeled dyes might be present in high concentrations. Osmolality depends only on the number of solute particles. Specific gravity depends on their number as well as their molecular weight.

Exchanges Between Body Fluid Compartments

- **Exchange of fluids**

  — ***Across the cell membrane*** Net fluid movement occurs instantaneously in response to an osmotic gradient.

— ***Across the capillary endothelium*** Net fluid movement occurs in response to a gradient in hydrostatic pressure or a gradient in oncotic pressure.

- **Exchange of electrolytes**

— ***Across the cell membrane*** Net electrolyte movement across the cell membrane occurs in response to changes in one or more of the following:

  - The gradient in electrical potential
  - The concentration gradient
  - The rate of active transport processes

— ***Across the capillary endothelium*** There is free diffusion of all ionic species except protein^{n-}.

- **Whole-body balance and distribution of fluids and electrolytes** Two principles determine content and distribution:

  - The kidneys determine how much of the water and electrolyte intake is to be left in the body.
  - Physico-chemical forces operate across compartment boundaries and distribute throughout the body the fluids and electrolytes which the kidneys have left behind.

Filtration, Reabsorption, and Secretion

Nephrons: The Basic Functional Units of the Kidney Fig. 7.1 shows typical cortical and juxtamedullary nephrons and their associated peritubular capillary network.

- Cortical nephrons are shorter than juxtamedullary nephrons. As a result they offer less total surface area for solute/solvent exchange.

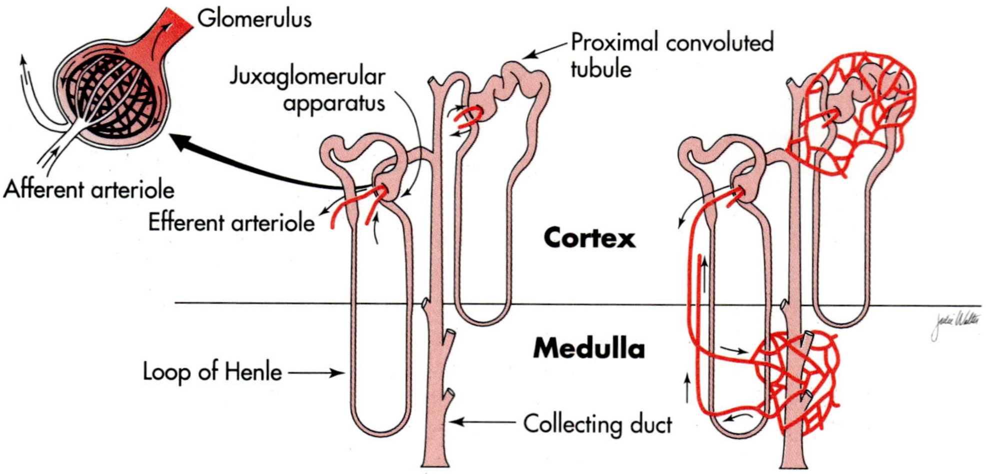

Fig. 7.1 Diagram of a typical cortical nephron and a typical juxtamedullary nephron. Their associated vascular networks are shown on the right in color. Efferent arterioles of cortical nephrons divide and form the peritubular capillary network of the proximal and distal convoluted tubules. In contrast, efferent arterioles of juxtamedullary nephrons form long, straight capillary vessels, the descending *vasa recta,* that run parallel to the loop of Henle into the medulla. At their target depth they form a capillary network that empties into the ascending *vasa recta.* Ascending *vasa recta* empty into a venular plexus.

- The early part of each distal convoluted tubule contains a region of special *macula densa* cells that make close contact with the *juxtaglomerular apparatus* in the afferent arteriole of the same nephron.
- Nephrons produce urine from blood by three mechanisms: *filtration, tubular reabsorption,* and *tubular secretion.*

From Renal Blood Flow to Glomerular Filtration Blood enters the nephron through the afferent arteriole of the glomerulus. It traverses the glomerular capillaries, the efferent arteriole, the peritubular capillaries, or vasa recta, and finally leaves through a venule.

- Glomerular filtration is driven by the same forces that drive capillary transepithelial exchange elsewhere: an imbalance between hydrostatic and protein-osmotic pressure gradients across the glomerular capillary wall. This imbalance is called the *net filtration pressure.*
  - Net filtration pressure is high at the beginning of the glomerular capillary and decreases progressively along its length as capillary albumin concentration rises with fluid loss to the space outside (Bowman's capsule).
  - Normally about 20% of the plasma entering as renal blood flow leaves the glomerular space as filtered fluid.
- Two functional kinds of glomeruli are distinguished: those showing a net filtration pressure of zero before the end of the glomerular capillary and those with a positive net filtration pressure at the end of the glomerular capillary.

Glomeruli with zero net filtration pressure before the end of the capillary are said to be in *filtration pressure equilibrium.* In such glomeruli the filtration rate is positively correlated with renal blood flow. A higher rate of flow "pushes" regions of higher intracapillary albumin concentration farther toward the end of the capillary and thereby moves the point of filtration pressure equilibrium farther toward the end of the capillary, allowing filtration to occur over a greater length.

Tubular Reabsorption Ninety-nine percent of the glomerular filtrate is reabsorbed as it moves along the nephron.

- Solute reabsorption occurs by both active and passive mechanisms.
  - Reabsorption of some solutes establishes electricochemical gradients across the tubular wall. Such gradients are then used for reabsorption of other solutes.
  - Reabsorption may also establish osmotic gradients across the tubular wall, and these may drive subsequent fluid reabsorption.
- There are two kinds of reabsorptive mechanisms: Those exhibiting a *transport maximum* (T_m-limited) and those exhibiting a *gradient-time maximum.* Both can be either active or passive.

T_m-limited transport

- T_m-limited transport occurs by means of substance-specific carriers that are limited in number. The reabsorption maximum (T_m) is reached when all carriers are occupied.
- If substance in excess of T_m is presented in the filtrate, all the excess will be excreted.
- Glucose and HCO_3^- are transported by T_m-limited mechanisms.

- **Gradient time–limited transport**
    - Epithelia can sustain only a finite transmural difference in concentration or electrical potential before backleakage becomes large enough to cancel active transport in the forward direction.
    - Whether the limiting gradient is reached depends on the time of contact between tubular fluid and tubular cell. As a result, the magnitude of the transepithelial gradient for any substance varies inversely with rate of flow of tubular fluid.
    - Na^+ is transported by a gradient time–limited mechanism.

Tubular Secretion Secretion is analogous to tubular reabsorption, but transport is into tubular fluid rather than out of it.

- Secretion may also exhibit a transport maximum.
- The most important examples of secreted substances in humans are creatinine, H^+, K^+, and NH_3.

Renal Clearance

Definition For any substance that is excreted in urine the total amount that is excreted in 1 minute can be considered to have come from a hypothetical volume of plasma that is now completely cleared of the substance. This hypothetical minute volume is defined as the *renal clearance* of the substance.

Calculation The clearance of any substance, *S*, is calculated as follows:

$$\text{Clearance}_S = \frac{\text{Urine concentration}_S}{\text{Plasma concentration}_S} \times \text{Urine flow rate}$$

Clearance of Inulin The concept of renal clearance is important mainly because there exists a substance, namely inulin, that is not secreted into the nephron but is freely filtered through the glomerular capillaries. Once inulin has entered renal tubules, it is not reabsorbed but is excreted in urine.

As a consequence of these features of the renal handling of inulin:

- At steady state the amount per minute of inulin appearing in the urine is equal to the amount per minute of inulin filtered through all glomeruli.
- The clearance of inulin (C_{In}) is identical to the rate of glomerular filtration (GFR), and C_{In} is used as a measure of GFR.
- The clearance of any substance can be used as an indication of the manner in which the substance is handled by the nephron; that is,
    - if $C_s > C_{In}$, then *S* is both filtered *and* secreted;
    - if $C_s < C_{In}$, then *S* is first filtered and then reabsorbed.

Clearance of Para-Aminohippuric Acid (PAH) PAH is freely filtered at the glomerulus and is also actively secreted into the nephron with such vigor that practically all of it is removed in one passage through the kidney and there is virtually no PAH in renal venous blood. As a result, clearance of PAH can be used to estimate renal plasma flow.

Renal Handling of Ions and Water

Although the energy turnover of the kidney is concerned mostly with reabsorption of Na^+, a variety of electrolytes and nonelectrolytes are transported across renal epithelia for the purpose of regulating body fluids, body electrolytes, and acid-base composition.

The nephron is divided into histologically and functionally different sections (see Fig. 7.1). Regional differences in function are a vital feature of nephron physiology.

The Proximal Convoluted Tubule

- **Epithelial structure** Several morphologic features of proximal tubular cells demonstrate that they have a large surface area for reabsorption and a high metabolic activity. These features are:
  - an extensive brush border on the luminal side
  - many infoldings on the interstitial side
  - large lateral intercellular spaces
  - many mitochondria
  - a high concentration of carrier proteins in the brush border
- **Solute reabsorption and secretion**

  • The cells of the proximal tubule form a leaky epithelium and a large fraction of fluid, and electrolyte reabsorption occurs by the *paracellular pathway,* through the gaps between adjacent cells.

  • The other route for reabsorption is called the *transcellular pathway.* Reabsorption by this route occurs in three steps: (1) from tubular lumen into tubular cells, (2) from tubular cells to the peritubular interstitial space, and (3) from the interstitial space to the peritubular capillaries.

— ***Transport of Na^+, K^+, Cl^-, or HCO_3^-*** Fig. 7.2 summarizes the mechanisms of proximal tubular reabsorption of the most significant ion species.

Sodium. The proximal tubule reabsorbs 40% to 45% of the filtered sodium. In the process, electrochemical gradients are established for reabsorption of Cl^- and HCO_3^-

• Na^+ enters passively on the luminal side, driven by a large electrochemical gradient.

• A high concentration of Na^+,K^+-ATPase in the membrane at the lateral intercellular spaces focuses active outward Na^+ transport in that region.

• Viewed globally, most Na^+ reabsorption from tubular lumen to blood occurs along with Cl^- reabsorption and subserves homeostasis of extracellular fluid volume. Some Na^+ reabsorption occurs along with HCO_3^- reabsorption and serves acid-base homeostasis.

Chloride

• Slightly more than 50% of proximal tubular Cl^- reabsorption occurs by the transcellular route in association with Na^+ (dashed arrows in Fig. 7.2). Both ions move together across the luminal and basolateral membranes of the tubular epithelium by mechanisms that are not yet fully understood.

• 35% to 45% of proximal tubular Cl^- reabsorption occurs by passive mechanisms through the junctions between adjacent tubular cells (wavy arrow in Fig. 7.2). These mechanisms include solvent drag as well as diffusion down an electrochemical gradient. The latter arises in the first instance from high tubular [Cl^-] in tubular fluid as a result of HCO_3^- reabsorption in the earliest tubule segment. Cl^- moves down this concentration gradient from tubular lumen to blood and thereby renders the proximal tubular lumen electropositive. This

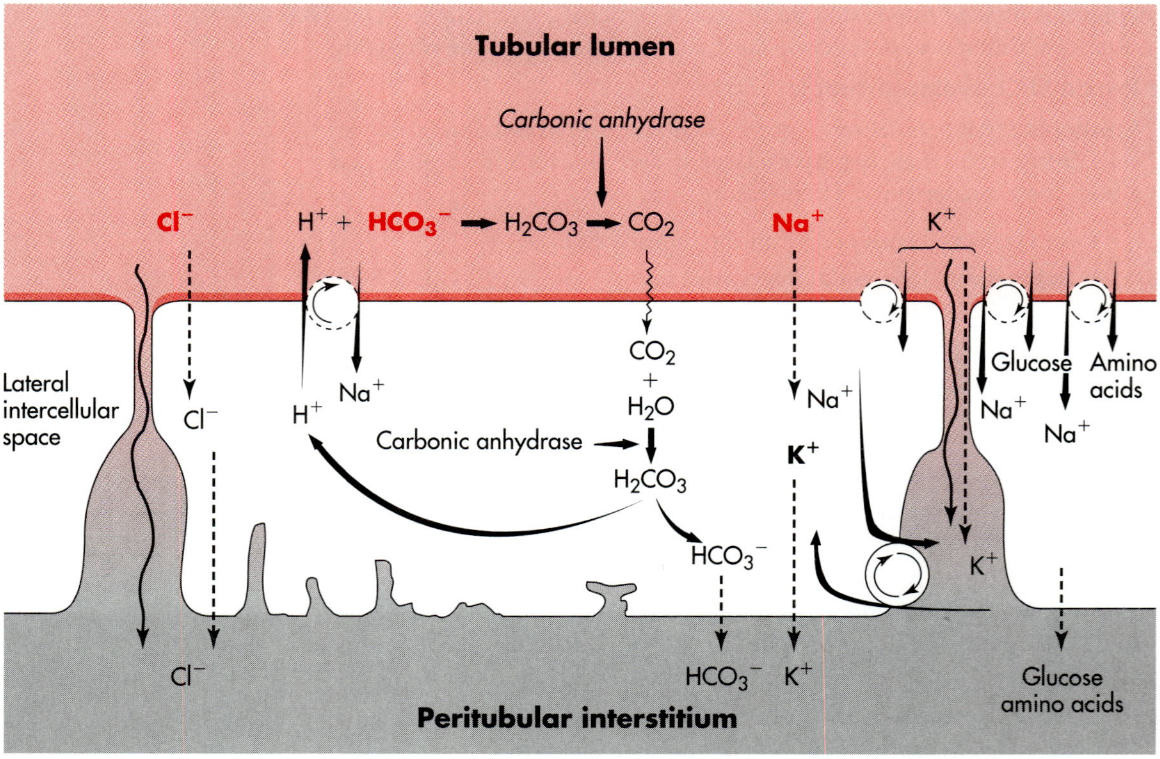

Fig. 7.2 Summary of mechanisms by which the major ions are reabsorbed from the tubular fluid to the renal interstitium. Cl^- is reabsorbed by passive mechanisms. HCO_3^- reabsorption occurs by intermediate formation of CO_2, which diffuses readily across the brush border. Na^+ and K^+ are reabsorbed by a combination of active and passive transport mechanisms. It is chiefly Na^+ reabsorption that creates electrical and osmolality gradients for reabsorption of other substances.

potential is important for the reabsorption of other ion species.

Bicarbonate. HCO_3^- is transported indirectly by means of CO_2 that is formed when H^+ is secreted into the lumen by a protein-based Na^+-H^+ exchanger located in the luminal membrane (Fig. 7.2). The presence of carbonic anhydrase in tubular fluid assures rapid dissociation of H_2CO_3 into CO_2 and H_2O. Similarly, carbonic anhydrase within tubular cells causes rapid formation of H_2CO_3 from CO_2 and H_2O. Subsequent dissociation of H_2CO_3 provides H^+ for further HCO_3^- reabsorption.

Potassium

- Three mechanisms participate in K^+ reabsorption by the proximal tubule: solvent drag (20%), passive diffusion (60%), and active transport (20%).
- Solvent drag (wavy arrow, Fig. 7.2) and passive diffusion (dashed arrow, Fig. 7.2) appear to occur primarily through the gaps between adjacent cells.
  - Solvent drag occurs at this site because the gaps form a ready pathway for water transport.
  - Passive diffusion occurs there because the Na^+-K^+ pumps set up a favorable concentration gradient for K^+ diffusion.

— ***Transport of glucose or amino acids***

• Glucose reabsorption in the proximal tubule is Na^+ dependent and occurs by means of a Na^+-glucose cotransporter.

• Almost all filtered amino acids are reabsorbed, and the reabsorption occurs almost exclusively in the proximal convoluted tubule in cotransport with Na^+.

● **Water reabsorption**

• Water reabsorption always follows passively down the osmotic gradients created by solute transport.

• As a consequence of water transport, interstitial hydrostatic pressure increases and interstitial protein osmotic pressure decreases. The resulting change in capillary transepithelial forces favors uptake of interstitial fluid in the peritubular capillary network.

■ **The Loop of Henle** Much of the NaCl escaping the proximal convoluted tubule is reabsorbed in the loop of Henle, particularly in the ascending thick segment.

● **Thin segment** Cells in this portion of the nephron are thin. They have no brush border and few mitochondria.

— ***Descending limb*** This region is highly permeable to water and moderately permeable to most solutes. Little solute transport occurs here.

— ***Ascending limb*** This portion is less water permeable than the descending portion. It shows a small amount of passive NaCl reabsorption and a moderately high permeability for urea.

● **Thick segment** Cells in this portion resemble those of the proximal convoluted tubule but lack the brush border.

- The paramount feature of this segment is its impermeability to water.
- The segment pumps Na^+, K^+, and Cl^- actively.

As shown in Fig. 7.3, the luminal cotransporter moves 4 ions simultaneously: two Cl^-, one Na^+, and one K^+.

- Na^+ leaves by means of the Na^+-K^+ pump at the basolateral membrane.
- Cl^- leaves probably by a Cl^- channel.
- K^+ diffuses back into the lumen, down its electrochemical gradient, and creates electropositivity in the lumen. Positivity inside the lumen creates a force for reabsorption of Na^+, K^+, Ca^{++}, or Mg^{++} into the renal interstitium, possibly by a transcellular route.

• The net effects of solute reabsorption in the ascending thick segment without accompanying water are the creation of (1) high osmolality in the interstitium of the corticomedullary region and (2) hypotonic tubular fluid at the beginning of the distal convoluted tubule.

■ **The Distal Convoluted Tubule** Only 10% to 20% of the filtered load reaches the distal convoluted tubule. This portion of the nephron is relatively impermeable to urea and water.

• The primary transport process is active Na^+ reabsorption. The resultant negativity in the lumen provides a driving force for passive Cl^- reabsorption into the interstitium as well as for secretion of K^+ and H^+ into the distal tubular lumen.

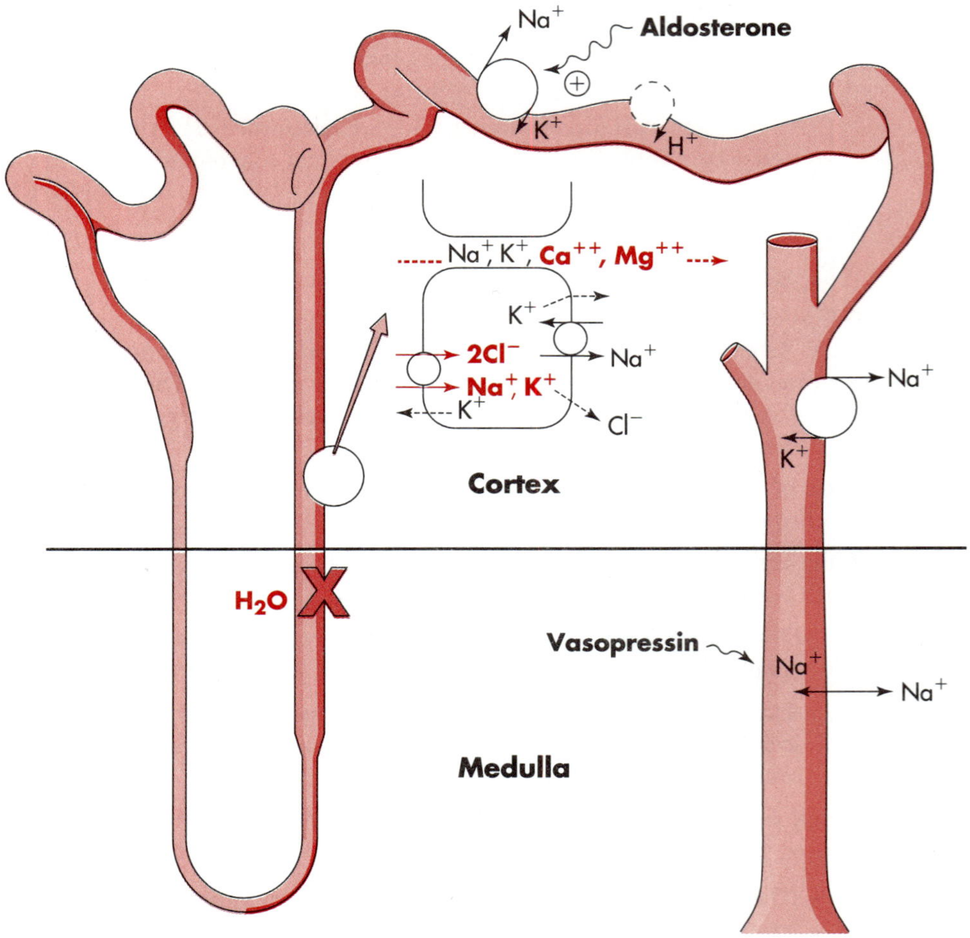

Fig. 7.3 Solute reabsorption or secretion in distal parts of the nephron. Sites of action of the hormones aldosterone and vasopressin are also shown.

- The K^+/H^+ secretion ratio depends on the acid-base status.
- Both Na^+ and K^+ transport rates are regulated by *aldosterone.*

The Cortical Collecting Tubule and Medullary Collecting Duct The fluid reaching the cortical collecting tubule is hypotonic because it comes from regions where solute is reabsorbed without accompanying water (ascending limb of the loop of Henle and distal convoluted tubule).

Na^+ and K^+ can be reabsorbed actively or secreted, depending on the status of body electrolytes. Water follows passively, with its rate of transport being determined by two factors:

- The osmotic gradient between lumen and medullary interstitium
- Epithelial water permeability, which is directly related to plasma levels of vasopressin (antidiuretic hormone)

Urinary Concentration and Dilution

Osmolality as the Driving Force Althouth the osmolality of plasma and of glomerular filtrate is remarkably constant at about 300 mOsm/kg, urine can be as dilute as 50 mOsm/kg in conditions of excess water intake or as concentrated as 1200 mOsm/kg in severe dehydration. Three factors are vital to the production of urine with such a range of osmolality:

- The presence of a very high osmolality in the medullary interstitium
- The anatomical routing of the water-permeable collecting duct through the region of high medullary interstitial osmolality
- The modulation of water permeability in the collecting duct by the hormone vasopressin (ADH)

Medullary Interstitial Osmolality Medullary interstitial osmolality increases progressively from renal cortex to renal medulla. Creation of highly concentrated fluid in this region requires:

- transport of osmotically active material (mostly Na^+, Cl^- and urea)
- exclusion of water

NaCl accumulation The progressive corticomedullary osmotic gradient is in part established by a *countercurrent multiplier* mechanism for NaCl. This mechanism requires:

- anatomical juxtaposition of the two limbs of the loop of Henle
- differentially selective permeabilities to NaCl, urea, and water in each limb
- active transport mechanisms that will take solute out of the ascending limb

— ***Countercurrent multiplier for NaCl*** As tubular fluid enters the descending limb of the loop of Henle.

- Active pumps, located in the luminal border of the thick ascending segment, add Na^+ and Cl^- to the interstitial space and to the fluid in the thin descending limb.
- Water does not follow the solute because the thick segment is water impermeable.
- The net result is that NaCl concentration increases slightly in this portion of the descending limb and at this level in the interstitium.
- A little deeper into the medulla, more NaCl is added by the active pumps, and the osmolality increases further.
- At steady state, each depth of the renal interstitium is characterized by a fixed concentration difference between ascending fluid and its surroundings.
- The longer the loop, the higher is the osmolality at its tip.

Urea accumulation Urea is formed during protein metabolism and is excreted mainly in urine where its presence is responsible for approximately half of the total urine osmolality. Urea is a normal constituent of glomerular filtrate. Its concentration in tubular fluid changes along the length of the nephron because urea permeability varies in the different sections (Fig. 7.4).

- Up to the thin descending segment of the loop of Henle, tubular urea concentration is low because to that point urea permeability of tubular cells is high and urea quickly moves down any concentration gradient that arises from water reabsorption.
- Tubular urea concentration increases slightly from the thin descending segment to the thin ascending segment partly because in this region

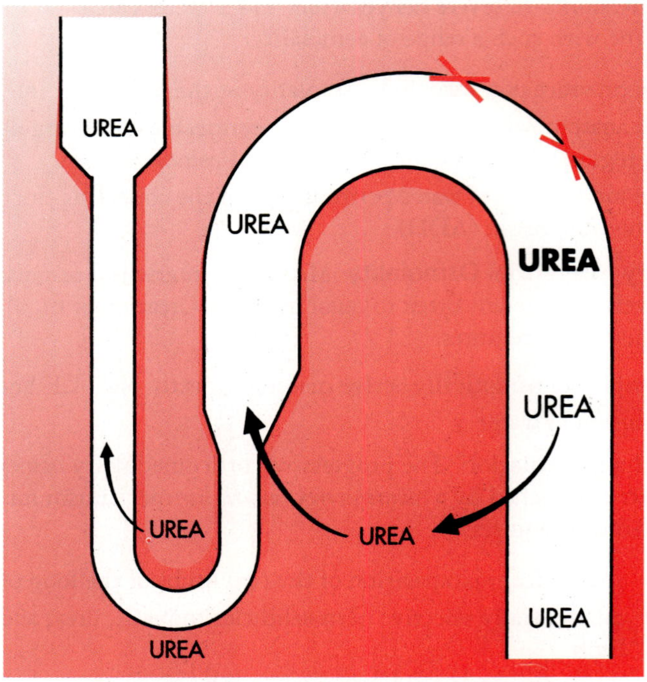

Fig. 7.4 Concentration, recycling, and excretion of urea. Different sections of the nephron vary in their permeability to urea. Local differences in the tubular permeability to water and urea allow urea concentration to change along the length of the nephron and allow urea to be an osmotically active agent in the renal interstitium.

water reabsorption occurs more readily than urea reabsorption and partly because urea is added to the tubular fluid from the interstitium in response to a concentration gradient.

- Urea becomes highly concentrated in the distal convoluted tubule and early collecting tubule because they are impermeable to urea. Any reabsorption of water will raise tubular fluid urea concentration progressively until the fluid reaches the inner medullary collecting tubule, which is highly urea permeable.
- As a result of the high urea concentration in the inner medullary collecting tubule and its urea permeability, urea is dumped into the renal interstitium at this site and contributes to the tonicity of interstitial fluid around any segment of tubule whose urea permeability is lower than its water permeability.

Regulation of Extracellular Volume and Osmolality

Extracellular osmolality and extracellular fluid volume give the appearance of being regulated. For example,

- Under normal conditions of diet, physical activity, and ambient temperature, plasma osmolality is maintained at a level about halfway between its threshold for vasopressin secretion and its (higher) threshold for thirst.

• Isotonic expansion of the extracellular fluid volume or isohemic changes in blood volume are quickly followed by appropriate changes in water intake and in renal excretion.

In pathological states, particularly those involving edema, the kidney continues to retain Na^+ and water even though the extracellular fluid compartment is already expanded. This observation perpetuates the question of which volume is being regulated: extracellular fluid volume, *effective circulating volume,* or *fullness of the arterial circulation.*

Sensory Mechanisms

Pressure

• Nervous receptors that respond with changes in firing frequency to changes in ambient stretch are located diffusely throughout the cardiovascular system. They are concentrated in the aortic arch, carotid sinus, cardiac ventricular wall, and cardiac atrial wall. Their action potentials are conveyed to the central nervous system either via sympathetic afferents or via the vagus and glossopharyngeal nerves.

• An additional stretch-sensitive mechanism resides in juxtaglomerular cells of the renal afferent arteriole. These cells synthesize the proteolytic enzyme *renin* and release it when afferent arteriolar stretch is diminished. Fig. 7.5 shows the cascade that is initiated by interaction of renin and angiotensinogen. Its final products are the biologically active peptides angiotensin II and III.

Volume

• The walls of the cardiac atria contain stretch-sensitive neuronal receptors that fire predominantly during the v wave of the atrial pressure tracing (see Chapter 6, Fig. 6.3). In view of the association of the v wave with atrial filling (as opposed to atrial contraction) these receptors are often described as atrial volume receptors. They are significantly involved in the modulation of vasopressin secretion by nonosmotic stimuli.

• Atrial muscle cells contain secretory granules filled with the immediate precursor to *atrial natriuretic factor,* a small peptide that is released mostly in response to atrial wall stress. Its actions include increased rate of glomerular filtration as well as inhibition of Na^+ reabsorption, particularly at the collecting tubule site.

Flow

• Cells in the macula densa region of the distal convoluted tubule are sensitive to the rate of distal Na^+ delivery. These cells are in intimate contact with the juxtaglomerular cells of the afferent arteriole.

• Changes in Na^+ delivery to the macula densa are directly related to renin secretion from the juxtaglomerular cells. Thus, increased distal Na^+ delivery leads to increased renin release with subsequent formation of angiotensin II, constriction of the afferent arteriole, and decreased glomerular filtration rate. This process by which distal Na^+ delivery in a nephron modulates its own glomerular filtration rate is called *glomerulotubular feedback.*

Osmolality Cells within the portal venous circulation of the liver and within the hypothalamus respond to changes in extracellular osmolality. Neurons near to but separate from the supraoptic nucleus of the hypothalamus are of greatest importance.

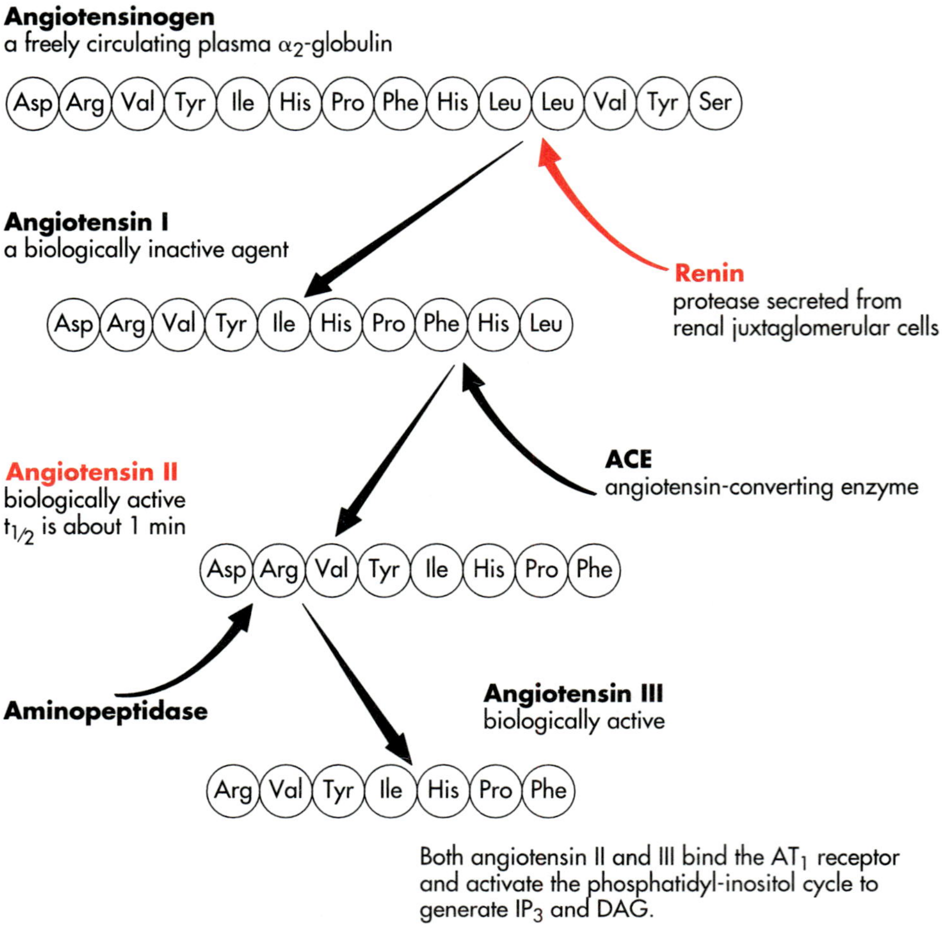

Fig. 7.5 The biochemical cascade by which renin leads to formation of the biologically active peptides angiotensin II and angiotensin III.

The functional properties of these receptors are that, at plasma osmolalities below a certain threshold, plasma vasopressin is suppressed below detectable levels. Above the threshold, plasma vasopressin concentration rises steeply with small increases in extracellular osmolality.

Reflex Centers

- **Hypothalamus** The central nervous area that is most significantly involved in the regulation of body fluid balance and its integration with cardiovascular function is the anteroventral region of the third cerebral ventricle (AV3V). It includes structures such as the *preoptic nuclei* and the *organum vasculosum of the lamina terminalis.*

Effector Mechanisms

Effector Mechanisms Thirst and renal excretion are the two major components of the systems that regulate body fluid volume and osmolality.

- **Thirst**

  - Water intake is regulated by the sensation of thirst. It emanates from neurons close to but not identical with those regulating vasopressin secretion.
  - The level of angiotensin II in the AV3V region has been identified as an important stimulant of drinking behavior.

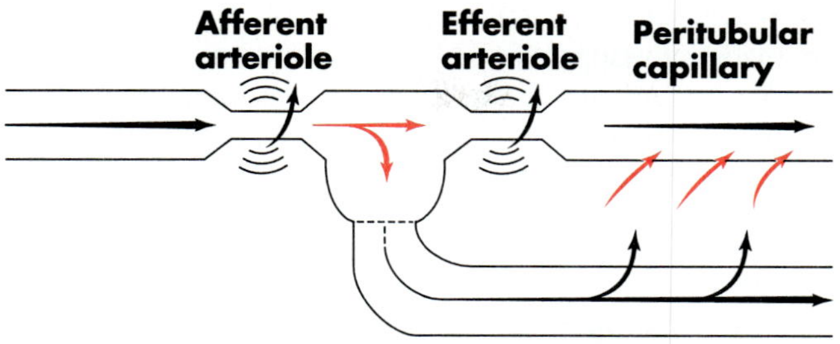

Fig. 7.6 Changes in afferent arteriolar resistance affect renal plasma flow (RPF) and glomerular filtration rate (GFR) equally and cause no change in filtration fraction (GFR/RPF). Changes in efferent resistance relative to afferent resistance will alter filtration fraction and with that the protein concentration in peritubular capillary plasma. As a result, changes in filtration fraction will change the rate of tubular reabsorption.

- **Renal excretion of water and sodium** Renal excretion of water and Na^+ are important in the regulation of body fluid volume and composition. Their excretion is modulated by physical factors, hormones, and nerves.

— ***Physical factors*** The tone in afferent and efferent renal arterioles influences glomerular, tubular, and peritubular capillary hydrostatic and oncotic pressures as well as local rates of flow. Such factors have a great influence on reabsorptive mechanisms, especially in the proximal convoluted tubule.

• Efferent and afferent arteriolar resistances together determine both renal blood flow at a given renal arterial pressure and hydrostatic pressure in glomerular capillaries, relative to a given renal arterial pressure.

• When efferent and afferent arteriolar resistances are changed individually, their relative settings determine the partitioning of flow between glomerular filtrate and peritubular capillaries. This partitioning is called the *filtration fraction.* It is normally 20% of renal plasma flow (Fig. 7.6).

- A relative increase in efferent resistance increases filtration fraction and thereby increases oncotic pressure in peritubular capillary plasma. This favors reabsorption of interstitial fluid and hence promotes reabsorption from the proximal convoluted tubule. Conversely,
- A relative decrease in efferent resistance decreases filtration fraction and decreases reabsorption from the proximal convoluted tubule.

— ***Hormones*** Both glomerular filtration rate and tubular reabsorption are influenced by hormones. In humans the most significant contributions to alteration of glomerular filtration rate are made by:

• circulating epinephrine, angiotensin II, and atrial natriuretic factor and by

• locally produced angiotensin II, prostaglandins, and endothelium-derived factors.

• Significant contributions to modulation of tubular reabsorption are made by vasopressin (ADH), aldosterone, atrial natriuretic factor, and kinins. At present the actions of vasopressin and aldosterone are understood best. Their importance derives from their ability to affect salt excretion and water excretion directly, not secondarily by primary effects on renal blood flow or glomerular filtration rate.

Vasopressin

• The plasma concentration of vasopressin is regulated by input from osmoreceptors and cardiac atrial stretch receptors.

• Vasopressin has two major actions in the kidney: (1) It acts as a ligand for V_2 receptors in the basolateral membrane of the collecting tubule and causes migration of water channel–containing vesicles (called *aggrephores*) toward the luminal membrane and subsequent insertion of water channels into that membrane. This causes increased water reabsorption down the steep osmotic gradient. (2) It promotes Na^+ reabsorption in the cortical collecting duct, partly by enhancing passive Na^+ entry through amyloride-sensitive channels on the luminal side and partly by increasing turnover of Na^+,K^+-ATPase on the basolateral side.

Aldosterone. As summarized in Fig. 7.7,

• plasma aldosterone levels can be elevated by one of two mechanisms:

- By increased cardiovascular hardship, which promotes aldosterone synthesis by increased synthesis of angiotensin II or
- by direct influence of the plasma $[K^+]/[Na^+]$ ratio on adrenal zona glomerulosa cells, which are the site of aldosterone synthesis.

• Onset of aldosterone action is measured in hours.

- It acts initially (during the first 5 to 6 hours) by increasing the number of conducting Na^+ channels in the luminal membrane of the cortical collecting tubule (Fig. 7.8).
- Its later phase of action (>10 hours) requires (1) aldosterone binding to a receptor in the cytosol; (2) migration of the ligand-receptor complex to the nucleus, and (3) induction of Na^+,K^+-ATPase synthesis. The outcome is an increase in the number of Na^+,K^+-ATPase pumps on the basolateral side (Fig. 7.8). As a result of these effects,

• aldosterone stimulates Na^+ reabsorption and K^+ transport from blood into tubular cells. Enhanced Na^+ reabsorption increases the electrical driving force for K^+ secretion into tubular fluid. In addition, aldosterone may promote insertion of K^+ channels into the luminal membrane and thereby further promote K^+ secretion (Fig. 7.8).

— ***Nerves*** Activity in renal nerves:

• Modulates the tone of vascular smooth muscle in afferent and efferent arterioles and thereby modulates the physical factors that determine renal plasma flow and glomerular filtration rate.

• Stimulates renin release from juxtamedullary cells.

• May stimulate Na^+ reabsorption in the proximal tubule.

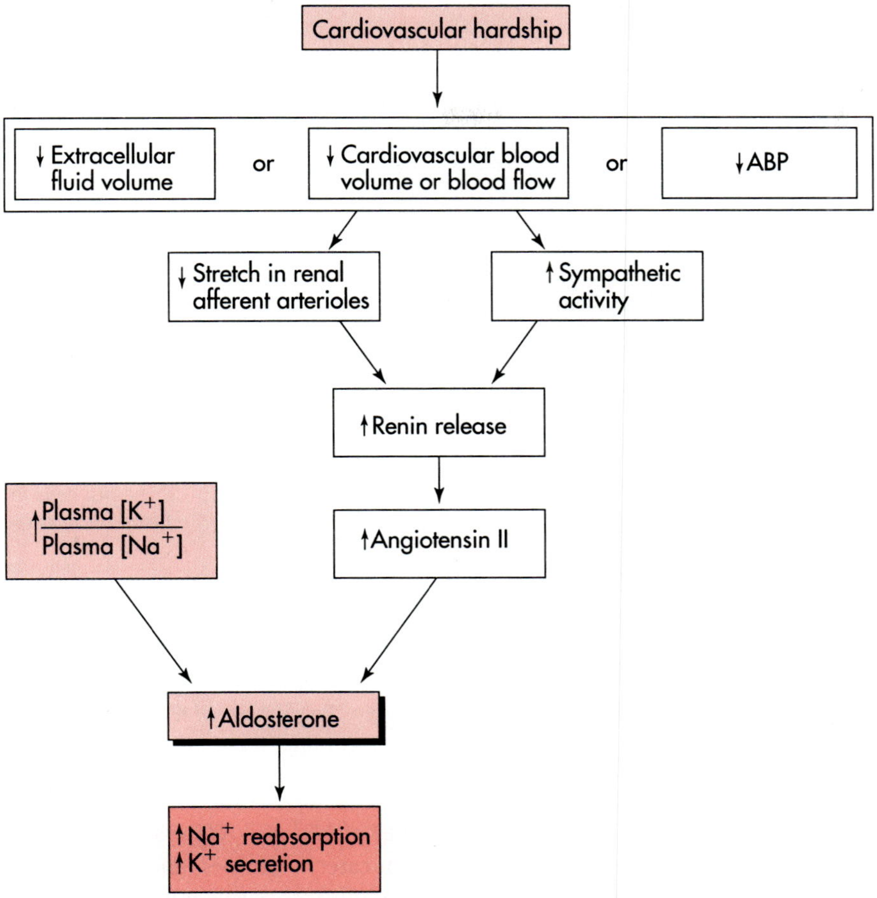

Fig. 7.7 Factors influencing the release of aldosterone. The three major influences are an increase in cardiovascular hardship, an increase in plasma K^+ concentration, or a decrease in plasma Na^+ concentration.

Regulation of Potassium Balance

Whole-body K^+ balance is regulated by factors governing temporary storage in (or release from) the intracellular depot as well as by factors that modulate permanent elimination of the ion, chiefly by the kidney. Fig. 7.8 summarizes the relationships among K^+ intake, internal sequestration, and renal excretion.

- **Internal Distribution** Most body K^+ resides in the intracellular compartment (about 3500 mmol) to which it gains access by active transport via the Na^+,K^+ pump.
  - Na^+,K^+-ATPase
    - *Biochemistry*

      • Na^+,K^+-ATPase is integrated into the lipid core of the plasma membrane and has access to both the extracellular and intracellular domains.

      • In most human cells the stoichiometry of the Na^+,K^+ pump is that the hydrolysis of one ATP molecule causes 3 Na^+ ions to be pumped out of the cell and 2 K^+ ions to be pumped in. Because of this inequality, the pump is electrogenic, with each "turn" causing a net loss of one positive charge from the cell interior.

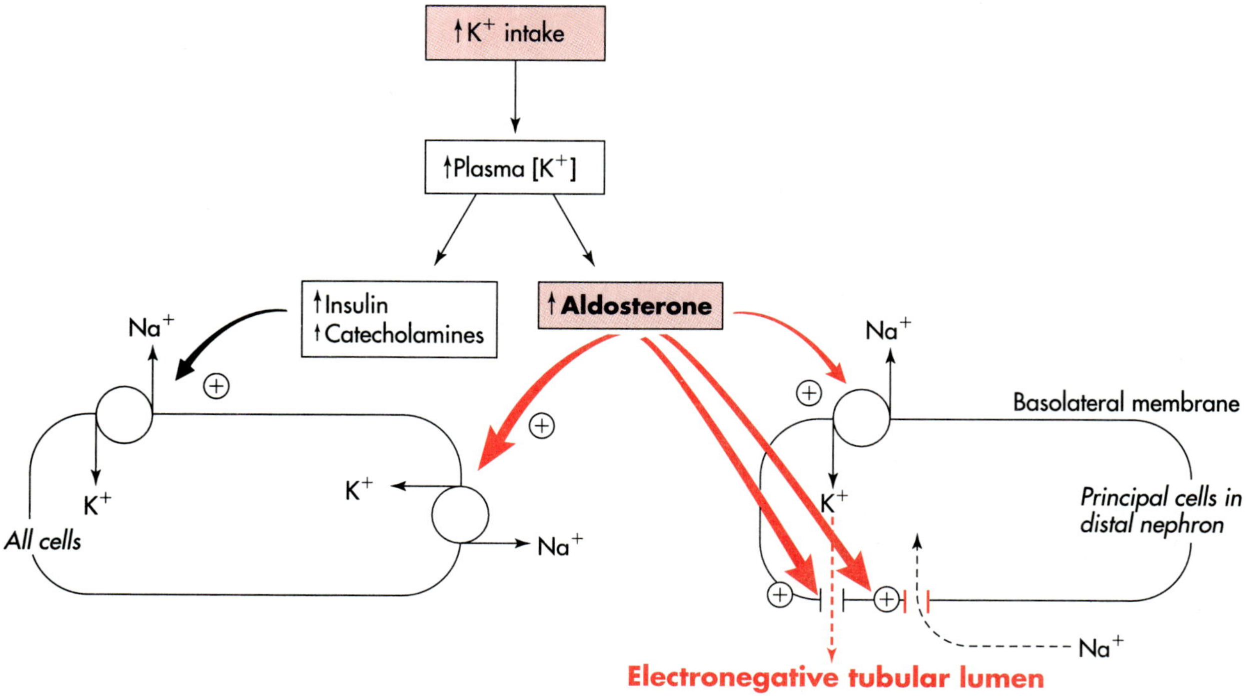

Fig. 7.8 The regulation of body K^+ distribution is dominated by the hormone insulin. Regulation of K^+ excretion is dominated by effects of the hormone aldosterone.

— ***Physiological regulation*** Activity of Na^+,K^+-ATPase is directly related to plasma levels of insulin, catecholamines, and aldosterone. As explained elsewhere, their rates of synthesis or release are governed by plasma K^+ concentration.

Insulin. Insulin is a major controller of extracellular K^+ concentration. It promotes K^+ uptake into cells by a receptor-mediated stimulation of Na^+,K^+-ATPase.

Catecholamines. Both β_2-adrenergic receptor agonists, like epinephrine, and α-adrenergic receptor agonists, like norepinephrine, influence K^+ uptake into cells.

- β_2-agonists enhance Na^+,K^+ pumping and hence K^+ uptake.
- α-agonists depress Na^+,K^+ pumping and promote K^+ loss from cells.
- The β_2-agonist effect normally dominates in a healthy, resting human.

Aldosterone. Aldosterone both increases the rate of activity of Na^+,K^+-ATPase and induces synthesis of new Na^+,K^+-ATPase and its insertion into the plasma membrane.

■ **Renal Excretion of K^+** Both sweat and stool contain small amounts of K^+, but renal excretion is the major route of K^+ elimination from the body and is regulated to suit homeostatic needs. The dominant site of regulation is the distal nephron, and the hormone aldosterone is the major vehicle for regulation.

● **Glomerulus and proximal tubule**

- K^+ enters the nephron by means of glomerular filtration, and about 50% of this load is reabsorbed in the proximal convoluted tubule in response to concentration gradients that are created by reabsorption of Na^+ and water.

- In the thin descending segment of the loop of Henle the K^+ concentration gradient generally favors diffusion of K^+ into the nephron. High interstitial K^+ concentration is created by reabsorption in the thick ascending segment and the cortical collecting tubule.

- **Thick ascending segment of the loop of Henle** K^+ is reabsorbed in this segment by a combination of active and passive processes (Fig. 7.3). The extent of reabsorption is such that virtually all K^+ that is presented to this nephron segment is reabsorbed. Only 5% to 15% of filtered K^+ remains as the tubular fluid enters the distal convoluted tubule.

- **Distal tubule and collecting duct**

  - The distal nephron contains a variety of cell types, including *principal cells,* which are located in the late distal tubule, the collecting tubule and the early collecting duct and are the main sites of renal K^+ secretion. Fig. 7.8 shows the steps involved in K^+ secretion by these cells as well as their modulation by aldosterone.

  - The first step is K^+ entry into the cytosol of principal cells by means of Na^+,K^+-ATPase that is preferentially located in the basolateral membrane. The second step, transfer from the cytosol to the tubular lumen, occurs through K^+ channels and in response to the electrical gradient that is created by Na^+ reabsorption.

    - The number and conductivity of luminal membrane K^+ channels is positively correlated with plasma aldosterone concentration.
    - Na^+ reabsorption by principal cells can alter the voltage of the tubular lumen, relative to blood, over the range 0 to −40 mV.

  - Quite apart from any influences on active Na^+,K^+ pumping, the rate of K^+ secretion is critically influenced by conditions in the tubular lumen:

    - Tubular concentration of K^+ will influence the epithelial concentration gradient for that ion.
    - Tubular concentration of Na^+ drives passive entry of Na^+ and determines cytosolic availability of that ion for Na^+,K^+ pumping, which is a requisite step in K^+ secretion.
    - Ion composition of tubular fluid determines which ions will be available to move in response to the electrical gradient established by Na^+ reabsorption. If, for example, there is an ample supply of readily reabsorbed ions like Cl^- available, relatively less K^+ will be used to eliminate the electrical gradient and high tubular $[C1^-]$ will depress K^+ secretion.
    - The flow rate of tubular fluid determines the extent to which local secretion-promoting gradients can be maintained. High rates of tubular flow prevent a local buildup of secreted K^+ and therefore promote K^+ secretion.

Renal Handling of Calcium, Phosphate, and Magnesium

- Extracellular Ca^{++} represents only 0.1% to 0.2% of total body calcium and exists as *bound* Ca^{++} (40%), *complexed* Ca^{++} (10%), and *free* Ca^{++} (50%).

- Bound Ca^{++} exists in association with plasma proteins, mostly albumin.
- Complexed Ca^{++} exists in association with various anions.
- Free Ca^{++} is available for physiologic activity, most prominently in the actions of nerves and muscles.

• Phosphorus-containing intracellular compounds serve vital roles in cell function. The element is stored mostly in bone and reaches cells via the extracellular space. Extracellular phosphorus constitutes 1% of total body phosphorus stores and exists mainly as the ionic form, *phosphate.* Phosphate exists as both *organic* (70%) and *inorganic* (30%) phosphate. Only the inorganic component is included in the routine measurement of plasma phosphate.

- Organic phosphate is mostly bound to lipids or albumin.
- Most plasma inorganic phosphate exists as $HPO_4^{=}$ and is freely available for physiologic activity.

• Magnesium is chiefly an intracellular ion and is an important cofactor and activator for many proteins and enzymes. Extracellular Mg^{++} represents about 2% of total body magnesium. Plasma Mg^{++} is present as *bound* Mg^{++} (25%; to plasma proteins), *complexed* Mg^{++} (15%; mostly with HCO_3^-) and *free* Mg^{++} (60%).

Daily intake of calcium, phosphorus, and magnesium is much less than the amount that is filtered at the glomerulus. Therefore, negative balances are prevented by avid tubular reabsorption of each of them.

Reabsorption and Secretion in the Nephron

Proximal tubule The proximal tubule reabsorbs about 60% of the filtered Ca^{++} load, 70% of the filtered $HPO_4^{=}$ load and 40% of filtered Mg^{++}.

• The major route for Ca^{++} reabsorption is the paracellular pathway. The mechanisms are passive, driven by the gradients in charge and concentration that result from active reabsorption of Na^+ and from the electrical gradient that is established by Cl^- reabsorption. Plasma levels of parathyroid hormone (PTH) have negligible influence on proximal Ca^{++} reabsorption.

• $HPO_4^{=}$ enters proximal tubular epithelial cells in association with Na^+ and leaves on the basolateral side, driven by a concentration gradient. Proximal tubular $HPO_4^{=}$ reabsorption is negatively correlated with intracellular cAMP. The plasma level of PTH is the most important regulator of proximal epithelial cell cAMP.

• Mg^{++} is poorly reabsorbed in the proximal tubule. Its transport mechanisms are coupled to Na^+ reabsorption.

Thick ascending segment of the loop of Henle

• This portion of the nephron is the major site for reabsorption of Ca^{++} and Mg^{++}. Both use the paracellular route and are driven by electropositivity in the lumen. That positivity is created by the Na^+,K^+,Cl^- cotransporter and the associated diffusion of K^+ back into the lumen.

• There is little $HPO_4^{=}$ reabsorption in the thick ascending limb.

Distal convoluted tubule and early collecting tubule

• Although these sites reabsorb only a small fraction of filtered Ca^{++}, they are important for modulating the amount of Ca^{++} that is finally excreted. This modulation resides in factors that influence the behavior of a $3Na^+/Ca^{++}$ antiporter, located in the basolateral membrane. This antiporter is similar to that found in cardiac muscle cells. The greatest effect on its rate of operation at these nephron sites is exerted by the influence of intracellular [Na^+] on the antiporter reversal potential.

- PTH stimulates Ca^{++} reabsorption in the distal convoluted tubule, presumably by cAMP-dependent Ca^{++} pumping.
- There is little reabsorption of Mg^{++} or $HPO_4^{=}$ at these nephron sites.

Regulation of Excretion Parathyroid hormone (PTH) and vitamin D are the two most important regulators for renal excretion of Ca^{++} and $HPO_4^{=}$. Renal Mg^{++} excretion is not under hormonal control.

Parathyroid hormone (PTH)

- PTH secretion from the *chief cells* of the parathyroid glands is stimulated by a decrease in the plasma concentration of free, ionized calcium.
- PTH stimulates renal $HPO_4^{=}$ excretion, suppresses renal Ca^{++} excretion and stimulates vitamin D production.

Vitamin D

- The active form of vitamin D is $1,25(OH)_2D_3$. Mitochondria in proximal tubular cells are the major locus of the enzyme 1α-hydroxylase, which converts a mildly bioactive precursor into the biologically potent form.
- $1,25(OH)_2D_3$ production is stimulated by a decrease in plasma [$HPO_4^{=}$] as well as increased plasma levels of PTH.
- $1,25(OH)_2D_3$ stimulates renal reabsorption of Ca^{++} and $HPO_4^{=}$.

The Role of the Kidney in Acid-Base Balance

The kidney plays three parts in maintaining the H^+ concentration of extracellular fluid within its normal, narrow limits: It

- reclaims any HCO_3^- that enters the nephron in glomerular filtrate;
- generates new HCO_3^- to replenish body buffer stores;
- excretes fixed acids.*

Reclaiming of Filtered HCO_3^- Bicarbonate ions are the major extracellular buffer. They are filtered freely through the glomerular membrane and would be lost in the urine if they were not reabsorbed.

- Reabsorption of HCO_3^- occurs mostly in the proximal convoluted tubule because the presence of carbonic anhydrase in proximal tubular fluid (but not in the more distal luminal fluids) and in proximal tubular cells allows the required chemical reactions to proceed rapidly.
- Fig. 7.9 shows the essence of the reabsorption mechanisms. It is the combination of filtered HCO_3^- with H^+ to yield CO_2. CO_2 diffuses into the cells and forms intracellular HCO_3^- and H^+.
- Although a great deal of H^+ is secreted in the process of HCO_3^- reabsorption, this process does not eliminate H^+ from the body.

Generation of New HCO_3^-

- New HCO_3^- is generated in the proximal and distal tubules as well as in

*Fixed acids are those that cannot be converted to CO_2 and H_2O.

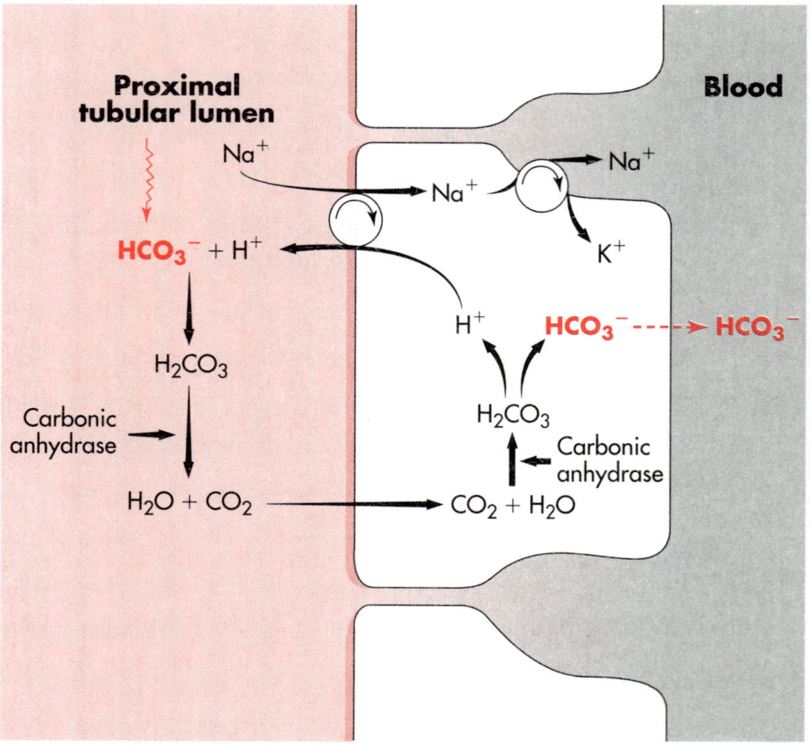

Fig. 7.9 Reclaiming of filtered HCO_3^- by proximal tubular cells. The Na^+/H^+ antiporter provides H^+ to the tubular lumen for the formation of H_2CO_3 from filtered HCO_3^-. Carbonic anhydrase catalyzes H_2CO_3 dissociation, and the resulting CO_2 diffuses into the epithelial cell where it eventually forms H^+ and HCO_3^-.

the collecting duct by a mechanism that depends on intracellular hydrolysis of body CO_2 and subsequent transfer of H^+ into the tubular lumen (Fig. 7.10).

- Depending on the chemical partner with which secreted H^+ combines in the luminal fluid, new HCO_3^- can be generated by the formation of either *titratable acid* (Fig. 7.10, *A*) or *ammonium* (NH_4^+; Fig. 7.10, *B*).

Excretion of titratable acid

- Buffered H^+ that are excreted as titratable acid appear mostly as $H_2PO_4^-$ because $HPO_4^=$ is the most readily available buffer anion.
- $H_2PO_4^-$ is termed *titratable acid* because it will liberate its H^+ if the urine were titrated back to plasma pH. It should be noted that NH_4^+ would give up little of its H^+ during titration to plasma pH because of the high pK of the ammonia-ammonium system.

Excretion of NH_4^+ Renal cortical cells increase glutamine metabolism in response to low extracellular pH. This yields ammonia (NH_3), a diffusible gas. In the tubular lumen secreted H^+ ions are used to form NH_4^+ (Fig. 7.10, *B*), and the removal of each H^+ leaves one HCO_3^- to diffuse into the body buffer stores.

Excretion of H^+ Although the kidney does excrete some acid in the form of free H^+, the properties of the distal nephron are such that it cannot maintain,

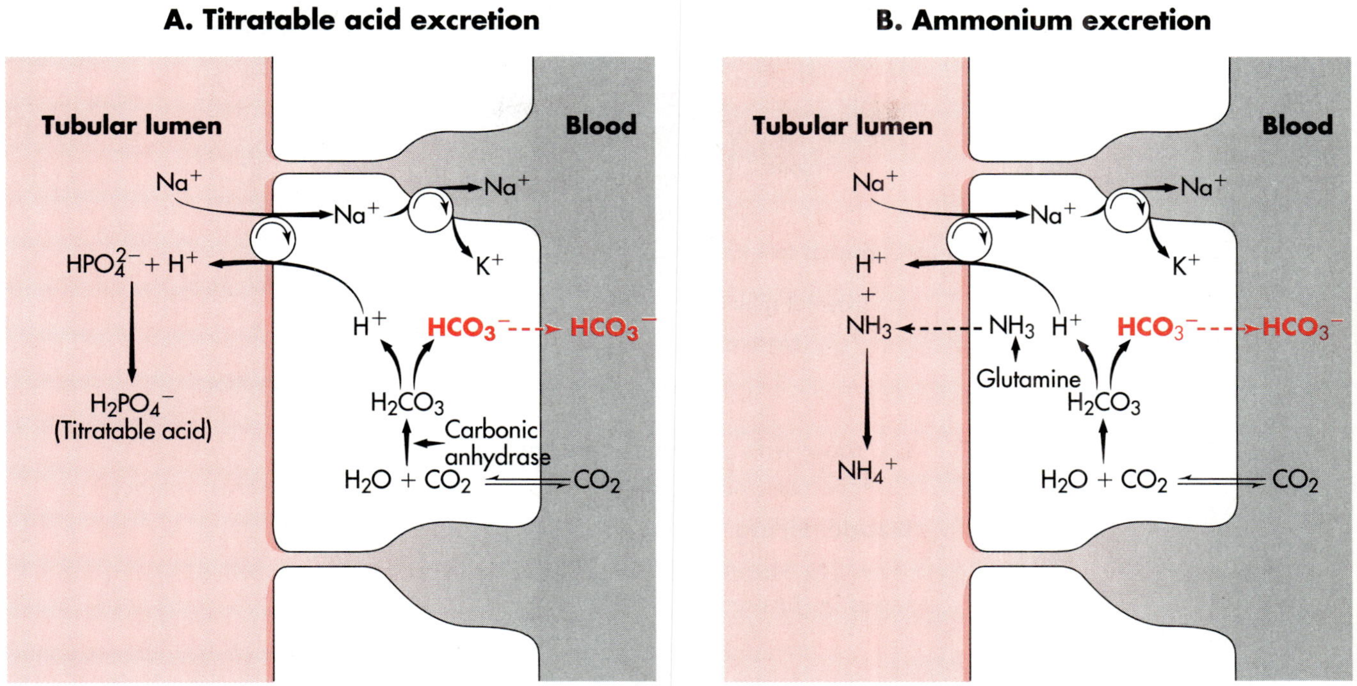

Fig. 7.10 Formation of new HCO_3^- by the kidney involves both the excretion of titratable acid, mostly in the form of $H_2PO_4^-$ and the excretion of NH_4^+. **A,** Titratable acid is mostly in the form of $H_2PO_4^-$ because of the relative abundance of $HPO_4^=$ in extracellular fluid. **B,** NH_4^+ (ammonium) is formed from NH_3 (ammonia), which derives from the metabolism of glutamine in renal cortical cells. Such metabolism is increased when there is an increase in extracellular H^+ concentration above normal values. The process takes about 2 days to be fully active.

across the tubular cell, a gradient in H^+ concentration of sufficient magnitude to meet the body needs for H^+ excretion. As a result, transported H^+ diffuses back from the tubular lumen into the cells and most acid excretion takes place in buffered form, the H^+ appearing in urine either as titratable acid or as NH_4^+.

MICTURITION

Anatomy of the Bladder and Urinary Tract The body of the bladder is a smooth-muscle chamber formed by the detrusor muscle (Fig. 7.11).

- The ureters and the urethra connect to the bladder through the *trigone,* a triangular area of fine smooth muscle fibers located near the bladder neck. This arrangement creates a functional internal sphincter that can be relaxed only by contraction of the detrusor muscle.
- The urethra is surrounded near its origin by the external sphincter. It is composed of striated muscle.
- Sympathetic efferents supply blood vessels in the bladder wall and the portion of the detrusor muscle surrounding the bladder neck. They have little influence on normal bladder function. Their major role may be closure of the internal sphincter during orgasm.
- Parasympathetic nerves are the major innervation of the bladder wall and urethra. Afferent fibers, arising from endings that respond to stretch and pain,

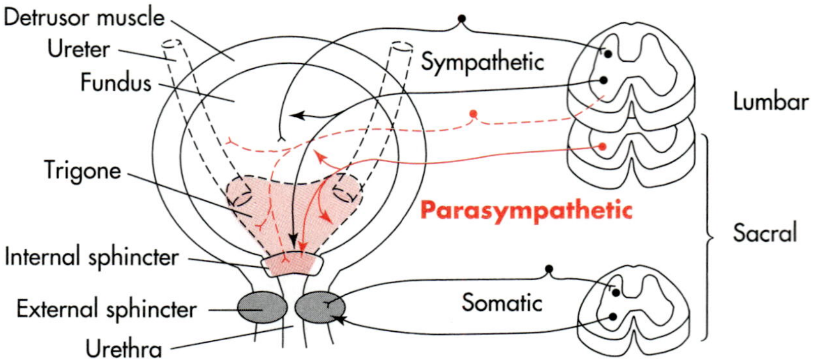

Fig. 7.11 Anatomy of the bladder and its innervation.

also travel with parasympathetic nerves.

- Somatic nerves control the striated muscle of the external sphincter.

Bladder Filling During the filling phase:

- The bladder fills passively, with the urine being propelled by peristaltic waves in the ureters.
  - Reflux into the ureters is prevented by the nature of the ureter-trigone junction; the oblique angle of entry of the ureters creates a sphincter-like junction.
  - Escape into the urethra is prevented by tonic constriction of the external sphincter by somatic input from the sacral spinal cord.
- As wall stretch increases, the distension excites stretch receptor afferents projecting to the brainstem. When about 300 mL of urine has been collected and the chamber pressure reaches about 20 mm Hg, stretch receptor activity is sufficient to elicit a conscious desire to empty the bladder. At 400 to 500 ml the desire becomes very strong.
- The urge to void can be suppressed for a while by inhibitory activity from hypothalamic and cortical centers to prevent emptying at unsuitable times. Such suppression is termed *continence.* If the urge is not suppressed, the *voiding reflex* is initiated.

Bladder Emptying (Voiding) Voiding begins with contraction of the detrusor muscle in response to parasympathetic nerve activity and contraction of abdominal muscles (somatic control).

- The resulting increase in bladder pressure further excites stretch receptor afferents, and when the pressure approaches 40 mm Hg, somatic input to the external sphincter is reduced and the sphincter relaxes.
- Once the bladder has begun to empty, the process accelerates explosively until emptying is completed.

Central Nervous Influence on Bladder Function Transection of the spinal cord above the sacral level may remove sympathetic influences and does remove higher nervous contributions to bladder control.

- Individuals with such injuries can learn to control bladder evacuation on the basis of local, spinal reflex paths.
- Such control involves initiation of detrusor contractions at suitable intervals by momentary elevation of bladder pressure above a tension threshold and can be accomplished, e.g., by slight tapping of the abdomen.

Multiple Choice Review Questions

1. Which one of the following statements is correct?
   a. Erythrocytes neither swell nor shrink in a solution isosmotic with the contents of the cell.
   b. Water moves from a hypertonic to a hypotonic solution.
   c. Erythrocytes neither swell nor shrink in a solution isotonic with the contents of the cell.
   d. If solutions are isosmotic, they must also be isotonic.

2. Administering 1 L of hypertonic saline to a 70 kg human will:
   a. inhibit vasopressin secretion.
   b. increase intracellular fluid volume.
   c. decrease interstitial fluid volume.
   d. lower plasma protein concentration.

3. Two substances are simultaneously injected into a patient. They are *Diodrast* and *inulin.* Their renal clearances are measured and are found to be $C_{Diod} = 1270$ mL/min $C_{inulin} = 110$ mL/min. On the basis of this observation it may be concluded that Diodrast is:
   a. filtered at the glomerulus but then reabsorbed.
   b. too large a molecule to be filtered.
   c. filtered at the glomerulus and secreted into the tubules.
   d. filtered but then neither reabsorbed nor secreted into the tubules.

4. A solution is infused intravenously into a patient. As a result extracellular fluid volume is increased, and intracellular fluid volume is decreased in the patient. The plasma water compartment is increased proportionately more than total ECF. These results are compatible with one of the following. Which one?
   a. The infusate is hypertonic and hyperoncotic.
   b. The infusate is isotonic, isoncotic.
   c. The infusate is hypotonic, hyponcotic.
   d. The infusate is hypotonic, hyperoncotic.

5. Which of the following could increase the total solute concentration in the urine of a normal, healthy person?
   a. Decreased plasma sodium-concentration.
   b. Increased NaCl transport in the ascending limb of the loop of Henle.
   c. Increased extracellular fluid volume.
   d. Decreased afferent arteriolar resistance.

6. When the secretion of PAH reaches the T_m, further increase in plasma PAH concentration causes its clearance to:
   a. increase in proportion to its plasma concentration.
   b. approach glucose clearance asymptotically.
   c. remain constant.
   d. approach inulin clearance asymptotically.

7. Afferent arteriolar constriction would probably:
   a. increase both RBF and GFR.
   b. decrease both RBF and GFR.
   c. increase filtration fraction.
   d. decrease filtration fraction.

8-10. Match each description with the most appropriate morphologic segment of the nephron. Lettered items may be used once, more than once, or not at all.
   a. Proximal convoluted tubule
   b. Thin descending limb of the loop of Henle
   c. Thick ascending limb of the loop of Henle
   d. Collecting tubule

8. Major site of vasopressin-mediated water reabsorption

9. Site where largest fraction of water is reabsorbed

10. Concentrating segment of the nephron

Chapter 8

Gastrointestinal Physiology

Gastrointestinal Motility

Motility is determined by the mechanical activity of intestinal smooth muscle. It has three functions:

- To transport food along the GI tract
- To mix ingested food with internal secretions that aid digestion
- To remove undigestibles and waste to the rectum

Contractions are evident in all GI muscle units, those surrounding the lumen as well as those forming sphincters within it; they are initiated by spontaneously generated smooth muscle action potentials. The consequent mechanical activity is coordinated by central nervous motor programs and modulated by local mechanical, chemical, neural, or hormonal influences.

■ **The GI Muscle Coat** Almost all the GI muscle coat is smooth muscle, arranged in two layers around the lumen.

- The inner layer encircles the lumen.
- The outer layer is arranged longitudinally.
- The region between the two layers is occupied by an extensive network of neurons, called the *intrinsic plexus* (myenteric plexus; Auerbach's plexus).

■ **Innervation of the GI Tract** GI innervation is supplied mostly by the intrinsic plexus. Some sections also show direct innervation by sympathetic postganglionic fibers, bypassing the plexus, but such innervation is sparse. The intrinsic plexus:

- modulates the activity of all components of the GI tract.
- is modulated by the autonomic nervous system.
  - Parasympathetic fibers arise from the vagus or pelvic nerves.
  - Sympathetic fibers arise from the splanchnic network.

■ **Motility in Different Regions** The movement of food through the GI tract begins with oral ingestion, chewing, and swallowing.

● **Chewing** Chewing reduces the size of solid food and mixes it with saliva. At suitable intervals a bolus to be swallowed is separated by the action of the tongue against the hard palate and is propelled into the oropharynx. Once there, its transport along the entire tract proceeds as a coordinated involuntary reflex, beginning with the processes of swallowing.

● **Swallowing** Swallowing conveys food into the stomach within a few seconds. Its initial phase requires complex central nervous coordination of respiration, speech, and upper esophageal relaxation.

- Subsequent transport of food toward the lower esophageal sphincter is caused by sweeping waves of peristalsis:
  - A primary peristaltic wave is initiated by distension of the pharynx and coordinated by neurons in the pons and medulla regions of the brainstem.
  - A secondary peristaltic wave occurs as a reflex response to afferent signals from esophageal stretch receptors, activated by food that was not propelled by the primary wave.
  - During the interdigestive period an all-encompassing peristaltic wave, called the *migrating motor complex,* sweeps remaining food through the system. It begins in the esophagus, moves toward the colon, occurs every 60 to 90 minutes, and lasts about 10 minutes.
- Peristalsis results from sequential timing of vagal impulses to progressively more distal segments of the esophagus.
- Swallowing also involves the fundus region of the stomach and adjacent portions of its body.

● **Stomach** The stomach is divided into three regions: fundus, body, and antrum. The role of the stomach is to store food temporarily.

— ***Reception and temporary storage of food*** As part of the swallowing reflex, the fundus relaxes in synchrony with relaxation of the lower esophageal sphincter. This allows food to enter the stomach. It is temporarily stored in the upper portion of the fundus because motility in that area is inhibited by vagal activity. This phenomenon is called *receptive relaxation.*

— ***Production and release of chyme*** While food resides in the stomach, it is blended with gastric secretions to produce *chyme.*

- The antrum and adjacent portions of the stomach body show greater motile activity than the fundus region.
  - Motile activity originates in gastric pacemaker cells that are located near the midpoint of the greater curvature.
  - The velocity of its travel toward the pyloric sphincter is greater than the velocity at which chyme travels toward the sphincter. As a result only a small portion of chyme is propelled into the duodenum while the remainder is thoroughly mixed in the stomach.
- The rate of stomach emptying is influenced by the nature and chemical composition of its contents:
  - Liquids leave more rapidly than solids.
  - Chyme that is high in lipids, osmolality or H^+ leaves at a slower rate.

● **Small intestine (duodenum, jejunum, ileum)** Semiliquid chyme is received from the stomach, is mixed with bile and digestive secretions, and is moved longitudinally by spontaneous, rhythmic smooth muscle activity (modulated by nerves and hormones).

Two kinds of contractions occur: localized circumferential contractions and longitudinal (peristaltic) contractions.

— ***Local circumferential contractions*** Most contractions in the small intestine are brief, localized events.

- They involve only 1 to 4 cm of bowel at a time.
- They last less than 5 seconds and occur in sets that are spaced 5 to 10 seconds apart.
- Their primary purpose is local mixing and circulating of bowel contents. In addition, they cause division of the contents into segments and are therefore described as segmenting contractions.

— ***Longitudinal contractions*** The small intestine is also capable of longitudinally coordinated, peristaltic contractions. These occur over distances of a few centimeters and serve to propel food toward the colon.

- **Colon** Segmental (nonpropulsive) and peristaltic (propulsive) activities in the colon resemble those in the small intestine, but they are of longer duration.

  - Local segmental contractions occur independently from those at adjacent sites and may be partly responsible for the *haustra* that characterize the colon.
  - Most propulsion of the colonic contents occurs during occasional characteristic sequences called *mass movement.* During mass movement segmental activity ceases, the haustra disappear, and the length of the colon is swept by a peristaltic contraction that moves feces into the rectum.

- **Rectum**

— ***Filling of the rectum*** The rectum fills by peristaltic contractions of the descending colon. Each wave stimulates stretch receptors in pudendal and pelvic nerve afferents (Fig. 8.1). Once the rectum contains about 2 L of feces such stretch receptor activity leads to conscious perception of an urge to defecate.

The arrival of each bolus in the rectum is also accompanied by transient relaxation of the internal sphincter (a reflex governed by the enteric nervous system). As a result, continence can be maintained only by tonic contraction of the external sphincter. The governing reflex is mostly independent of higher function and involves segmental afferents and efferents in the pudendal nerve (see Fig. 8.1).

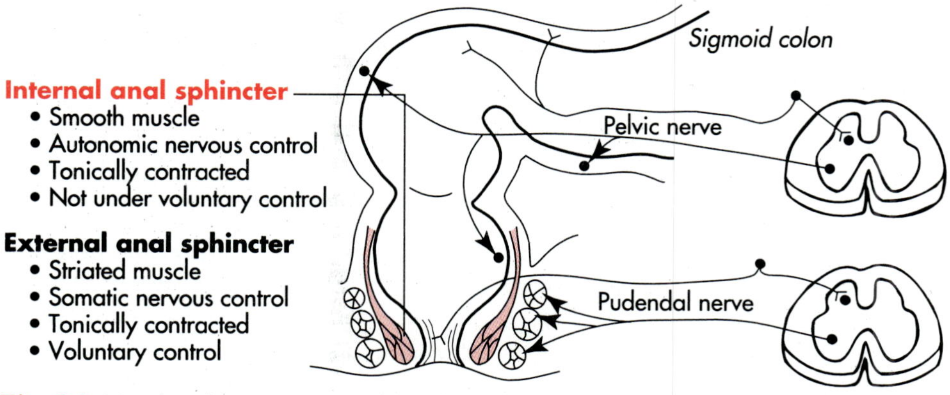

Fig. 8.1 Muscle sphincters involved in the control of bowel evacuation. The pelvic nerve carries both afferent and autonomic cholinergic efferent fibers. The pudendal nerve also carries both afferent and efferent fibers. The efferents belong to the somatic nervous system.

Table 8.1 *Summary of the Physiology of Gastric Regulatory Peptides*

Peptide	Influence of Release Mechanisms					
	H^+	Somatostatin	Parasympathetic Activity		Products of Digestion	Stretch
			ACh	GRP		
Gastrin	–	–		+	+	+
Cholecystokinin (CCK)	+				+	
Secretin	+					
Gastric inhibitory polypeptide (GIP)					+	
Somatostatin	+		–			
Motilin			+	+	+	
Pancreatic polypeptide			+		+	
Vasoactive intestinal peptide (VIP)	+		+		+	

Soon after the arrival of each bolus the rectum accommodates to its increased volume, wall stretch is reduced, the internal sphincter contracts, and the external sphincter is no longer the sole guardian of dignity.

— ***Emptying of the rectum*** At a suitable time evacuation of the rectum is initiated by voluntary effort This involves four components:

1. Intra-abdominal and intrarectal pressures are increased by voluntary contraction of abdominal muscles.
2. Rectal contents are moved distally by reflex contraction of longitudinal musculature in the colon and rectum. This increases intrarectal pressure further.
3. The internal sphincter relaxes as each pressure wave arrives in the rectum.
4. The external sphincter is relaxed by voluntary effort.

Central nervous influence. Transection of the spinal cord above the sacral level abolishes the voluntary motor patterns that assist defecation. As a result paraplegics must learn special techniques for relaxing the external anal sphincter.

Vomiting and Retching Vomiting and retching are complex reflexes of motility. They are initiated by stimuli that include:

- activation of chemosensitive neurons located centrally or peripherally;
- conflicting visual and vestibular sensory inputs;
- emotional input from cerebrocortical centers.

Vomiting and retching require coordination of three physiologic systems:

- Smooth muscle activity in the GI tract shows spastic contractions of the gastric antrum but complete relaxation of all structures from the gastric body to the upper esophagus.
- The respiratory pattern is changed to slow, deep inspirations. The effect is to decrease intraesophageal pressure.
- Skeletal muscle in the abdominal wall is contracted. This raises intra-abdominal pressure and creates a gradient between intra-abdominal pressure and esophageal pressure. The pressure gradient forces gastric contents upward into the relaxed esophagus.

Major Actions on . . .

H^+ Secretion	Gastric Motility	Gastric Emptying	Pancreatic Enzymes	Pancreatic HCO_3^-	Bile HCO_3^-	Gallbladder Contraction	Intestinal Motility	GI Peptide Release	Insulin Release	Growth Promotion
+	+	+								+
				+		+				+
	−	−		+	+					
−	−	−							+	
−	−							−		
	+	+					+			
			−	−						
−								−		

ACh, Acetylcholine; *GRP,* gastrin-releasing peptide.

The difference between vomiting and retching, once gastric contents have been pushed into the esophagus, is the state of the upper esophageal sphincter.

• If the upper esophageal sphincter is opened, evacuation into the mouth takes place (= vomiting).

• If the upper esophageal sphincter remains closed, distension of the esophagus will initiate a wave of secondary peristalsis that sweeps the gastric contents back from the esophagus into the stomach. This is perceived as retching.

Regulatory Peptides of the GI Tract

The GI system is particularly rich in peptides. They originate from secretory cells of the gut mucosa (endocrines and paracrines) or from nerves (neurocrines). Their major release mechanisms and actions are summarized in Table 8.1.

The most familiar of the gastrointestinal regulatory endocrines and paracrines can be divided into two families of structural homology:

- Gastrin and cholecystokinin (CCK), which have the same 5 amino acids at their C-terminuses.
- Secretin, vasoactive intestinal peptide (VIP), gastric inhibitory peptide (GIP), and glucagon, which show several regions of homology.

Gastrin

Synthesis and secretion Gastrin is synthesized and released from mucosal G-cells, which are found within pyloric glands in the gastric antrum, duodenum, and jejunum. Its synthesis or release are influenced by a variety of chemical and nervous factors (Fig. 8.2) including *gastrin-releasing peptide (GRP).*

- Antral cells produce the 17–amino acid form (G17) in large quantities just after a meal.
- The small intestine produces the equipotent but longer-lasting G34.
  - During the basal, nondigestive state most serum gastrin in humans is of the 34–amino acid form (G34).
  - Although G34 can be split by trypsin to yield G17 and an ad-

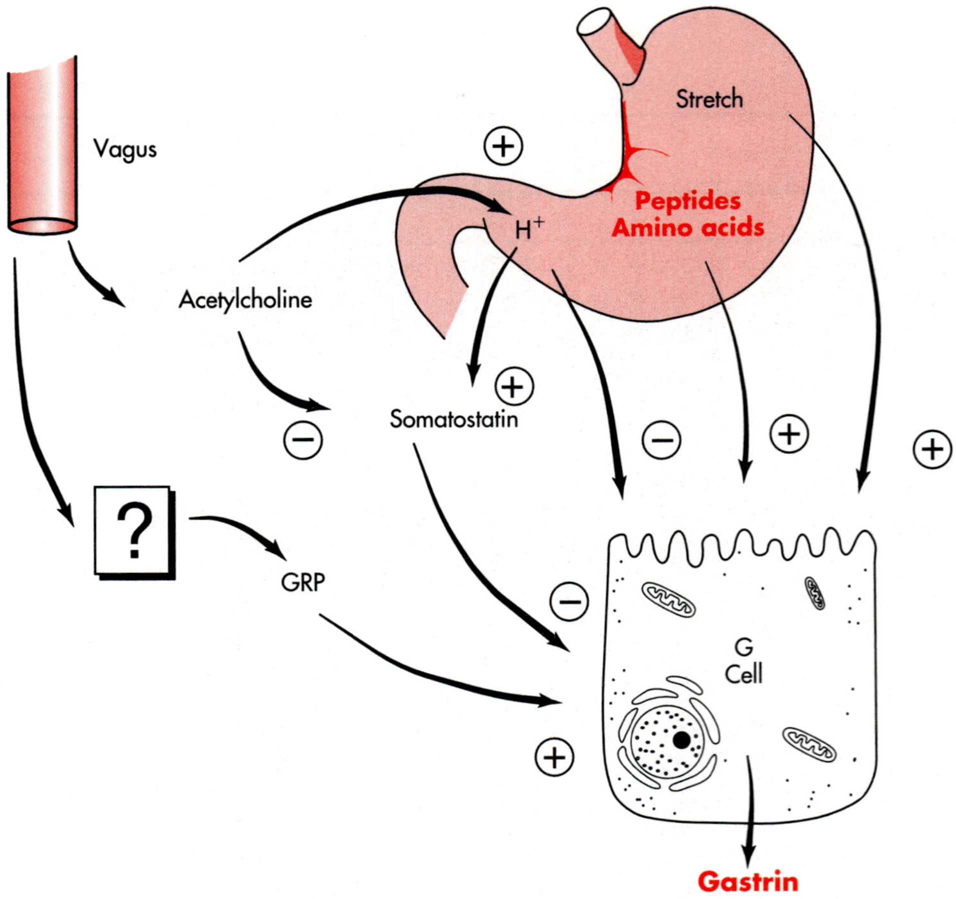

Fig. 8.2 The control of gastrin synthesis and release from G-cells within pyloric glands. Peptides and amino acids are the most important stimuli. Physical stretch and action potentials in vagus nerves will also promote gastrin release. The mechanism of vagally mediated gastrin release is believed to be related to a concomitant elevation of *gastrin-releasing peptide* (GRP). Gastrin release is inhibited by somatostatin and by elevated [H^+] (no gastrin release occurs when the pH is less than 2).

ditional 17-residue peptide, G34 is not the normal precursor of G17.

- **Actions** The major actions of gastrin are to:
  - stimulate parietal cells in the oxyntic glands of the stomach mucosa to secrete H^+;
  - stimulate pancreatic acini to secrete enzymes;
  - enhance gastric motility;
  - promote mucosal growth in the stomach and small intestine.

Cholecystokinin (CCK)

- **Synthesis and secretion** CCK is synthesized and secreted by specialized cells in the mucosa of the duodenum, jejunum, and, to some extent, the ileum.
  - It has 33 amino acids and the last 5 at the C-terminus are identical to the last 5 in gastrin. As a result they are partial agonists of each other.
  - Peptides belonging to the gastrin-CCK family and having

an unsulfated tyrosyl residue in position 6 or 7 from the C-terminus display the gastrin pattern of activity: pronounced stimulation of gastric H^+ secretion and weak contraction of the gallbladder. On the other hand,

- peptides belonging to the gastrin-CCK family and having a sulfated ($—SO_3H$) tyrosyl residue in position 7 from the C-terminus display the CCK pattern of activity: weak stimulation of gastric H^+ secretion and strong contraction of the gallbladder.

• The major stimulus for CCK release is the presence, within the intestinal lumen, of products of fat or protein digestion. Duodenal H^+, Mg^{++}, Zn^{++}, and Ca^{++} also cause release.

• Somatostatin is the major inhibitor of both CCK release and CCK actions.

- **Actions** The major CCK actions are:

  - stimulation of pancreatic HCO_3^- secretion;
  - potent stimulation of gallbladder contraction (by receptor-activated Ca^{++} channels) and relaxation of the sphincter of Oddi;
  - inhibition of gastric emptying;
  - promotion of pancreatic growth.

Secretin

- **Synthesis and secretion** Secretin is a 27–amino acid peptide synthesized and secreted by the S-cells of duodenal and jejunal mucosa. The strongest stimulus for secretin release is duodenal H^+. In humans plasma secretin increases postprandially and coincidentally with increases in duodenal $[H^+]$.
- **Actions** The biologic function of secretin is to reduce $[H^+]$ in the duodenum. This is accomplished mostly by:

• stimulating secretion of HCO_3^--rich fluid from cells located in the neck of secretory units in the pancreas and bile ducts. In addition, secretin:

• inhibits gastrin-stimulated H^+ secretion from parietal cells in the gastric mucosa. Although this is probably not a physiologically significant effect at the normally low levels of secretin, it may be important in some disease states.

- inhibits gastric motility and gastric emptying;
- stimulates growth of the exocrine pancreas.

Secretin and CCK are synergistic in their effects on pancreatic secretion and pancreatic growth.

Gastric Inhibitory Polypeptide (GIP)

- **Synthesis and secretion** GIP is a 42–amino acid peptide produced mostly in the jejunum but also in the duodenum and upper ileum. GIP release is stimulated most prominently and in a dose-dependent fashion by glucose and fatty acids, but also by high levels of extracellular Ca^{++} or after intraduodenal infusion of amino acids or bile. The group of amino acids that will cause release is different from that which promotes release of gastrin or CCK.
- **Actions** The major function of GIP is that of a potent stimulus for insulin release, but by virtue of its inhibitory effect on gastric motility, H^+ secre-

tion, and gastric emptying it acts as a regulator of gastric activity relative to the capacity of the upper small intestine.

Somatostatin

- **Synthesis and secretion** Somatostatin is found in the central and peripheral nervous system, but 70% of body stores is contained in the D-cells of the antral, duodenal, and pancreatic mucosa. Somatostatin release and plasma levels are influenced by many metabolic, humoral, and nervous factors:

  - Basal levels show a circadian rhythm and are decreased by vagal stimulation.
  - Postprandial peaks coincide with duodenal [H^+] and are depressed by cholinergic vagal activity.

- **Actions** Somatostatin is a strong suppressor of many gastrointestinal and pancreatic functions, including salivary secretion, gastrointestinal motility, gastric acid secretion, and peptide release (gastrin, CCK, secretin, insulin, and glucagon).

Motilin

- **Synthesis and secretion** This 22–amino acid peptide is most highly concentrated in duodenal and jejunal mucosal cells.

  - Its release is stimulated by the passage of nutrients, by vagal activity, and by gastrin-releasing peptide (GRP).
  - Motilin is inhibited by somatostatin.

- **Actions**

  - Motilin produces contraction of intestinal smooth muscle, including the gallbladder. As a result it enhances phasic contractions of the lower esophageal sphincter, increases gastric as well as intestinal motility, and accelerates gastric emptying.
  - Motilin enhances (and may initiate) the migrating motor complex.

Pancreatic Polypeptide (PP) PP derives from specific cells in the acini of the pancreas. The most potent stimulus for its release is intraduodenal protein, but vagal cholinergic stimulation also plays a major role in postprandial release. The most important physiologic action of PP is believed to be an inhibitory effect on pancreatic exocrine secretion.

Vasoactive Intestinal Peptide (VIP) VIP is found in endocrine cells throughout the entire length of the GI tract as well as in the urogenital tract, the brain, and certain neurons of the autonomic nervous system.

- VIP is released by cholinergic agonists, hypertonic fluids, H^+, ethanol, and fat.
- Its actions affect many physiologic systems. Those in the GI tract include inhibition of acid, gastrin, and pepsin secretion. It also is a weak inhibitor of pancreatic HCO_3^- secretion.

Nonpeptide Secretions of the GI Tract

Different portions of the GI tract secrete substances that have three basic functions:

- Mechanical modification of ingested food

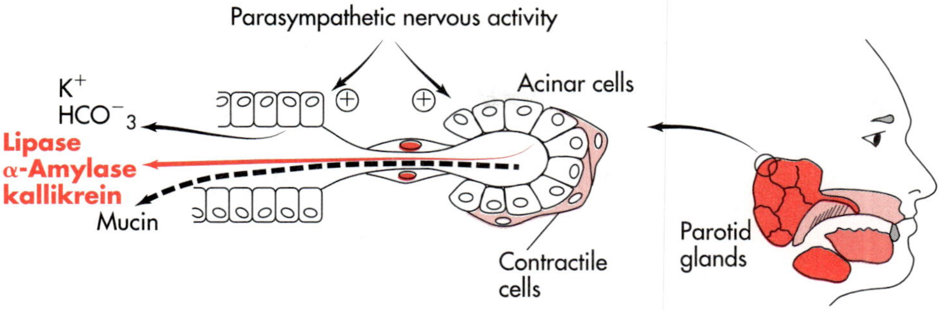

Fig. 8.3 Cells in the acini of salivons, the functional units of salivary glands, secrete a watery fluid having an ionic composition resembling that of plasma. As the fluid moves along the ducts, it is modified and emerges relatively rich in K^+ and HCO_3^-. Saliva also contains a few enzymes and the lubricant mucin.

- Digestion of food
- Protection of the GI tract from its own destructive actions

Secretions from the Mouth Each day the adult parotid, submandibular, and sublingual pairs of glands (Fig. 8.3) together secrete about 500 mL of saliva. The main purpose of saliva is lubrication and protection.

- Salivary glands consist of a system of ducts, each with a salivary acinus at its end, surrounded by a thin layer of contractile cells (see Fig. 8.3). They are innervated by both sympathetic and parasympathetic nerves.
- The composition of saliva varies among glands. It is a watery fluid, rich in K^+ and HCO_3^- and containing mucin, α-amylase, lipase, and kallikrein.
  - The acini secrete a fluid resembling plasma in its composition. As the fluid moves along the salivary duct, Na^+ and Cl^- are reabsorbed while K^+ and HCO_3^- are secreted into the lumen. Accordingly, as saliva flow rate increases and less time is available for compositional changes along the duct, the ionic composition of saliva more and more approaches that of plasma. However, even at high flow rates saliva has a higher concentration of K^+ and HCO_3^- than plasma and, correspondingly, lower concentrations of Na^+ and Cl^-. HCO_3^- serves a protective functions against the acids that are constantly being produced by oral microorganisms.
  - Mucin has buffering functions similar to those of HCO_3^-. In addition, mucin forms a protective, low-friction film in the mouth and around food particles.
  - Alpha-amylase and lipase provide token digestive functions that are quickly inactivated by gastric acidity.
  - Kallikrein is a substrate for a ubiquitous converting enzyme and forms bradykinin. This potent vasodilator acts to increase blood flow to the secreting glands.
- The autonomic nervous system is the major controller of salivary function. Parasympathetic influences dominate (see Fig. 8.3).

Gastric Secretion The proximal 80% of the gastric mucosa contains *oxyntic glands* (Fig. 8.4) as the major secretory unit, whereas the distal 20% of the stomach, near the pyloric sphincter, contains mostly gastrin-secreting *pyloric glands.* The oxyntic gland forms three types of secretions in large volumes of fluid:

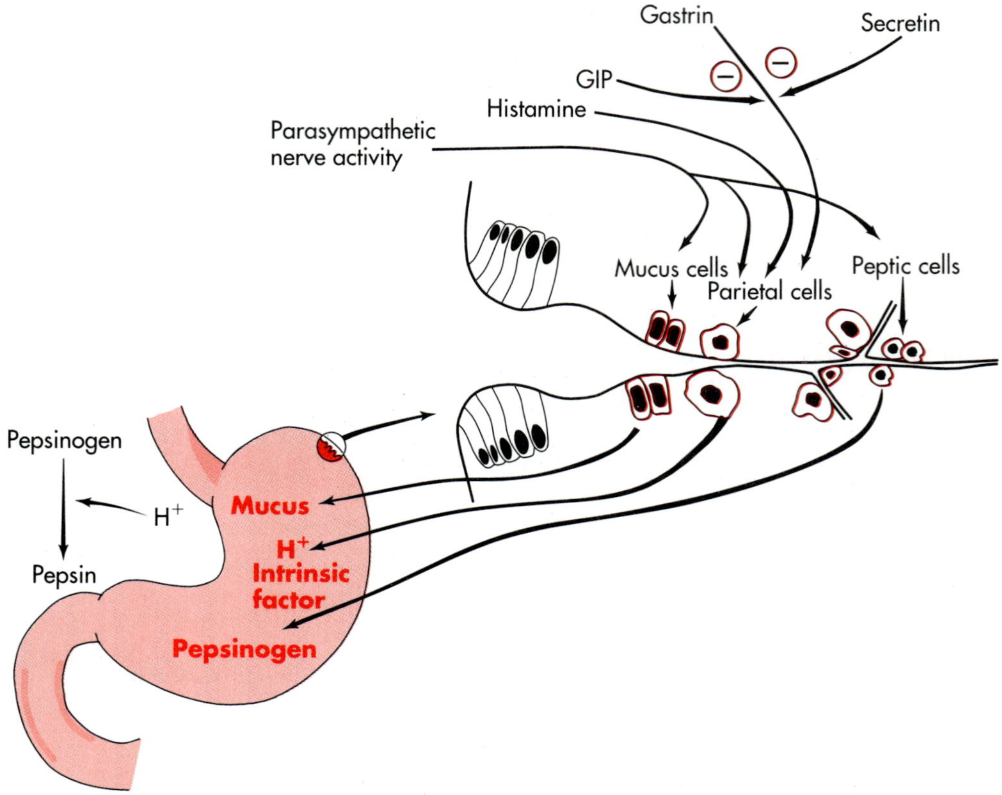

Fig. 8.4 Specialized cells within the oxyntic gland of the gastric mucosa are responsible for the elaboration of mucus, H^+, and pepsinogen. Mucus cells are stimulated by parasympathetic nerves to secrete soluble mucus. Parietal cells secrete H^+ when they are stimulated by parasympathetic activity, histamine, or gastrin. Peptic cells secrete pepsinogens when they are stimulated by parasympathetic activity. Pepsinogen is rapidly converted to the active enzyme pepsin in the presence of H^+. Enterokinase is an enzyme that is located on the brush border of duodenal cells. Its function is to activate trypsinogen.

- Mucus cells near the neck of the gland are stimulated by cholinergic agonists (normally acetylcholine from parasympathetic nerve terminals) to secrete *soluble mucus.* Soluble mucus is rich in mucoproteins and mixes with chyme for the purpose of lubrication. *Insoluble mucus* is a gel that is secreted from mucus cells at the surface of the mucosa for the purpose of protecting the GI mucosa from being digested itself.
- Parietal cells secrete H^+ as well as *intrinsic factor.*
  - H^+ is generated from CO_2 at a high rate because parietal cells are rich in carbonic anhydrase. The mechanisms are identical to those governing the handling of HCO_3^- by renal tubular cells and are described in chapter 7. H^+ is required to activate pepsinogen and to stimulate enzyme secretion from later portions of the GI tract.
  - Intrinsic factor is a glycoprotein that is required for absorption of vitamin B_{12}. Vitamin B_{12} must be supplied in the diet. Deficiency or malabsorption of B_{12} causes atrophy of parietal cells as well as megaloblastic anemia.
  - Parasympathetic nerve activity and gastrin are the most important physiologic stimulants of parietal cell secretion. In addition, histamine acts through H_2 receptors to stimulate parietal cells and to

potentiate the actions of cholinergic agonists and of gastrin on the secretory activity of these cells.

- Peptic cells secrete pepsinogen, the inactive precursor of pepsin. Pepsin breaks peptide bonds and thereby begins the process of protein digestion. Its contribution to total protein digestion is only 10% to 15%.
    - Acetylcholine from parasympathetic nerve terminals is the strongest stimulant for peptic cells. Its release may be initiated centrally or by an H^+-triggered local cholinergic reflex. Secretin and gastrin can also stimulate pepsinogen secretion but are not normally significant regulators of peptic cell secretion in humans.

Pancreatic Secretion Almost all the pancreas is devoted to the elaboration of digestive secretions from groups of acini that resemble salivary glands. This part is called the *exocrine pancreas* and produces about 1 L of fluid per day. Only 2% of the organ, labeled the *endocrine pancreas,* is the islets of Langerhans. They are dispersed throughout the pancreas and are concerned with glucose metabolism.

Pancreatic acini contain two types of secreting cells:

- Cells near the neck of the interlobular ducts secrete a fluid that is rich in HCO_3^-. Their secretions neutralize the high acidity of duodenal contents. The cellular mechanisms of HCO_3^- production depend on the presence of carbonic anhydrase and are similar to those described earlier for H^+ production in parietal cells of the gastric mucosa.
- Cells deep within the acini secrete digestive enzymes or enzyme precursors that can be categorized into four classes:

1. Amylolytic enzymes (starch breakdown)
   For example, *amylase*
2. Lipolytic enzymes (breakdown of lipids and cholesterol)
   For example, *lipase; phospholipase A, cholesterol esterase*
3. Proteolytic enzyme precursors* (protein breakdown)
   For example, *trypsinogen, chymotrypsinogen, procarboxypeptidase*
4. Nucleolytic enzymes (breakdown of RNA and DNA)
   For example, *ribonuclease, deoxyribonuclease*

- Regulation of pancreatic secretion is summarized in Fig. 8.5.
    - Some secretion occurs during the anticipatory (cephalic) and gastric phases of a meal. The main stimulus during these phases is parasympathetic nervous activity, and there is some contribution from gastrin.
    - The major stimulus to pancreatic secretion is the presence of H^+ and digestion products in the small intestine. In this phase of a meal parasympathetic nerve activity, secretin, and cholecystokinin (CCK) are of secondary importance as stimuli.

- Although secretin and CCK are synergistic in their effects (see Fig. 8.5), they each have primary target cells in the pancreatic acini:

*Proteolytic enzymes are secreted from acinar cells as inactive precursors (zymogens). Their activation begins with the formation of trypsin from trypsinogen. *Enterokinase,* a membrane-bound enzyme of the duodenal brush border, is necessary for this step. Trypsin, in turn, acts on other proteolytic zymogens (chymotrypsinogens, procarboxypeptidases, and proelastases) to form the respective fully active proteolytic enzymes.

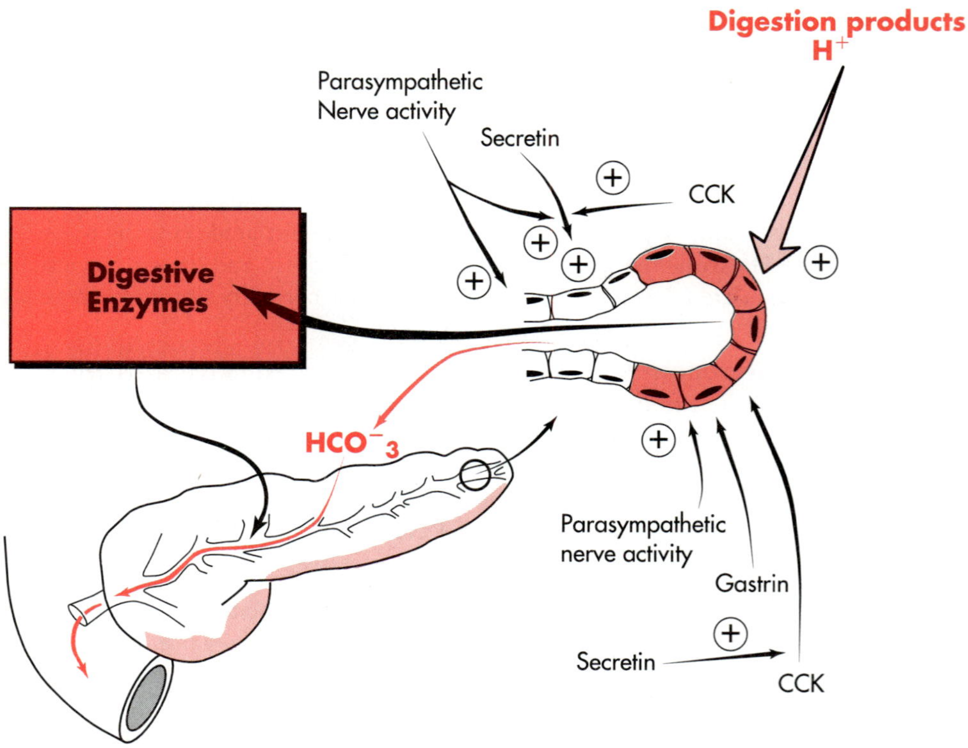

Fig. 8.5 Control of pancreatic exocrine secretions from acinar cells. The presence of digestion products and H^+ in the small intestine accounts for 70% to 80% of pancreatic stimulation. Of the remaining 20% to 30%, parasympathetic nerve activity (vagal activity) and secretin are the major nervous and endocrine stimuli for HCO_3^- secretion. Parasympathetic activity and cholecystokinin (CCK) are the significant stimuli for enzyme or enzyme precursor secretion. Gastrin is a relatively minor stimulus. It acts mostly during the anticipatory and gastric phases of food intake.

- HCO_3^- secretion is mediated by secretin.
- Secretion of digestive enzymes is mediated by CCK.

Secretions from the Liver and Gallbladder The liver continuously metabolizes cholesterol, producing the primary bile acids *cholic acid* and *chenodeoxycholic acid* as breakdown products. They are immediately conjugated with the amino acids taurine and glycine, thereby becoming water soluble and more resistant to precipitation in acid media. Even in conjugated form the bile salts will precipitate out of solution at pH <4. To maintain the pH above that level in the bile ducts, their epithelial cells actively secrete HCO_3^-.

- Conjugated bile salts, phospholipids (primarily lecithin), cholesterol, and bile pigments (primarily bilirubin) form the major components of *bile.*
- Between meals bile is stored and concentrated in the gallbladder, a distensible, muscular organ; during meals, when bile is required for the digestion and absorption of fats, it is expelled into the lumen of the duodenum. Once in the duodenum,
  - secondary bile acids (deoxycholic and lithocholic acid) are formed by the action of intestinal bacteria on the primary acids;
  - conjugated and unconjugated bile acids are passively absorbed along

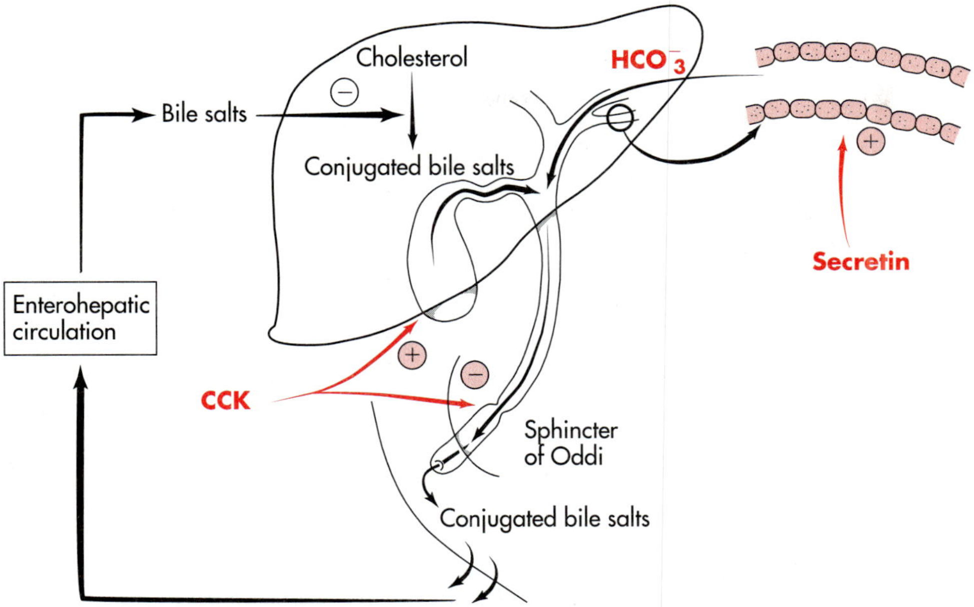

Fig. 8.6 Control of bile synthesis and expulsion. Synthesis of bile salts in the bile canaliculi occurs continuously but at a rate that is inversely related to the amount of bile salts returning to the liver through the portal vein. Secretin promotes secretion of HCO_3^--rich fluid for the purpose of preventing bile salts from precipitating, and cholecystokinin (CCK) promotes bile expulsion by causing simultaneous gallbladder contraction and sphincter of Oddi relaxation.

the entire gut. In addition, conjugated bile acids are actively transported from the lumen of the distal ileum.

- Reabsorbed bile acids reach the liver through the portal vein. This loop for recycling and conserving bile acids is called the *enterohepatic circulation* (Fig. 8.6).
- The control of bile synthesis and secretion involves two components. They are described as the bile acid–dependent and bile acid–independent portions.
  - As summarized in Fig. 8.6, the amount of primary bile acids synthesized in hepatocytes depends inversely on the amount of bile acid returned to the liver by means of the enterohepatic circulation.
  - The bile acid–independent portion resides mostly in the hormones secretin and cholecystokinin and their actions on bile ducts, the gallbladder, and the sphincter of Oddi.

Integrated Regulation of GI Function

Fig. 8.7 traces the events that link anticipatory vagal activity and the subsequent arrival of food in the stomach to release of gastrin, secretin, GIP, and other peptides so that they can trigger the secretion of digestive or catalytic enzymes from the pancreas and other areas.

Digestion and Absorption

The primary function of the digestive system is absorption of nutrients. Although electrolytes, trace metals, and vitamins can be absorbed directly by passive or specific active transport mechanisms, carbohydrates, proteins, and fats first undergo enzymatic degradation into smaller and chemically simpler units.

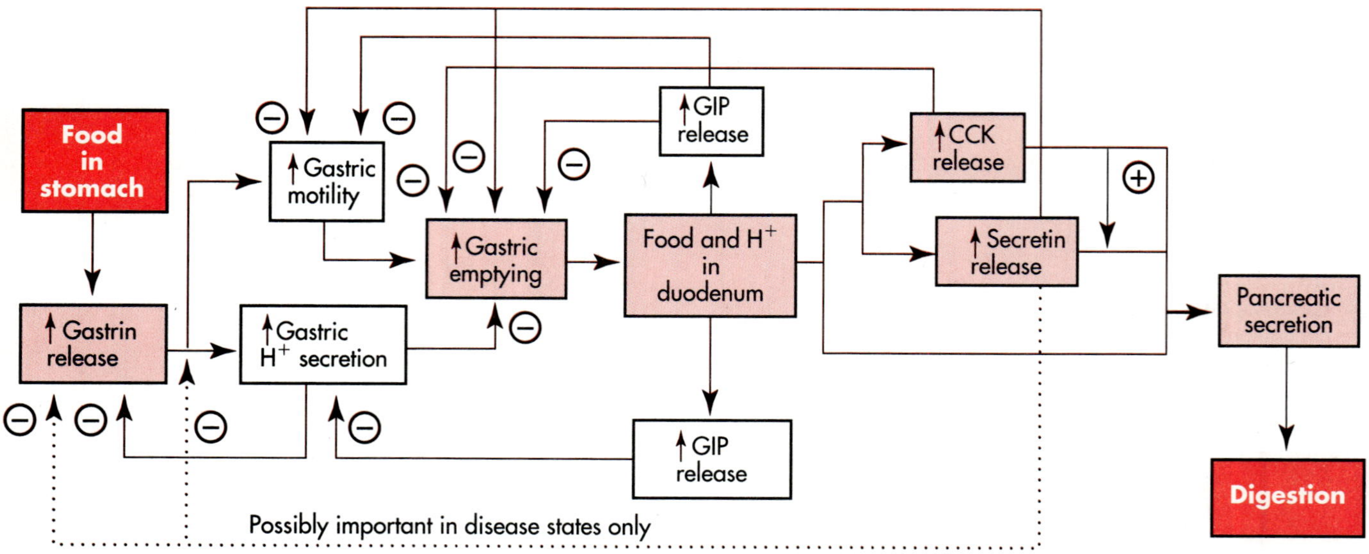

Fig. 8.7 Integrated regulation of GI secretions. The principal hormones controlling GI function are gastrin, cholecystokinin (CCK), secretin, and gastric inhibitory peptide (GIP). Their actions can be viewed as a sequence with a gastric phase and an intestinal phase. Gastrin is involved in the gastric phase and promotes events that result in the placement of food and H^+ into the small intestine. In the small intestine both H^+ and certain constituents in the ingested food will activate specialized cells to release secretin, CCK, and GIP; they in turn will promote pancreatic secretion and bile expulsion and at the same time decrease the rate of gastric emptying.

Carbohydrates Carbohydrates in human diet derive from starch or from disaccharides such as sucrose and lactose. All must be broken down to monosaccharides before they can be absorbed.

Disaccharides

- Sucrose and lactose are broken down to monosaccharides by *disaccharidases* in the brush border of the small intestine. For example, the action of the enzyme *sucrase-isomaltase* on sucrose yields the monosaccharides glucose and fructose.
- Monosaccharides are absorbed from the intestinal lumen either by passive transport down a concentration gradient (e.g., fructose) or by specific active membrane transport systems that couple monosaccharide absorption to Na^+ secretion.

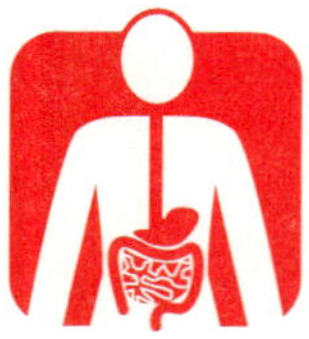

- Absence of one or more of the disaccharidases causes intolerance to certain dietary sugars. For example, congenital lack of lactase (β-galactosidase) causes lactose intolerance. In this common disorder dietary lactose (contained in milk products) is not broken down to reabsorbable monosaccharides. As a result, it holds water in the intestine by osmotic forces and causes diarrhea. In addition, some of the ingested lactose is degraded by intestinal bacteria to organic acids and CO_2. These cause the symptoms of bloating, belching, flatulence, and cramping that accompany lactose intolerance.

Starch

- Dietary starch consists largely of two polysaccharides, *amylopectin* (75%) and *amylose.* They are broken down to smaller polysaccharides mostly by pancreatic α-amylase (Fig. 8.8), and they in turn are quickly digested by brush-border enzymes to yield fructose, glucose, and galactose.

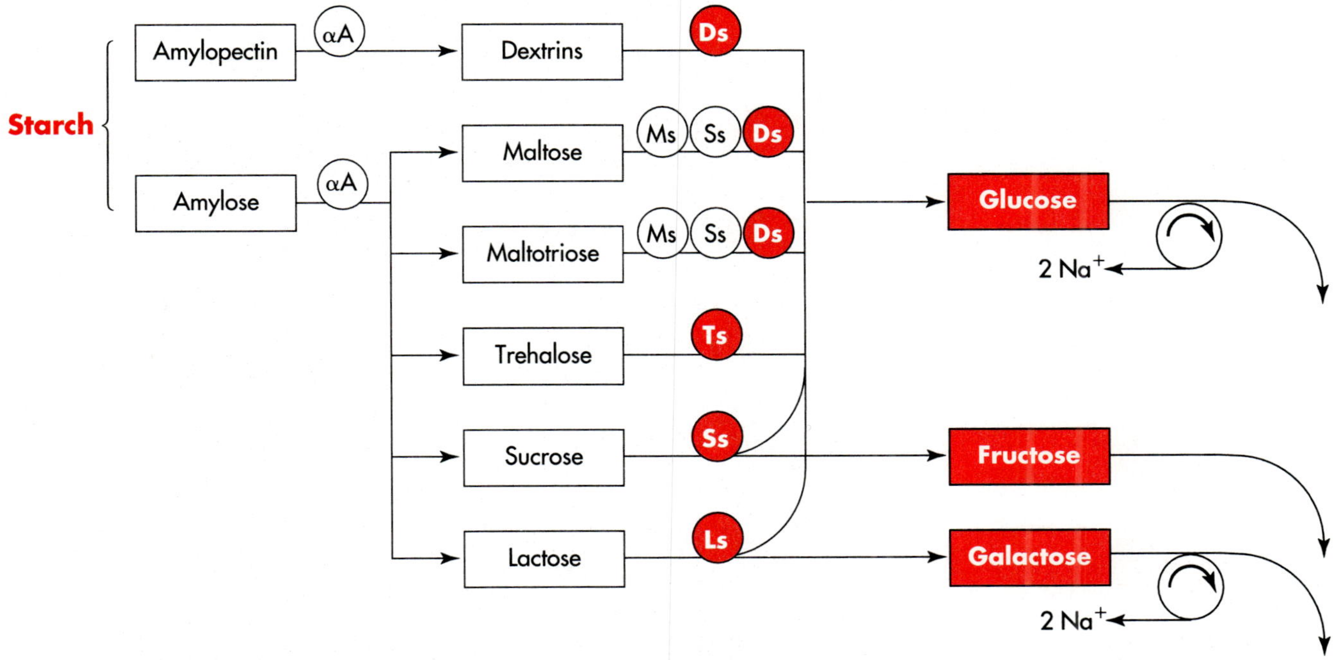

Fig. 8.8 Digestion of starch involves several enzymes and finally yields mostly glucose as well as some fructose and galactose. Maltose and maltotriose are hydrolyzed by three different enzymes: maltase and sucrinase, each contributing about 25%, whereas dextrinase contributes 50%. Glucose and galactose are reabsorbed across the mucosa by a common, Na^+-dependent active transporter. Fructose is absorbed passively by facilitated transport. αA, α-amylase; *Ds*, dextrinase; *Ls*, lactase; *Ms*, maltase; *Ss*, sucrinase; *Ts*, trehalase.

- Glucose is absorbed by the time the chyme has traveled 20 cm into the jejunum.
- Indigestible carbohydrates such as cellulose are not reabsorbed and are excreted in the feces. They are commonly called "bulk," or "fiber," and are believed by some to be healthier than sliced bread.

■ **Proteins** Protein in the intestinal lumen derives from three sources: the diet, sloughed mucosal cells, and secreted enzymes.

- Some digestion of protein occurs in the stomach (as a result of the enzymes pepsin 1 and 2), but this accounts for no more than 15% of total protein digestion.
- Most protein digestion occurs by pancreatic peptidases that belong to one of two classes, *endopeptidases* or *exopeptidases.*
    - Endopeptidases attack the interior peptide bond between neighboring amino acids. They differ from one another in their specificity for amino acids adjacent to a cleavage site. Trypsin, chymotrypsin, and elastase are prominent examples.
    - Exopeptidases act on the last peptide bond at the C-terminus of protein molecules.
- Digestion by pancreatic peptidases yields amino acids and small peptides whose final digestion occurs at the duodenal-jejunal brush border as the result of local and specific surface enzymes such as aminopeptidases and enterokinase (Fig. 8.9).
- The end products are reabsorbed by specific transport mechanisms.

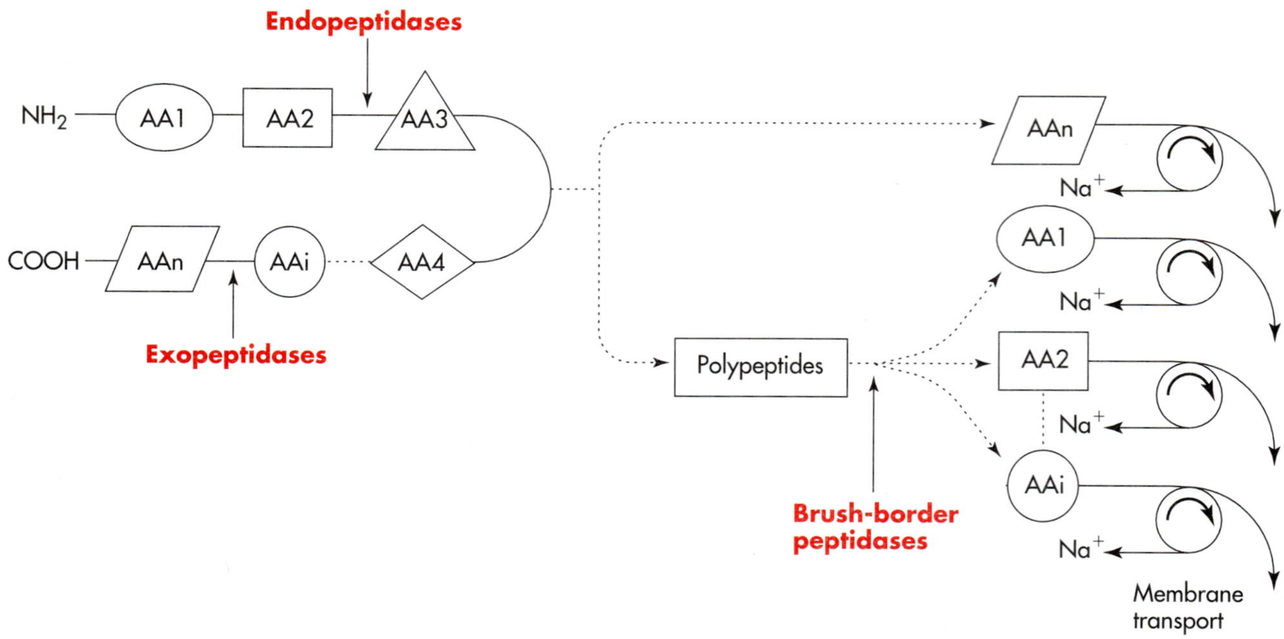

Fig. 8.9 Digestion of proteins. The action of endopeptidases or exopeptidases yields single amino acids or shorter polypeptides that are amenable to further enzymatic digestion. A variety of Na^+-coupled transporters exists for absorption of amino acids, dipeptides, or tripeptides.

- **Fats** A small portion of fat in the diet is in the form of lecithin, cholesterol esters, and medium-chain triglycerides (with 6 to 12 carbon atoms), but the majority is long-chain triglycerides (esters of glycerol and long-chain fatty acids of 16 to 18 carbon atoms).

  - **Long-chain triglycerides** Digestion and absorption of these fats involves *emulsification, micelle formation* in the lumen of the duodenum, and formation of *chylomicra* within mucosal epithelial cells. The processes are summarized in Fig. 8.10. They begin in the stomach.

    - The churning action of the stomach creates large fat drops within the aqueous chyme. They are reduced to even smaller drops by mechanical activity in the duodenum (emulsification) and are subjected to the actions of lipases (from the pancreas) and bile salts (from the gallbladder).
    - The outcome of lipase action is β-monoglycerides and free long-chain fatty acids.
    - Long-chain fatty acids and β-monoglycerides are poorly soluble in water and therefore difficult to absorb. They are brought into solution by a special property of bile salts:
      - Bile salts have a water-soluble portion and a fat-soluble portion. When they aggregate, they form *micelles* (French from Latin 'small morsel') in which the water-soluble portion of each molecule faces outward while the lipid-soluble portion faces inward.
      - Long-chain fatty acids and β-monoglycerides dissolve into the structure of bile acid micelles to form *mixed micelles* and are transported in that form to the brush border of the small intestine.
    - Mixed micelles package monoglycerides and long-chain FFA (free fatty acids) in water-soluble form for presentation to the unstirred layer near

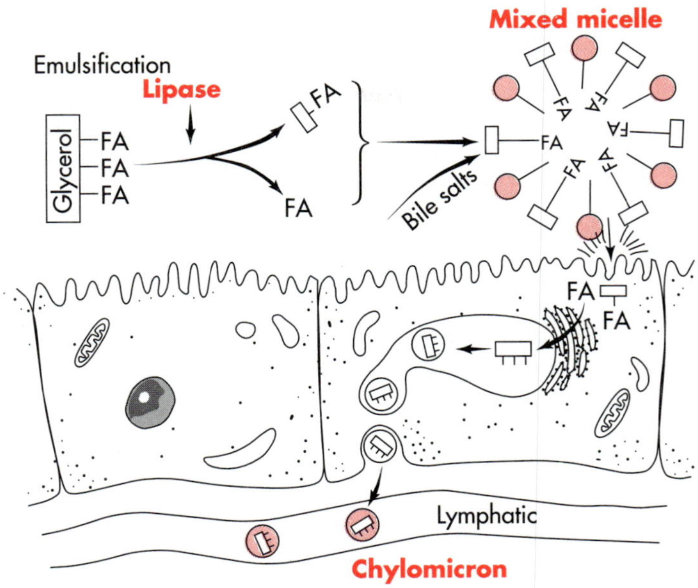

Fig. 8.10 Initial steps in the digestion of dietary fats. Emulsification is a mechanochemical process by which large fat drops are reduced to smaller drops. Mixed micelles are water-soluble structures of monoglycerides and long-chain free fatty acids *(FA)* with bile salts aggregated on their surface. The water-soluble portion (carboxyl end) of the bile salts faces outward.

the brush border of mucosal cells. They permit gradual release of lipids to the unstirred layer so that they can be gradually absorbed by mucosal cells.

- Within the endoplasmic reticulum of the mucosal cell, absorbed fatty acids and β-monoglycerides are reassembled with endogenously synthesized glycerol to form triglycerides again. These and dietary cholesterol are formed into *chylomicra* (droplets coated with protein and phospholipid) and are delivered in that form to the intestinal lymphatic system and from there to the blood.

- **Medium-chain triglycerides** Such smaller triglycerides are digested and absorbed by a much simpler process than long-chain triglycerides because they are rapidly and completely hydrolyzed by lipases and isomerases in the intestinal lumen. The resulting fatty acids:

  - require no micelle formation for absorption into epithelial cells because they have better water solubility than long-chain FFA. They also:
  - do not form chylomicra within the epithelial cells. Instead,
  - they are absorbed directly into the portal blood without being reconstituted as triglycerides.

- **Cholesterol** Cholesterol is required for cell membrane formation and for synthesis of steroid hormones. It can be obtained from the diet or synthesized in the liver and intestine from acetyl CoA. Dietary cholesterol is:

  - digested by cholesterol esterase (secreted by the pancreas),
  - incorporated into micelles, and

• absorbed from the small intestine in *chylomicra,* the primary means of transporting dietary lipids and cholesterol from the intestine to the liver.

Transport systems for lipids and cholesterol Lipids and cholesterol are insoluble in water and therefore require a special transport system. This is provided by a family of particles that differ in size but resemble one another in their components:

• Four classes of lipoprotein particles exist in fasting humans: high density lipoproteins (HDL), intermediate density lipoproteins (IDL), low density lipoproteins (LDL) and very low density lipoproteins (VLDL). They differ in diameter, density, molecular weight, fractional composition, and protein makeup.

• Immediately after a meal, in the postabsorptive period, there exists a fifth type of blood lipoprotein, the chylomicron.

• Blood lipoproteins consist of a core of triglycerides and cholesterol esters. The core is surrounded by a phospholipid/lipid monolayer shell containing nonesterified cholesterol and *apolipoproteins.* Each class of blood lipoprotein particles has a characteristic apolipoprotein composition. Apolipoproteins have two functions: They solubilize lipids, and they serve receptor functions.

• The transport mechanisms for lipids and cholesterol are summarized in Fig. 8.11.

- The endothelium of capillaries in adipose tissue and muscle contains lipoprotein lipase, an enzyme capable of rapidly hydrolyzing the triglycerides contained in chylomicra. The enzyme is activated by apo C-II, one of the apolipoproteins found in the shell of chylomicra. The cholesterol-rich residue, called *chylomicron remnants,* is taken up by the liver by means of an Apo E–mediated receptor process. The liver uses cholesterol from a variety of sources and combines it with locally synthesized triglycerides to form and secrete VLDL.
- VLDL reach fat and muscle tissue through the circulation and there, at the capillary wall, the triglycerides are removed from the VLDL by the same lipases that act on chylomicra. The remnants are rich in cholesterol esters and are called intermediate density lipoprotein (IDL) particles.
- Some of the IDL are immediately taken up by the liver. The others are mostly converted to LDL, the major carrier of cholesterol in blood. The role of LDL is to transport cholesterol to peripheral tissues and to regulate synthesis of cholesterol in them.
- When a cell requires cholesterol, it synthesizes receptors for LDL and inserts them into the plasma membrane within specialized surface pits. The apo B-100 component of LDL binds to these receptors and initiates a sequence whereby LDL are internalized by endocytosis, fuse with lysosomes, and are broken down to cholesterol and amino acids. The LDL receptor is recycled to the plasma membrane.
- The released cholesterol can be used for biosynthesis (of membrane components or steroids) or be esterified for storage inside the cell. When cholesterol is abundant inside the cell, new LDL receptors are not synthesized.

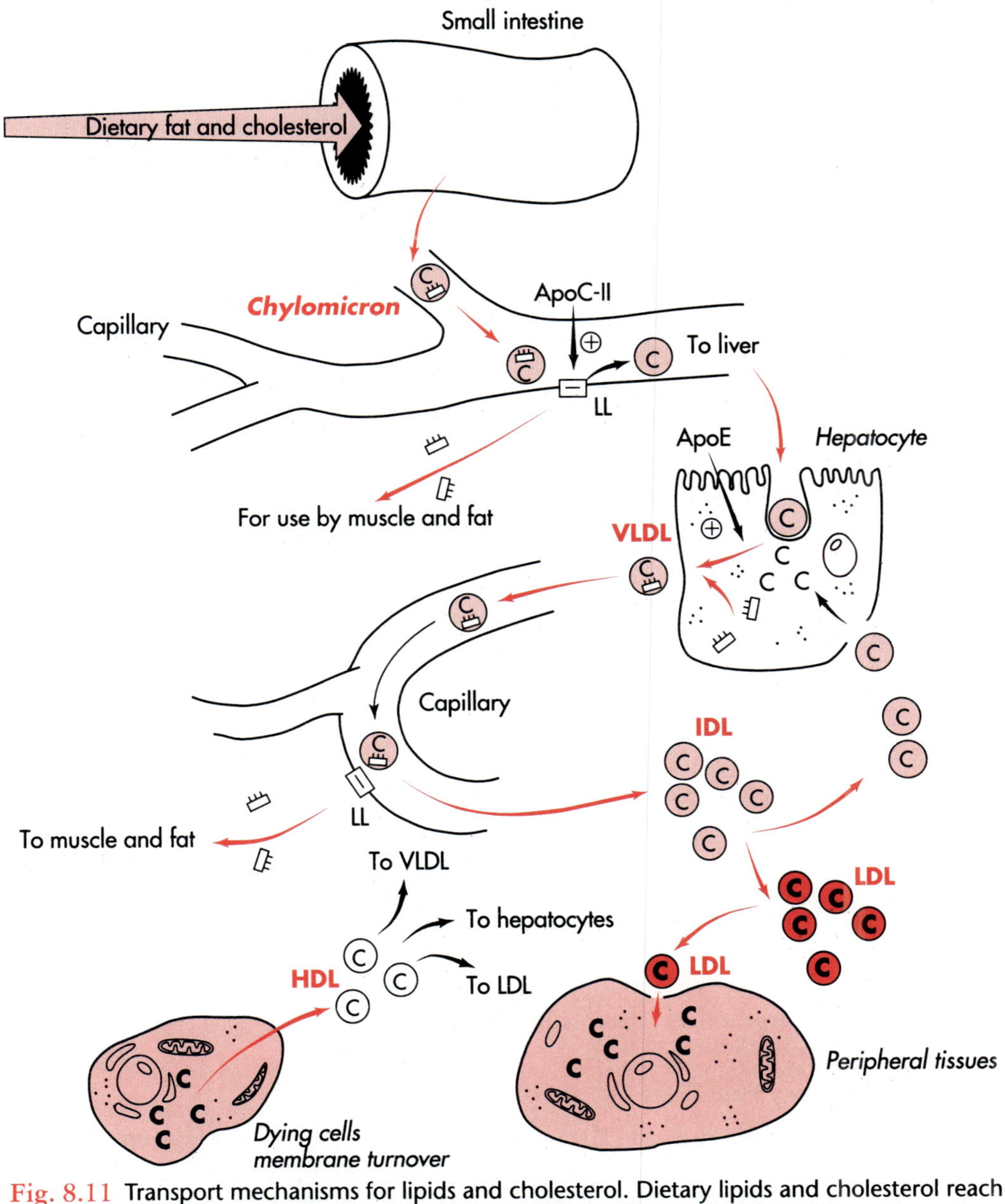

Fig. 8.11 Transport mechanisms for lipids and cholesterol. Dietary lipids and cholesterol reach the peripheral circulation in chylomicra. Lipoprotein lipase hydrolyzes their triglyceride component, and the chylomicron remnant is taken up by hepatocytes. Locally synthesized triglycerides and cholesterol from several sources, including chylomicron remnants, are used to form and secrete VLDL. In fat and in skeletal muscle the triglycerides are removed from the VLDL by lipoprotein lipase, and cholesterol remains behind in IDL, some of which quickly form LDL particles. LDL are used by cells as the source of cholesterol for their biosynthetic activities. Cholesterol from dismantled cellular components is carried in the blood by HDL. *ApoC-2,* apolipoprotein C-II; *ApoE,* apolipoprotein E; *C,* cholesterol; *HDL,* high density lipoprotein; *IDL,* intermediate density lipoprotein; *LDL,* low density lipoprotein; *LL,* lipoprotein lipase; [E, triglycerides; *VLDL,* very low density lipoprotein.

- Absence or ineffectiveness of LDL receptors leads to hypercholesterolemia and atherosclerosis. There is a strong positive correlation between serum LDL levels and ischemic heart disease.

- Free cholesterol from dying cells or from membrane turnover is picked up by HDL particles and is quickly esterified. Cholesterol esters can be exchanged to VLDL and LDL if specific transfer proteins are available or transported in HDL to the liver where the cholesterol content is disposed, partly as bile acids.
- There is a strong negative correlation between serum HDL levels and ischemic heart disease.

Liver Functions

Most substances that enter the body by mouth reach the liver via portal venous blood after they have been absorbed by the intestinal mucosa. The metabolic functions of the liver determine their subsequent disposition.In addition to these crucial functions, the liver performs several other vital tasks.

Metabolic Functions

- **Gluconeogenesis and glycolysis** When plasma glucose concentration is low, the liver forms glucose:

  - from amino acids (gluconeogenesis),
  - from glycogen (glycolysis), or
  - by converting fructose or galactose to glucose.

  When plasma glucose concentration is high, the liver:

  - stores glucose in the form of glycogen
  - breaks down glucose to amino acids
  - converts glucose to fatty acids
- **Protein synthesis** Plasma albumin, fibrinogen, and most globulins are synthesized by the liver from amino acids that are delivered to it via portal blood.
- **Fat metabolism** Phospholipids and lipoproteins are synthesized from fatty acids.

Circulatory Functions

- **Energy dissipation** The structure of the liver sinusoids forces incoming hepatic arterial blood to flow in a circular pattern. This contributes to dissipation of hepatic arterial blood pressure before mixing occurs with portal venous blood at a considerably lower pressure.
- **Clotting factors** The liver synthesizes most of the clotting factors and thereby ensures a functioning hemostatic system.
- **Volume storage organ** The vascular bed of the liver is a large, spongy organ. Its high compliance permits storage of blood volume in an accessible reservoir.

Secretory Functions In addition to a variety of peptide secretions, the liver secretes cholesterol, lecithin, bile salts, and bilirubin.

- **Cholesterol** Cholesterol is synthesized from acetyl coenzyme A and is either secreted within VLDL or broken down to form bile salts.

- **Bilirubin** Bilirubin is an end product of heme metabolism. It derives from the hemoglobin of degraded erythrocytes and is carried in plasma as a bilirubin-albumin complex. Bilirubin reaches hepatocytes through large fenestrations in the liver sinusoids and is transported into the cytosol. There a fraction of it is returned to the plasma. The remainder is conjugated mostly to glucuronate and secreted into the bile canaliculi.

Conservation Functions The liver conserves bile salts and amino acids after reclaiming them from the portal blood.

Excretion Functions The liver clears blood of a variety of substances including urea and other products of metabolism. The liver is also a major pathway for the clearance of drugs.

- **Metabolic filter** The liver acts as a filter for a variety of substances and uses conjugation, oxidation, reduction, methylation, and other processes to detoxify and excrete potentially harmful agents.
- **Urea** The liver is the major site of urea formation, removing in the process the ammonia that results from catabolism of amino acids.

Multiple Choice Review Questions

1. Which one of the following motility patterns in the small intestine is primarily responsible for the mixing of chyme with digestive juices and exposing the products of digestion to the intestinal wall?
   a. The migrating motor complex
   b. Peristaltic waves
   c. Haustrations
   d. Segmentation

2. Which one of the following choices correctly describes the changes in electrolyte composition as the secretion of the acinar cells in salivary glands flows from the acini along the ductules and ducts?

	K^+	HCO_3^-	Cl^-
a.	Decreased	Decreased	Increased
b.	Increased	Increased	Decreased
c.	Decreased	Increased	Decreased
d.	Increased	Increased	Increased
e.	Decreased	Decreased	Decreased

3. Gastric acid secretion can be stimulated by:
   a. Histamine only.
   b. Phospholipids.
   c. Histamine, gastrin, and acetylcholine.
   d. Cimetidine and atropine.
   e. Angiotensin II.

4. Which one of the following is the major factor stimulating HCO_3^- secretion from the pancreas?
   a. Cholecystokinin
   b. Secretin
   c. Glucagon
   d. Insulin
   e. Histamine

5. When the enterohepatic circulation of bile acids is interrupted by ileal disease or resection, what is the usual consequence?
   a. Very low fecal bile acid loss
   b. Reduction of the bile acid concentrations in the upper small bowel, resulting in disturbed fat absorption
   c. Increased serum cholesterol concentration
   d. Reduced hepatic bile acid synthesis

6. Which one of the following substances shows the greatest degree of digestion in the stomach?
   a. Fats
   b. Carbohydrates
   c. Proteins
   d. Vitamins
   e. Minerals

7. The end products of carbohydrate and protein digestion have one of the following characteristics in common. Which one?
   a. They require chylomicra in order to leave the mucosal cell.
   b. They enter the bloodstream after being transported through the lymphatic vessels of the intestine.
   c. They require Na^+ for active transport.
   d. They require mixed micelles for efficient absorption.

8. Which of the following secretions is most dependent on vagal stimulation?
   a. Saliva
   b. HCI
   c. Pepsin
   d. Pancreatic juice
   e. Bile

9. Which one of the following is *not* part of the swallowing sequence:
   a. The tongue pushes a bolus of food into the pharynx.
   b. A central reflex results in closing of the nasopharynx by the soft palate.
   c. The upper esophageal sphincter is relaxed.
   d. The respiratory center induces exhalation to prevent food from entering the pharynx.
   e. The swallowing center stimulates efferent vagal activity to initiate a wave of contraction down the esophagus.

10. Na^+-dependent transport is responsible for the absorption of all of the following *except:*
    a. Vitamin E
    b. Amino acids
    c. Glucose
    d. Bile salts

Chapter 9

Endocrinology

Mechanisms of Chemical Communication

- **Forms of Chemical Communication** Chemical communication among cells is named *autocrine, endocrine, paracrine,* or *neurocrine,* depending on the physical relationship between secreting cells and target cells or on the mode in which the chemical agent is delivered (Table 9.1).
- **Cellular Mechanisms of Hormone Action** Hormonal interaction with target cells begins with reversible binding to highly specific receptors. Such receptors are located in the plasma membrane, in the cytosol, or in the cell nucleus.
  - **Interactions with membrane receptors** The general sequence, described in detail in Chapter 1, is that the hormone-receptor complex activates a coupling mechanism (commonly a G protein) that will influence a catalytic enzyme whose role it is to transform a precursor into a second messenger.

See Appendix on p. 275 for abbreviations.

Table 9.1 *Forms of Chemical Signaling Among Mammalian Cells*

Form of Signaling	Relationship between Secreting and Target Cells	Mode of Effector Delivery
Autocrine	They are one and the same	Exocytosis
Endocrine*	Physically separated Connected by the vascular system	Bloodstream
Neurocrine	Physically separated Connected by the vascular system	1. Overflow into the capillaries near the primary target cell and subsequent transport to other target cells† or 2. Drainage of the extracellular space surrounding the secreting neuron by a portal circulation‡§
Paracrine	They are neighbors	Diffusion through extracellular space

*Agents in this category have traditionally been called "hormones."

†*Example:* Circulating norepinephrine secreted from sympathetic nerves.

‡*Example:* Vasopressin is synthesized in the supraoptic nucleus of the hypothalamus and is released into the portal circulation of the posterior pituitary. Its primary target cells are vascular smooth muscle and renal collecting tubule cells.

§The term "portal circulation" is used to describe a unique anatomical pattern of the circulation. The pattern consists of a capillary network that drains into a vein that subsequently divides into a second capillary network in another organ and then drains into yet another vein.

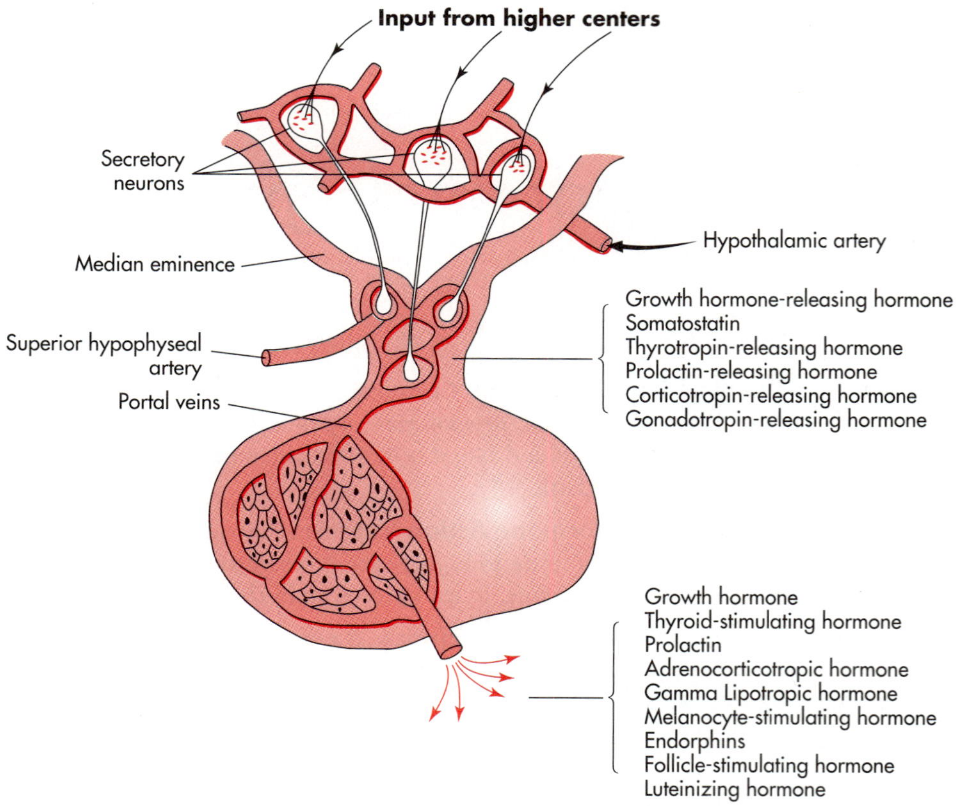

Fig. 9.1 Secretory neurons in the hypothalamus synthesize trophic factors and transport them along their axons to the region of the median eminence. They are released from the axon terminals at a rate determined by action potentials in the secretory neurons. This electrical activity is governed partly by higher nervous centers and partly by hormonal feedback reaching the neurosecretory cells through the hypothalamic artery.

The second messenger, in turn, activates an intracellular effector (commonly a kinase) to bring about the target cell response.

- The known catalytic enzymes are adenylate cyclase, guanylate cyclase, and phospholipase-C.
- The known second messengers are cAMP, cGMP, diacyl glycerol (DAG), inositol trisphosphate (IP_3), and Ca^{++}.

- **Interactions with nuclear receptors** Hormones that can penetrate the cell membrane and form intracellular hormone-receptor complexes (steroid and thyroid hormones) bind to DNA and alter the expression of specific genes. As a result, the time scale of action of these hormones is in the range of a few days rather than in the range of several minutes.

Hypothalamus and Anterior Pituitary

- **Anatomy** Special secretary neurons in various hypothalamic regions synthesize agents that are transported down their axons and are released from the axon terminals into a portal system that links the median eminence to the anterior pituitary (Fig. 9.1). This vascular system is a significant feature of the hypothalamus-pituitary unit; because of it the anterior pituitary receives blood from two sources:

Hypothalamic trophic factor	GHRH SS	TRH SS	PIF (mostly DA) PRF (TRH, VIP)	CRH ⊕	GnRH ⊕
Anterior pituitary hormone	⊕ ⊖ **Growth hormone**	⊕ ⊖ **TSH**	⊖ ⊕ **Prolactin**	**POMC** → **ACTH MSH END**	**FSH LH**
Target	Circulating precursors; Muscle Fat	Thyroid	Lacteals	Adrenal cortex	Ovaries; Testes
Target organ response	Somatomedins; ↑ Fatty acid mobilization metabolism	Thyroxine (T_4)	Lactation	Cortisol	Progesterone Estrogen; Testosterone

Fig. 9.2 Summary of anterior pituitary hormones; their actions and their control by hypothalamic trophic factors. *ACTH,* Adrenocorticotropic hormone; *CRH,* corticotropin–releasing hormone; *DA,* dopamine; *EN,* endorphins; *FSH,* follicle-stimulating hormone; *GHRH,* growth hormone–releasing hormone; *GnRH,* gonadotropin-releasing hormone; *LH,* luteinizing hormone; *MSH,* melanocyte-stimulating hormones; *PIF,* prolactin-inhibiting factor; *POMC,* pro-opiomelanocortin; *PRF,* prolactin-releasing factor; *SS,* somatostatin; *TRH,* thyrotropin-releasing hormone; *VIP,* vasoactive intestinal peptide.

- Arterial blood from the superior hypophyseal artery provides nutrients. This portion of the circulation is not shown in Fig. 9.1.
- Venous blood from the portal system that originates in capillaries of the median eminence (see Fig. 9.1) provides releasing factors for a variety of anterior pituitary hormones.

Secretions from Hypothalamic Neurons The hypothalamus coordinates whole-body function. The chemical aspects of this role are accomplished by synthesis and secretion of several trophic factors from specialized neurons. These are shown in Fig. 9.1 and their physiologic roles are summarized in Fig 9.2.

Anterior Pituitary Hormones

Growth hormone (GH)

— ***Synthesis and release of growth hormone*** Human growth hormone (GH) is a single polypeptide chain of 191 amino acids. It is synthesized in pituitary cells, called *somatotrophs.* They make up about a half of all anterior pituitary cells. All factors affecting its release use growth hormone–releasing hormone (GHRH) and somatostatin (SS) as final common pathways.

- GH release requires stimulation by GHRH. SS inhibits GH release.
  - GHRH is a 44-residue peptide, produced mostly in the arcuate nucleus of the hypothalamus but also in the mucosa of the small intestine. It promotes both GH synthesis (by enhancing gene transcription) and release (by a Ca^{++}-dependent mechanism that is initiated by receptor-mediated activation of adenylate cyclase).
  - Somatostatin exists as both a 14– and a 28–amino acid peptide. It is:
    - widely distributed in the central nervous system and functions there as a neurotransmitter. It is also

  - synthesized in cells of the gastrointestinal mucosa where it serves a variety of inhibitory functions (see Chapter 8,) and
  - synthesized in D-cells of the pancreas, where it is a paracrine modulator of insulin and glucagon secretion.
- The 28–amino acid form is the more potent inhibitor of GH and TSH (thyroid-stimulating hormone) than the 14–amino acid form. It has no effect on GH mRNA but inhibits GH secretion. The mechanism is receptor-mediated inhibition of adenylate cyclase.

- GH release is pulsatile and its half-life in plasma is 20 to 30 minutes. In any one person, plasma levels of GH rise after a meal and during sleep. Over the course of a lifetime, GH levels are highest during puberty.

— ***Functions of growth hormone*** Growth hormone is required for normal linear growth. It also has metabolic effects. The balance between these two functions is determined by the nutritional state of the body. When the nutritional state is adequate and insulin is active, GH promotes growth. On the other hand, when the nutritional state is inadequate and when insulin levels are low, the metabolic effects of GH dominate.

Growth effects. Growth effects are not attributable to GH directly but to the *somatomedins,* also called *insulin-like growth factors* (IGF). IGFs are synthesized under the control of GH.

- Somatomedin C (a.k.a. IGF-1) is the most important somatomedin for postnatal growth. It is produced in several tissues, but the liver is the most important source. IGF-1 circulates, bound to IGF-binding proteins, and thereby acquires a half-life of up to 20 hours. The effect of this prolongation of half-life is a relatively constant biologic effect of GH even in the face of pulsatile release.
- All tissues that are capable of growing are affected in cell size and rate of cell division.
  - Before puberty, while androgen levels are low, the major effects are on the longitudinal growth of cartilage and bone.
  - After puberty, when androgens have caused ossification and closure of the epiphyseal growth plates of the long bones, the somatomedins continue to stimulate circumferential growth in all bones but promote longitudinal growth only in membranous bones, which have no epiphyses.

Metabolic effects. GH raises blood glucose by two mechanisms:

- It increases insulin output from the liver and
- it inhibits glucose uptake and glucose utilization by muscle.

In addition GH increases lipolysis and thereby increases the plasma level of free fatty acids.

- **Thyroid-stimulating hormone (TSH)** TSH is the major regulator of thyroid function and thyroid size. It is synthesized in and secreted from pituitary thyrotrophs, which constitute about 5% of the cells in the anterior pituitary.
  - TSH release is increased by thyrotropin-releasing hormone (TRH)

and decreased by somatostatin (SS). The TRH receptor uses the IP_3 pathway of increasing cytosolic Ca^{++}. The SS receptor inhibits adenylate cyclase.

• The pituitary thyrotroph is also controlled by feedback inhibition from the plasma levels of thyroxine (T_4) or triiodothyronine (T_3), the active form of thyroid hormone.

- **Prolactin** Prolactin is synthesized in lactotrophs. They constitute 15% to 20% of the normal pituitary cell mass and increase to 70% during pregnancy. The predominant form of prolactin contains 198 amino acids.

• The major hypothalamic influence on pituitary lactotrophic cells is inhibition by prolactin-inhibiting factors (PIF, Fig. 9.2), mostly dopamine. This inhibitory action is exerted via D_2 receptors. Their predominant effector mechanisms are inhibition of adenylate cyclase, inhibition of Ca^{++} channels and activation of K^+ channels. Prolactin-releasing factors (PRF, Fig. 9.2) appear to be required for the prolactin increase that occurs during breast feeding. Thyrotropin-releasing hormone (TRH) and vasoactive intestinal peptide (VIP) may act in that capacity.

• The principal role of prolactin is to act on the mammary epithelium to increase milk production. This action is inhibited by high estrogen levels.

- **Adrenocorticotropic hormone (ACTH)**

— ***Synthesis and release*** ACTH is synthesized by corticotrophs. They constitute about 15% of the anterior pituitary cells.

• ACTH is part of a large, 265–amino acid precursor molecule, called *pro-opiomelanocortin* (POMC, Fig. 9.2).

• POMC is cleaved in the anterior pituitary to yield ACTH (39 amino acids), β-lipotropin (91 amino acids) and γ-melanocyte-stimulating hormone (γ-MSH). Further cleavage of β-lipotropin yields β-MSH and several endorphins, of which β-endorphin is the most abundant.

• The major regulator of ACTH synthesis and release is corticotropin-releasing hormone (CRH).

  - CRH is a single-chain polypeptide that is produced by neurons in the paraventricular nucleus of the hypothalamus as well as by other areas of the brain and the viscera. The placenta has the highest concentration of CRH outside the nervous system.
  - CRH increases mRNA for POMC and promotes ACTH release by a receptor-mediated, cAMP- and Ca^{++}-dependent mechanism.

• ACTH release is pulsatile. It shows a circadian rhythm that peaks a few hours before awakening and troughs about an hour before normal bedtime.

— ***Functions of ACTH*** ACTH controls the adrenal cortex (Fig. 9.3). This involves primarily the release of cortisol, but ACTH also promotes release of mineralocorticoids.

• Under normal circumstances the hypothalamic-pituitary-adrenal axis is regulated by negative feedback: cortisol inhibits both

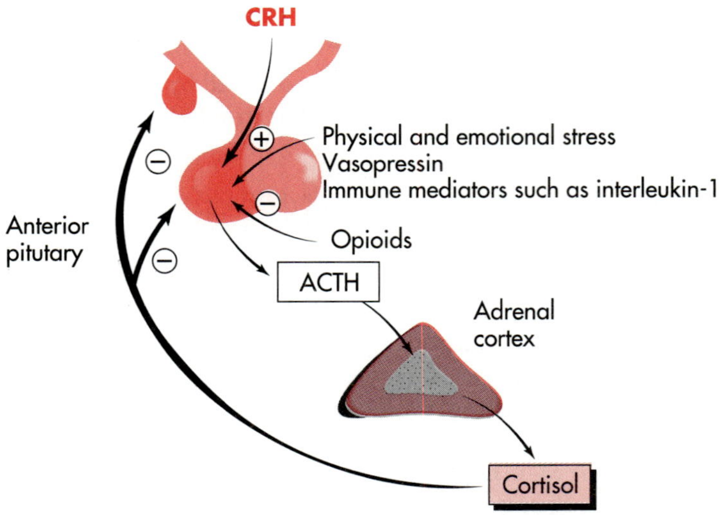

Fig. 9.3 Control and function of adrenocorticotropic hormone (ACTH). ACTH release is promoted by corticotropin-releasing hormone (CRH) and several nervous or humoral factors. ACTH release is inhibited most prominently by cortisol, which is itself an agent whose release is promoted by ACTH.

CRH release and the ACTH-releasing function of CRH (see Fig. 9.3). However,

- the axis is a significant component in the mechanisms that allow us to deal with stress. As a result it is modulated by many nervous, hormonal, and immunologic factors (see Fig. 9.3).

Endogenous opioid peptides The opioid peptides include *endorphins, enkephalins,* and *dynorphins.* They are chemically related but are produced by different biosynthetic pathways. The pituitary synthesizes endorphins as part of POMC, and β-endorphin is the most abundant form. The endogenous opioid peptides may play a role in a variety of complex physiologic patterns associated with pain perception, learning, behavior, and addiction.

Follicle-stimulating hormone (FSH) The gonadotrophs make up about 10% of all anterior pituitary cells. Most of them secrete both follicle-stimulating hormone (FSH) and luteinizing hormone (LH) for the purpose of regulating ovarian or testicular function.

- In women FSH induces growth of follicles in preparation for the next ovulation cycle. It also stimulates growth and estradiol synthesis in granulosa cells of the follicle (see Chapter 11 for the discussion of ovarian function and women's monthly rhythm).
- In men FSH stimulates secretory activity in Sertoli cells and thereby helps maintain the spermatogenic epithelium (see Chapter 11 for the discussion of testicular function).

Luteinizing hormone (LH) The glycoprotein luteinizing hormone is synthesized by pituitary gonadotrophs, along with FSH.

- In women LH stimulates the ovarian theca cells to produce androgens, which then diffuse to the granulosa cells where they are converted to estrogens (see Chapter 11 for the discussion of ovarian function). A surge in LH secretion at about day 10 to 12 of the cycle triggers ovulation from the dominant follicle (see Chapter 11 concerning women's monthly rhythm).

- In men LH is primarily responsible for controlling testosterone synthesis in Leydig cells of the testes (see Chapter 11 concerning testicular function).

Hypothalamus and Posterior Pituitary

Posterior Pituitary Hormones The posterior lobe of the pituitary gland produces no biologic responses to hypothalamic agents. However, it acts as a way station for transferring two hypothalamic neurosecretions to the bloodstream for transport to peripheral target organs, *vasopressin* and *oxytocin.* This function requires the portal network formed in the posterior pituitary by capillaries of the inferior hypophyseal artery. The network is a vital anatomic feature because it provides a point of contact with the terminal boutons of the secretory neurons.

Synthesis, transport, and release of vasopressin and oxytocin Vasopressin and oxytocin are small peptides (9 amino acids). They are synthesized in the magnocellular portion of the supraoptic and paraventricular nuclei of the hypothalamus as parts of precursor molecules:

- Vasopressin derives from *prepropressophysin.*
- Oxytocin derives from *prepro-oxyphysin.*
- Cleavage of the precursor molecules occurs during transport down the neurons toward the posterior pituitary. The storage granules in the neuron terminals contain free vasopressin or oxytocin and the corresponding neurophysin.
- Each hormone is released along with its neurophysin in response to nerve action potentials.

Functions The physiologic functions of neurophysins are unknown.

— ***Vasopressin*** Vasopressin exerts its effects via G protein–coupled receptors:

- V_1 receptors are located mostly in vascular smooth muscle. Their activation causes vasoconstriction (by an IP_3-mediated increase in cytosolic Ca^{++}).
- V_2 receptors are found on the basolateral side of cells in the renal collecting tubule. When they are activated, they cause insertion of water channels (aggrephores) into the luminal side of these cells by a cAMP-dependent mechanism. This increases water permeability of such cells, increases water reabsorption at this site, and thereby regulates body fluid osmolality and body fluid volume. Alcohol inhibits vasopressin secretion. This causes increased urine flow and is responsible for the observation that "You never buy beer; you only borrow it."

— ***Oxytocin*** Oxytocin acts specifically on uterine smooth muscle and on the smooth muscle cells that surround the distal portion of the mammary gland duct system. Its biologic effect is to elevate intracellular Ca^{++} and therefore to cause muscle contraction.

- In the mammary glands muscle contraction results in transport of milk to the lactiferous sinuses and subsequent milk ejection.
- Responsiveness of the uterus to oxytocin is dependent on many factors including the presence of estrogen (which enhances contrac-

tion) and progesterone (which inhibits contraction). The sensitivity of the uterus to oxytocin increases in late pregnancy as a result of an increase in the number of oxytocin receptors.

• Oxytocin may also be responsible for orgasm-related contractions of the uterus in women and the vas deferens in men. The purpose of such contractions is sperm transport.

The Thyroid Gland

Anatomy The thyroid elaborates two classes of hormones from histologically distinct cell populations:

- Parafollicular cells (also called C-cells) synthesize calcitonin.
- Follicular cells synthesize a mixture of thyroxine (T_4) and triiodothyronine (T_3).

Calcitonin

Synthesis and secretion C-cells, the source of calcitonin, are found in other tissues, but they are located chiefly in the thyroid gland where they are also called *parafollicular cells.* The rate of synthesis and secretion of calcitonin increases linearly with serum levels of ionized calcium.

Biologic effects Calcitonin appears to act mostly as a short-term regulator of serum Ca^{++} concentration:

• Its immediate effect is to decrease osteolytic activity in osteocytes (see Chapter 10). This decrease favors calcium deposition in the rapidly exchanging pool of bone calcium salts ($CaHPO_4$).

• Although calcitonin inhibits the formation of osteoclasts (large bone-reabsorbing cells that are normally active over a small portion of bone surfaces), its long-term effects on serum Ca^{++} are small in adult humans because such effects trigger compensatory changes in osteoblastic activity and parathyroid hormone secretion.

Thyroxine (T_4) and Triiodothyronine (T_3) T_4 and T_3 are iodinated derivatives of the amino acid tyrosine. They are synthesized in the cells that form the shell of thyroid follicles, are stored in droplets within the colloid core of the follicles, and are secreted from the core in response to stimulation by thyroid-stimulating hormone (TSH).

Dietary iodine is essential for synthesis of thyroid hormones. Most of it is reduced to iodide (I^-) in the GI tract before it is absorbed. It reaches the thyroid through the circulation. Fig. 9.4 shows the subsequent steps in the synthesis and secretion of thyroid hormones.

Synthesis of T_3 and T_4

• The external membrane of follicular cells absorbs iodide (I^-) by an active, Na^+- coupled transport mechanism. This process is called *iodide trapping.* I^- is oxidized by thyroid peroxidase and hydrogen peroxide to an unstable intermediary that is quickly incorporated into tyrosine moieties (TYR) of thyroglobulin (TGL), a glycoprotein that is synthesized by follicular cells. This forms monoiodotyrosine (MIT) or diiodotyrosine (DIT) within TGL.*

---

*Thyroglobulin (TGL) is a large molecule, containing many tyrosines. Only a few of them are in sites that are amenable to iodination.

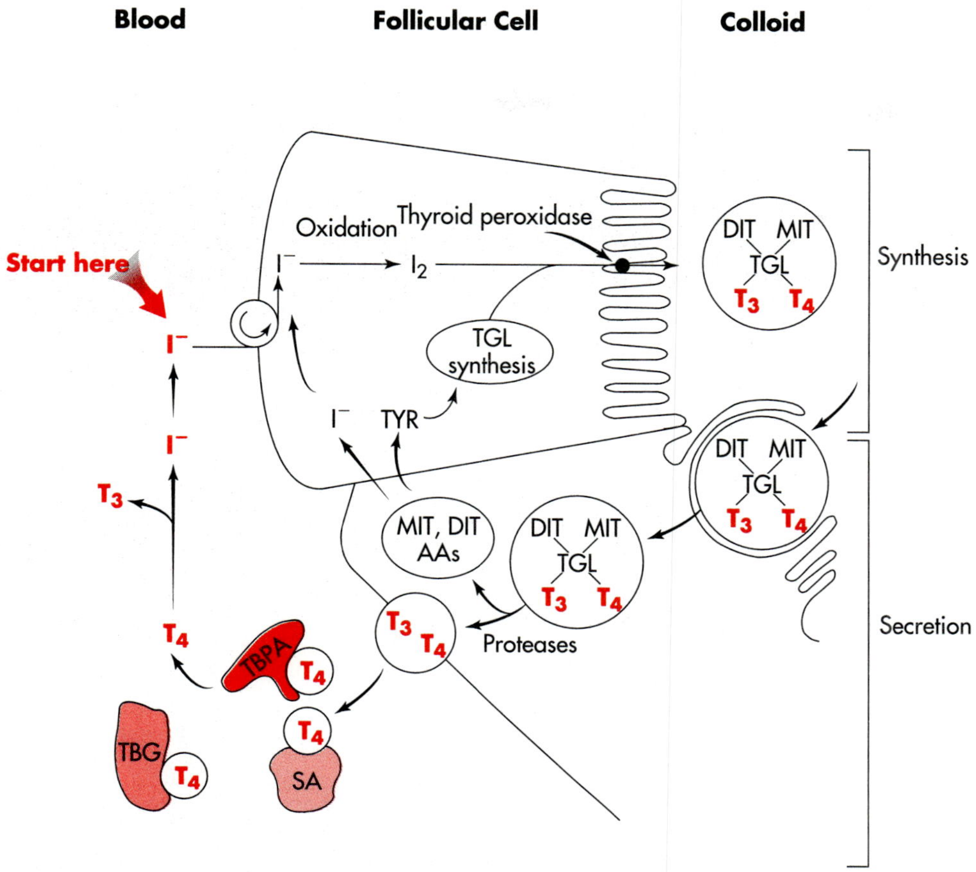

Fig. 9.4 Synthesis and secretion of thyroid hormones. Iodide (I^-)derives from dietary sources and enters the follicular cell by a Na^+-coupled transporter. Thyroid peroxidase, located at the colloidal border, promotes iodination of the amino acid tyrosine (TYR) within thyroglobulin (TGL). The outcome is complexes of TGL containing MIT, DIT, T_3, and T_4, and these complexes are stored within vesicles in the colloid.

Exocytosis of the secretory vesicles is promoted by thyroid-stimulating hormone. Only the T_3 and T_4 portions are finally secreted into the blood. The remaining components are recycled within the follicular cells. T_3 and T_4 are carried in the blood in association with three different proteins. *AAs,* Amino acids; *DIT,* diiodothyronine; *I^-,* iodide; *MIT,* monoiodothyronine; *SA,* serum albumin; *T_3,* triiodothyronine; *T_4,* tetraiodothyronine; *TBG,* thyroxine-binding globulin; *TBPA,* thyroxine-binding prealbumin; *TGL,* thyroglobulin; *TYR,* tyrosine.

The peroxidase that catalyzes the oxidation is located mainly in the apical portion of the follicular cell; as a result the iodination of TGL is restricted to the cell-colloid interface.

- A certain proportion of these iodinated TYR residues within the TGL molecule will couple to form triiodothyronine (T_3; a coupling of MIT with DIT) and tetraiodothyronine (T_4; a coupling of DIT with DIT). T_3, T_4, DIT, and MIT are all attached to TGL, and this complex is stored as colloid in the core of the thyroid follicles.

- **Secretion of T_3 and T_4** Secretion begins with reabsorption of colloid droplets into the follicular cell, where they fuse with lysosomes and are digested by proteolytic enzymes. This yields MIT, DIT, T_4, T_3, and a remnant of amino acids (see Fig. 9.4).

  - MIT and DIT are deiodinated, and the resultant I^- and tyrosine

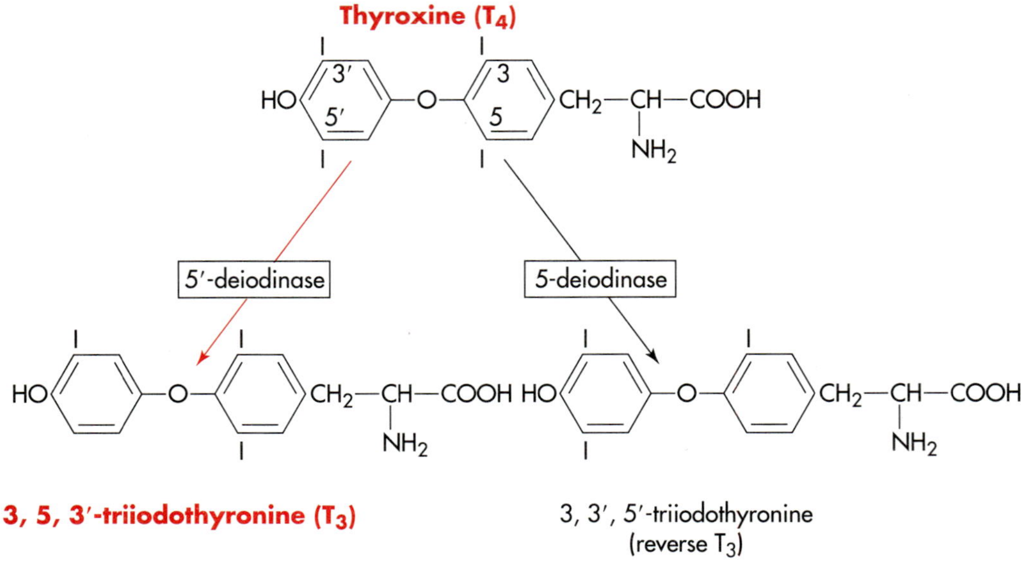

Fig. 9.5 Deiodination of T_4 to yield the active form of thyroid hormone. Peripheral tissues contain two deiodinases. Only 5′-deiodinase yields biologically active T_3. Production of reverse T_3 increases in old age. It is biologically inactive.

as well as other amino acid remnants remain in the cell to be recycled in thyroglobulin synthesis (see Fig. 9.4).

• T_4 and T_3 are secreted to the extracellular space in a ratio of 9:1. Once they reach the circulation they are reversibly bound to three carrier proteins:

- Thyroxine-binding globulin (TBG; highest binding affinity);
- Thyroxine-binding prealbumin (TBPA) and
- Serum albumin (SA; lowest binding affinity but present in large concentration).

• Free T_3 is the active form of thyroid hormone. Most of it is produced in the periphery by deiodination of T_4 (Fig. 9.5). There are two deiodinases in peripheral tissues. One produces T_3, which is biologically active; the other produces *reverse* T_3, which has no biologic activity.

- **Regulation of secretion**

• Thyroid-stimulating hormone (TSH) is the primary regulator of thyroid growth and function. Its effects are mediated by an adenylate cyclase–activating membrane receptor in follicular cells.

• In turn, circulating levels of T_3 and T_4 are the principal negative-feedback regulators of TSH and TRH.

- **Biologic effects of T_3** All biologic effects of thyroid hormone depend on the binding of free T_3 to nuclear receptors and are ultimately caused by changes in gene expression. Many systems and functions are affected:

— ***Growth and development*** T_3 is of special importance during fetal life and in the postnatal period. When it is absent, *cretinism* follows. When it is present, it promotes body growth as well as normal development of nervous tissue. These actions result from its:

- ability to promote protein synthesis

- synergism with growth hormone–releasing hormone in the secretion of growth hormone and synthesis of somatomedins
- synergistic influence on somatomedins at the epiphyseal growth plate in bone

— ***Energy metabolism*** In adult life the main physiologic role of T_3 is the regulation of energy metabolism, presumably within the framework of body temperature control. The mechanisms of these effects are primarily enhanced protein synthesis and increased mitochondrial size and number. As a result, the rate of formation of ATP and several cellular enzymes is increased, leading to increased metabolism of fats and carbohydrates. In nearly all tissues,

- T_3 stimulates basal rate of metabolism, oxygen consumption, and heat production. This effect is so strong that the metabolism of fat and carbohydrates can be increased to such an extent that caloric intake cannot be increased sufficiently to maintain body weight.
- Hypothyroidism, on the other hand, is associated with high serum levels of triglycerides and cholesterol, a reduced rate of lipid degradation, and a pronounced reduction in basal metabolic rate.

— ***Cardiovascular and respiratory effects***

- T_3 induces synthesis of β-adrenergic receptors and may promote expression of the more active isoenzyme form of cardiac myosin ATPase. As a result, hyperthyroid individuals show:
  - increased heart rate,
  - increased cardiac performance, and
  - cardiac hypertrophy
- T_3-mediated increases in metabolic activity, heat production, and oxygen consumption as well as associated decreases in peripheral vascular resistance (arising from both enhanced β-adrenergic activity and cutaneous vasodilatation) are stimuli for reflex changes in cardiorespiratory activity, leading to additional increases in:
  - heart rate,
  - cardiac performance, and
  - ventilation.
- In addition, hyperthyroidism is often associated with tachyarrhythmias such as atrial fibrillation.
- Individuals with low serum T_3 show reduced cardiac contractility and reduced resting heart rate. The latter observation suggests that T_3 has a cardiac chronotropic effect.

— ***Central nervous effects*** Central nervous system effects of T_3 are indicated by several observations:

- T_3 is required during infancy for normal intellectual development.
- Hypothyroid individuals show:
  - decreased mental performance,
  - impaired memory, and
  - personality changes.

- Hyperthyroid young individuals show these symptoms:*
    - Diffuse anxiety
    - Emotional lability
    - Extreme nervousness
    - Frequent movement

— ***Endocrine effects*** In normal physiologic amounts, T_3 potentiates the actions of insulin in the promotion of glycogenesis and glucose utilization. Its involvement in other endocrine systems is suggested by certain consequences of thyroid imbalance:

- Altered menstrual patterns:
    - Lack of thyroid hormone is associated with excessive and frequent menstrual bleeding.
    - Excess thyroid hormone causes reduction or cessation of menstrual bleeding.

- Disordered regulation of vasopressin secretion. Hypothyroidism can be accompanied by hyponatremia resulting from impaired renal water excretion.

- Altered sensitivity to catecholamines. Induction of β-adrenergic receptors in hyperthyroidism causes increased sensitivity to β-adrenergically mediated metabolic and vascular effects of catecholamines.

The Parathyroid Glands

■ **Anatomy** The parathyroid glands are four pill-sized structures embedded in the upper and lower thyroid poles. They are composed mainly of "chief" cells, which synthesize and secrete parathyroid hormone, a polypeptide of 84 amino acids.

■ **Parathyroid Hormone (PTH)**

● **Synthesis and secretion** PTH is synthesized in response to decreased serum levels of ionized calcium and virtually all the synthesized product is secreted rather than stored.

● **Biologic activity** PTH plays a vital role in the homeostasis of body calcium and phosphate. Its target organs are bone and kidney (see Chapter 10).

— ***In bone*** The effects of PTH are:

- Rapid removal of calcium salts from bone crystals.
- Activation of existing osteoclasts and formation of new osteoclasts. The effect of increased osteoclast activity is to dissolve bone and to increase serum [Ca^{++}].
- Depression of the rate at which osteoblasts lay down *osteoid,* the precursor of new bone.

— ***In the kidney*** The renal effects of PTH are:

- To enhance Ca^{++} reabsorption in the distal tubule.
- To decrease PO_4^{3-} reabsorption in the proximal tubule.

*In young thyrotoxic individuals nervous symptoms dominate the clinical picture. In older thyrotoxic individuals cardiovascular effects and symptoms of muscle weakness dominate.

- To promote the conversion of vitamin D to its biologically most active form, 1,25-dihydroxycholecalciferol [1,25-$(OH)_2D_3$]. This configuration of vitamin D acts on the intestinal epithelium to promote reabsorption of Ca^{++}. These aspects are described more fully in Chapter 10.

The Adrenal Cortex

Anatomy of the Adrenal Glands Approximately 90% of each adrenal gland is cortex, surrounding an inner core, called the *adrenal medulla.*

In the cortex three zones can be recognized on the basis of both microscopic anatomy and secretory products (Fig. 9.6).

- *Zona glomerulosa* cells synthesize aldosterone.
- Cells in the *zona fasciculata* and the *zona reticularis* synthesize cortisol and precursors for androgens.
- All three zones secrete corticosterone.

The anatomy of the vascular supply is an important determinant of the nature of the products that will be synthesized at different sites in the glands.

- Arteries enter the outer portion of the cortex.
- Successive branchings take the vessels toward the central medulla, where they unite in the collecting veins. As a result of this arrangement,
- synthesized hormones are transported toward the medulla, and their concentration increases progressively from cortex to medulla. As a result,
  - the specific inhibition or induction exerted by these hormones on selected steroidogenic or catecholaminergic enzymes increase progressively toward the core. Therefore,
  - the vascular anatomy ensures regional specificity in the synthesis of adrenal hormones.

Synthesis of Steroid Hormones Steroids are synthesized from cholesterol, which derives mostly from circulating low-density lipoprotein particles (LDL). Adrenocortical cells are especially rich in LDL receptors.

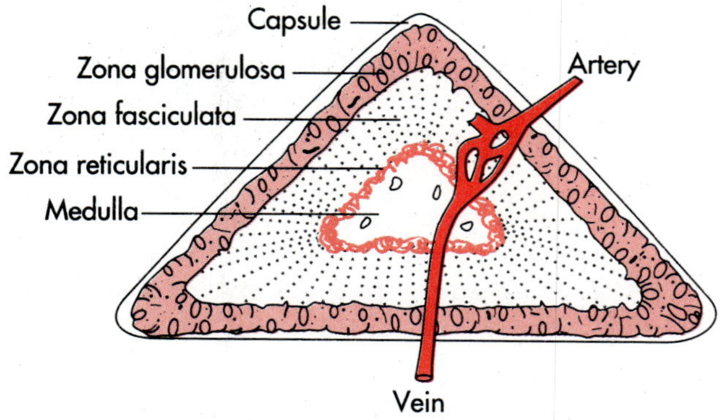

Fig. 9.6 Anatomy of the adult human adrenal glands. Three contiguous regions can be recognized in the cortex under the capsule. The vascular supply is arranged such that secretory products from the outer zones are carried toward inner zones and the medulla.

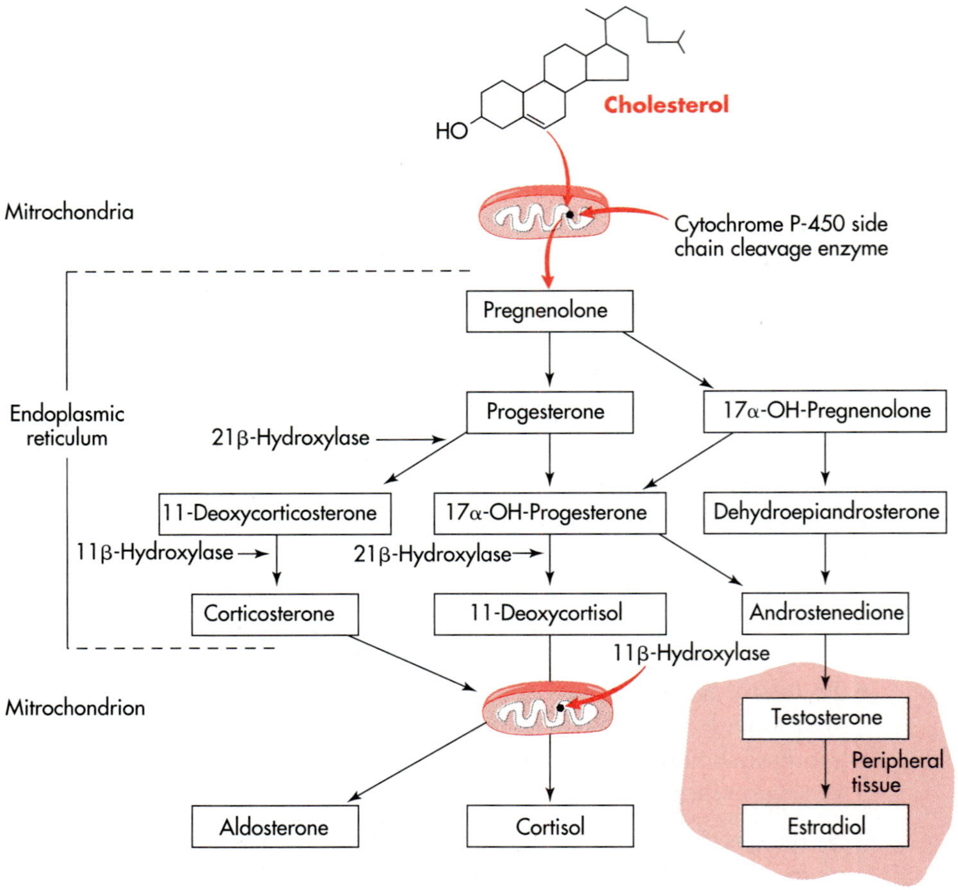

Fig. 9.7 Synthesis of steroid hormones. Cholesterol reaches the cytosol by means of low-density lipoprotein particles (LDL) and is transported to the inner membrane of mitochondria. One of the cytochrome P-450 enzymes removes a side chain from cholesterol and thereby produces pregnenolone. Successive enzymatic modifications eventually yield three classes of steroids, mineralocorticoids, glucocorticoids, and androgens. The sites of action of 11β-hydroxylase and 21β-hydroxylase are shown because genetic deficiencies of these enzymes are common. They lead to overproduction of adrenal androgens at the expense of the other two steroid classes.

• Steroid synthesis begins with transport of free cholesterol to the inner membrane of the mitochondria, where cytochrome P-450 side-chain cleavage enzyme removes a side chain and thereby produces *isocaproaldehyde* and *pregnenolone* (Fig. 9.7).

• Pregnenolone quickly leaves the mitochondria and is serially modified in the smooth endoplasmic reticulum and finally in the mitochondria again, to yield three classes of adrenocortical steroid hormones: mineralocorticoids, glucocorticoids, and androgens (see Fig. 9.7).

- **Mineralocorticoids** Deoxycorticosterone and aldosterone are normally secreted in equal amounts, exist mostly in the free form rather than being bound to plasma proteins, and are therefore quickly metabolized (in the liver). Their plasma half-life is only 10 to 20 minutes. Deoxycorticosterone has negligible mineralocorticoid potency compared to aldosterone. Therefore, aldosterone is the principal mineralocorticoid.

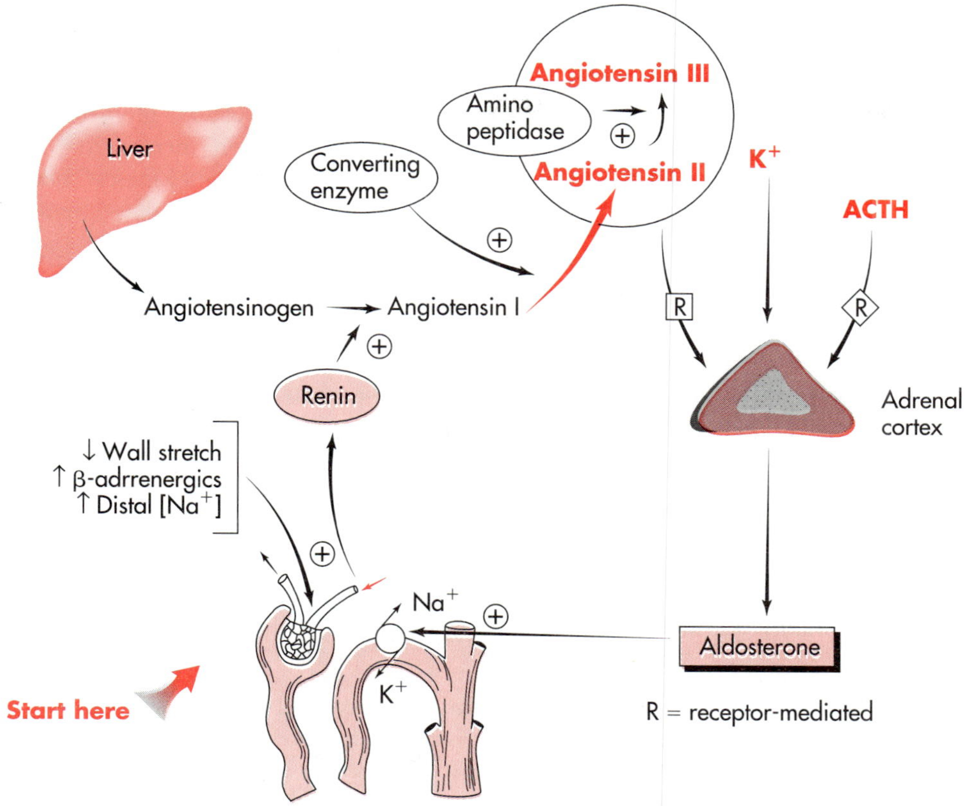

Fig. 9.8 Regulation of the renin-angiotensin-aldosterone system, an important regulator of arterial blood pressure, body electrolytes, and extracellular fluid volume. Angiotensin II and III, both acting by means of the AT_1 receptor, are the most important stimulus for aldosterone secretion. Hyperkalemia and adrenocorticotropic hormone (ACTH) are stimuli of less physiologic importance. The major biologic role of aldosterone is to synthesize and enhance the activity of Na^+,K^+-ATPase in the distal part of the nephron.

- **Glucocorticoids** Corticosterone possesses glucocorticoid activity, but its plasma concentration in humans is generally too low for significant biologic effects. Therefore, cortisol is the dominant glucocorticoid.

  Most plasma cortisol is bound to corticosteroid-binding globulin, the synthesis of which (by the liver) is stimulated by estrogen. Globulin-bound cortisol is biologically inactive but acts as a pool from which free cortisol can be drawn for biologic activity.

- **Androgens** In adult humans the adrenals represent a minor source of androgens when their output is compared with that of the ovaries or testes. The principal androgens secreted by the adrenals are:
  - dehydroepiandrosterone and
  - androstenedione.

Regulation of Steroid Synthesis

- **Mineralocorticoids** Aldosterone synthesis is governed by angiotensin II, angiotensin III, serum [K^+], and serum [ACTH] (Fig. 9.8).
  - Angiotensin II and III are the major promoters of aldosterone synthesis.

- Their plasma levels are driven by plasma levels of *renin,* a proteolytic enzyme that is released from juxtaglomerular cells of renal afferent arterioles.*
- Both bind the AT_1 receptor in the membrane of zona glomerulosa cells to activate phospholipase-C. The subsequent increase in cytosolic [Ca^{++}] and the production of activated protein kinase-C both enhance aldosterone synthesis.

• Serum K^+ levels affect aldosterone synthesis and secretion primarily by changes in membrane potential. Increased extracellular [K^+] tends to depolarize the cell and activate voltage-gated Ca^{++} channels, leading to increased [Ca^{++}] in the cytosol of zona glomerulosa cells.

• ACTH influences aldosterone synthesis when it binds to a G protein–coupled membrane receptor, causes elevation of intracellular cAMP, and leads to activation of protein kinase-A. The outcome is mostly increased glucocorticoid synthesis, but aldosterone synthesis is increased as well.

- **Glucocorticoids** Cortisol is an especially important adrenal product because it provides feedback inhibition of ACTH synthesis (Fig. 9.9) and because its presence is important for activation of phenylethanolamine *N*-methyltransferase which is required for epinephrine synthesis.

• Basal secretion and stress-induced increases in cortisol secretion are regulated by ACTH. ACTH, in turn, is secreted in bursts, most prominently early in the morning. Stress-induced increases in ACTH are driven by corticotropin-releasing hormone (CRH), whose synthesis and release from neurons in the median eminence is influenced by tracts from many central nervous nuclei and a variety of blood-borne agents (see Fig. 9.3).

• Free cortisol inhibits ACTH secretion at both the pituitary and the hypothalamic levels (see Fig. 9.9). The mechanism of inhibition is steroid receptor–mediated inhibition of protein synthesis.

- **Androgens** Since the adrenals normally supply only a minor portion of circulating androgens, regulation of this aspect of adrenal function assumes importance only in certain states of enzyme deficiency.

Deficiencies in 21β-hydroxylase and 11β-hydroxylase are common. Inspection of Fig. 9.7 shows that the effect of either of these is to shunt steroid synthesis toward overproduction of androgens.

Biologic Effects of Steroids

- **Effects of mineralocorticoids** The homeostatic role of aldosterone is to regulate the body balance of Na^+ and K^+. This is accomplished by its stimulatory effect on Na^+ and K^+ transport primarily in the distal nephron but also in sweat glands, in salivary glands, and in the GI mucosa.

The major target for aldosterone is Na^+,K^+-ATPase in the distal convoluted tubule portion of the nephron (Fig. 9.8). It increases both Na^+,K^+-ATPase synthesis and Na^+,K^+-ATPase activity so that its biologic effect is to increase Na^+ reabsorption from the tubule and K^+ secretion into the

*Renin release is increased during cardiovascular hardship by several specific influences on the juxtaglomerular cells of renal afferent arterioles. The function of renin is to cleave the decapeptide, angiotensin I, from angiotensinogen (a circulating glycoprotein that is synthesized by the liver). Angiotensin I is further processed by *converting enzyme* to yield angiotensin II. Angiotensin III is produced from angiotensin II when a terminal Asp is cleaved off by the action of an aminopeptidase.

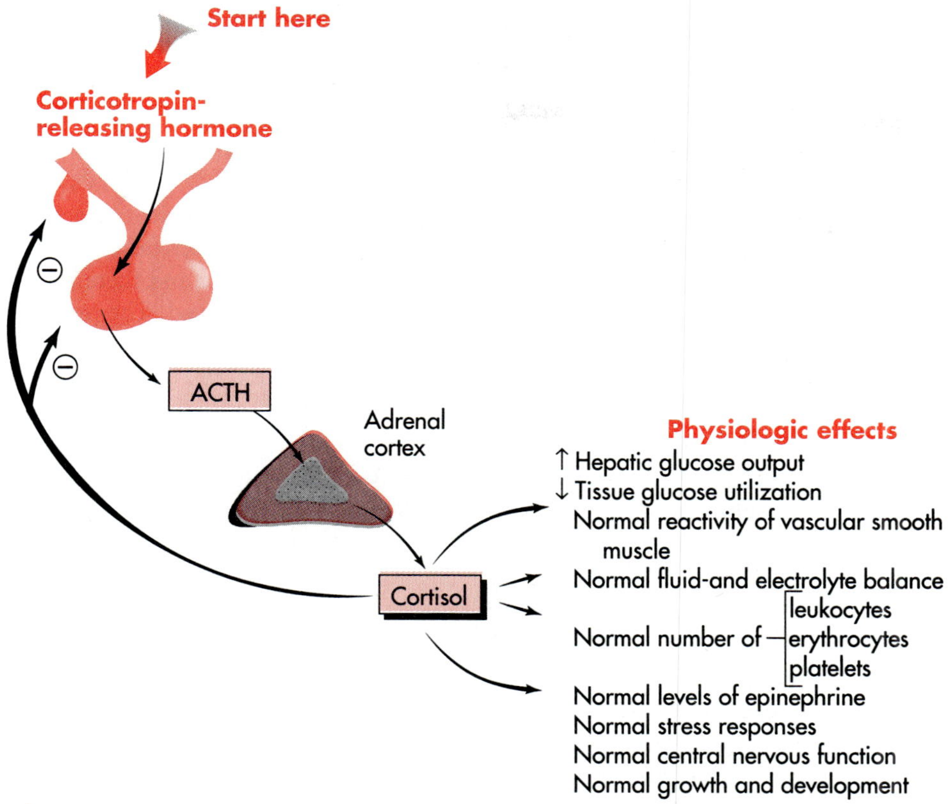

Fig. 9.9 The major function of adrenocorticotropic hormone (ACTH) is to stimulate cortisol synthesis. Cortisol has an amazingly large number of beneficial effects. Its levels increase, as a result of elevated ACTH, in response to physical or emotional stress. Increases in ACTH secretion to meet demands imposed by stress are mediated almost exclusively through the hypothalamus by release of corticotropin-releasing hormone (CRH).

tubule. (Control of Na^+,K^+-ATPase is described in greater detail in Chapter 7 concerning regulation of K^+ balance.)

- **Effects of glucocorticoids** Cortisol affects most aspects of body functions and affects them differently at physiologic or pharmacologic levels.

— ***Effects of cortisol at physiologic levels***

Metabolism of carbohydrate, protein, and fat. The overall effect is to help supply glucose to critical tissues when this is needed. To that end cortisol:

- Increases catabolism of proteins so that there is an increase in amino acid substrate for glucose production by the liver.
- Increases activity of several enzymes involved in the formation of glucose in the liver.
- Decreases utilization of glucose in peripheral tissues.
- Increases blood levels of free fatty acids and ketones provided that this action is not inhibited by elevated insulin levels. Such elevation would normally be triggered by the rise in plasma glucose that is caused by cortisol.

Control of ACTH secretion. Free cortisol inhibits ACTH secretion at both the pituitary and hypothalamic levels. The mechanism of inhibition is steroid receptor–mediated inhibition of protein synthesis.

Vascular reactivity. Glucocorticoids must be present for epinephrine and norepinephrine to affect the tone of vascular smooth muscle and for capillaries to maintain normal permeability.

Nervous system. In the absence of glucocorticoids there are personality changes (greater irritability, a feeling of unease, distractedness) and increased sensitivity to olfactory and gustatory stimuli.

Fluid and electrolyte balance. When glucocorticoids are absent, glomerular filtration rate is low, and nephron reabsorptive mechanisms are altered in such a way that hypertonic urine is excreted and there is a severe curtailment in the ability to excrete a water load. The impairment is so severe that "water intoxication" may be present, complete with cell swelling. When this occurs in brain cells, there may be disruption of temperature regulation.

Blood cells. Glucocorticoids increase the number of circulating neutrophils, platelets, and erythrocytes. They decrease the number of eosinophils and lymphocytes. They also inhibit production of interleukin-2 by T-lymphocytes and thereby stop lymphocyte proliferation.

Permissive actions. Glucocorticoids assist glucagon and the catecholamines epinephrine and norepinephrine in their metabolic effects.

Resistance to stress. Elevated levels of circulating glucocorticoids are necessary to withstand the physiologic impact of "stress." Their benefits derive largely from mechanisms that are still not known but may include maintenance of vascular responsiveness to catecholamines and permitting catecholamines to boost energy supplies by liberating free fatty acids.

Growth and development. In the fetus, cortisol is one of the factors that aids in the maturation of the surfactant system. Additional roles in growth and development are suggested by the observation that cortisol inhibits secretion of growth hormone.

— ***Effects of cortisol at pharmacologic levels*** Large doses of glucocorticoids, applied for short periods of time,

- inhibit the inflammatory responses to tissue injury (by suppressing cell-mediated immunity, inhibiting the mobility of leukocytes and macrophages, stabilizing lysosomes, and depressing the release of vasoactive and proteolytic enzymes such as kinins);
- inhibit phospholipase A_2, an enzyme that releases arachidonic acid from membrane phospholipids. Such inhibition reduces formation of arachidonic acid metabolites such as prostaglandins and leukotrienes. Leukotrienes are mediators of inflammatory responses, including those of rheumatoid arthritis;
- stabilize lysosome membranes and thereby prevent release of lysosomal enzymes in inflamed tissue;
- inhibit fibroblast activity and thereby prevent
  - encapsulation of foci of infection
  - formation of keloid or adhesions around surgical wounds;
- prevent histamine release and thereby alleviate the symptoms of allergic reactions;
- decrease growth hormone secretion and thereby inhibit growth.

Chronic excess of cortisol leads to *Cushing's syndrome* and is characterized by

- protein depletion (leading to poor muscle development, poor wound healing, fragile capillaries and thin skin);
- overdevelopment of fat cells in the upper back, face, and abdominal wall;
- hyperglycemia and decreased peripheral utilization of glucose;
- retention of water and Na^+ and depletion of K^+;
- inhibition of osteoblasts and subsequent osteoporosis; and
- signs of central nervous involvement (insomnia, euphoria, psychoses).

- **Effects of androgens** Dehydroepiandrosterone and androstenedione, the principal androgens secreted by the adrenal cortex, have little biologic potency. They become active only after peripheral tissues have converted them, chiefly to testosterone. Peripheral conversion of testosterone to estradiol also occurs. These conversion products are tightly bound in blood to specific globulins. Testosterone and other androgens have some biologic activity in most tissues at all stages of life.

— ***In fetal life*** Androgens determine the development of gender-linked features in anatomy and patterns of gonadotropin release. High levels of androgens have a masculinizing effect. Thus androgen concentration in fetal blood during the first 10 weeks determines whether:

- female or male genitalia (internal as well as external) develop;
- the hypothalamus will develop a cyclic pattern of gonadotropin release after puberty (female) or a noncyclic pattern (male).

— ***In adult life*** Two androgen effects are observed, depending on the target organ. Androgens

- stimulate protein synthesis (anabolic effects) or
- influence development and growth of male sexual characteristics such as:
  - pattern of body growth and muscle development,
  - maturation of external genitalia and accessory sexual organs (scrotum, seminal vesicles, prostate),
  - size of the larynx and vocal cords, and
  - patterns of hair growth and hair loss.
- Effects of androgens on gender-specific behavior have been asserted, but the evidence for such effects in humans is conflicting.

The Adrenal Medulla

The adrenal medulla contains neuronal cells *(chromaffin cells)* that have endocrine function. They are arranged in close relationship with preganglionic cholinergic fibers and with venules that drain the adrenal cortex. As a result of this anatomic arrangement

- both sympathetic nervous activity and adrenocortical humoral products influence the synthesis of catecholamines (norepinephrine and epinephrine).

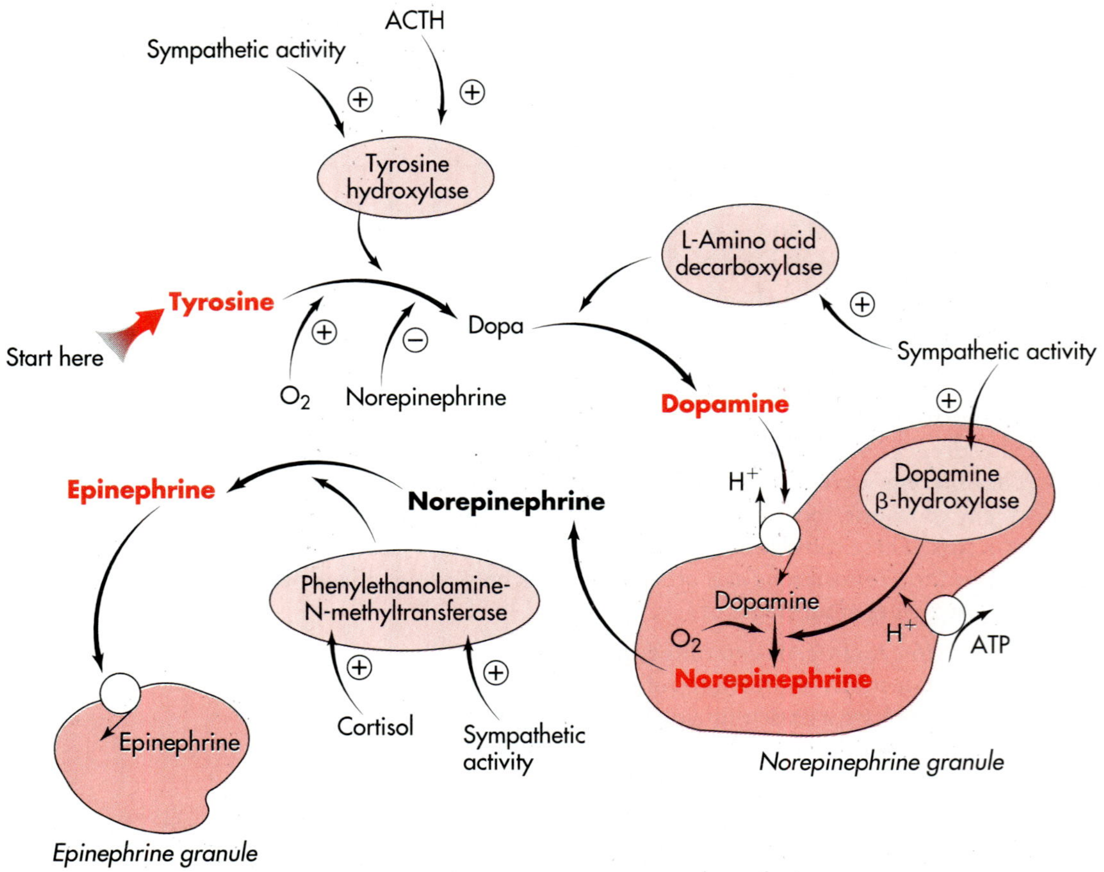

Fig. 9.10 Adrenal medullary synthesis of epinephrine from the amino acid tyrosine. The rate-limiting step is the activity of tyrosine hydroxylase and the formation of dopa (dihydroxyphenylalanine). Sympathetic nervous activity, ACTH, norepinephrine, and cortisol are important regulators of epinephrine synthesis. ATP-consuming active transport of H^+ into the norepinephrine storage granules provides the H^+ gradient required for the transport of dopamine by the amine/H^+ antiport.

- The enzyme phenylethanolamine *N*-methyltransferase is a unique feature of chromaffin cells. Its function is to convert norepinephrine to epinephrine. In healthy adult humans so much of this enzyme is present that
- mostly epinephrine is released into the circulation when the adrenal medulla is stimulated.

Epinephrine

- **Epinephrine synthesis and storage** The steps involved in epinephrine synthesis are summarized in Fig. 9.10.
  - The first step is the conversion of tyrosine to dopa.
    - This is the rate-limiting reaction in epinephrine synthesis.
    - It requires the presence of O_2.
    - It is catalyzed by *tyrosine hydroxylase,* an enzyme that is maintained by ACTH and elevated above normal levels mostly by sympathetic activity.
  - In the second step a carboxyl group is removed from dopa, and dopamine is formed.

- Dopamine is transported into vesicles by an amine/H^+ antiport and is further processed by *dopamine* β*-hydroxylase,* an enzyme that is present only within vesicles, not in the cytoplasm.

  - Dopamine β-hydroxylase is also under sympathetic control but to a lesser extent than either tyrosine hydroxylase or L-amino acid decarboxylase. As a result, high levels of sympathetic nervous activity, such as they may occur under stress, can alter the proportion of dopamine to epinephrine or norepinephrine released from sympathetic nerves and from the adrenal medulla.
  - In the presence of oxygen, dopamine β-hydroxylase catalyzes the formation of norepinephrine from dopamine.
  - Most of the norepinephrine diffuses out of the vesicles into the cytoplasm.

- *Phenylethanolamine N-methyltransferase* (PNMT), a cytosolic enzyme, converts norepinephrine to epinephrine. Two factors regulate PNMT:

  - Cortisol, draining from the adrenal cortex, is required in high concentration for activation of PNMT.
  - Sympathetic stimulation elevates PNMT above its resting level.

- Synthesized epinephrine is actively pumped into storage vesicles.

Epinephrine secretion

- Stored epinephrine is secreted into the circulation by exocytosis in response to cholinergic preganglionic nerve activity reaching the adrenals by means of the greater splanchnic nerve.
- Once released, epinephrine is metabolized within a few minutes by two key enzymes, *catechol O-methyltransferase* and *monoamine oxidase* (MAO).

Biologic effects of epinephrine

— ***Metabolic actions*** These arise chiefly from interactions with β-adrenoceptors. Two kinds of metabolic actions are observed: diabetogenic and thermogenic effects.

Diabetogenic effects of epinephrine. β-adrenoceptor activation in liver, skeletal muscle, pancreas, or adipose tissue causes increased production and decreased uptake of glucose and ketoacids.

Thermogenic effects of epinephrine. These effects occur diffusely in many tissues. They may arise from direct enhancement of metabolic rate or may happen secondarily to an increase in core temperature that is caused by epinephrine-induced constriction of cutaneous blood vessels and consequent decrease in heat loss (see Chapter 11 for a discussion of temperature regulation).

— ***Cardiovascular actions of epinephrine*** Epinephrine has two significant cardiovascular effects.

- It increases cardiac output by increasing heart rate and cardiac performance. Both result from activation of β_1-adrenoceptors.
- It influences the selective distribution of cardiac output. The fraction of cardiac output that is directed to a tissue is determined by the vascular resistance that is offered by that tissue in comparison to all other tissues. Net tissue vascular resistance, in turn, is determined by the degree of imbalance between local vasoconstrictor factors and

local vasodilator factors. Tissues that have a high proportion of vascular smooth muscle β_2-adrenoceptors (such as skeletal muscle) show a high potential for vasodilatation in the presence of epinephrine. A physiologically important example is given in the discussion of cardiorespiratory responses to exercise in Chapter 11.

The Endocrine Pancreas

- **Anatomy of the Islets of Langerhans** A minuscule portion (1% to 2%) of the pancreas consists of discrete bodies, the *islets of Langerhans,* that contain at least four distinct cell types. Each is responsible for synthesis, storage, and release of one of the hormones insulin (B-cells), glucagon (A-cells), somatostatin (D-cells), and pancreatic polypeptide (F-cells).

Insulin

- **Synthesis and secretion of insulin** Insulin, a small peptide, is synthesized in islet B-cells and is stored in secretory granules within these cells. Its secretion is driven by intracellular [Ca^{++}] and is therefore regulated by metabolic, nervous, endocrine, and ionic factors (Table 9.2).

  — *Metabolic influences*

  - The most important stimulus for insulin release is an increase in plasma glucose concentration above its normal level of 100 mg/dL.

    - Glucose is transported into B-cells by the GLUT 2 transporter, which does not require insulin for activation. Intracellular metabolism of glucose produces ATP, which inhibits $I_{K,ATP}$, the ATP-sensitive K^+ channel. Such inhibition restricts K^+ outflow, depolarizes the cell, and causes voltage-gated, L-type Ca^{++} channels to open. This and possibly additional sources of Ca^{++} raises intracellular [Ca^{++}], causes activation of Ca^{++}-dependent kinases, and leads to exocytosis of insulin-containing vesicles.

Table 9.2 *Factors Controlling Synthesis and Secretion of Insulin*

Synthesis and Secretion	
Increased by	**Decreased by**
Factors that increase [Ca^{++}]$_i$:	*Factors that decrease [Ca^{++}]$_i$:*
glucose	**somatostatin**
protein metabolites	insulin
fat metabolites	α_2-adrenoceptor agonists
β_2-adrenoceptor agonists	hypokalemia
muscarinic agonists	
hyperkalemia	
gastric inhibitory peptide	
gastrin	
secretin	
cholecystokinin	
glucagon	
cortisol	
growth hormone	

- Insulin release from individual B-cells is normally oscillatory. The reason is that Ca^{++} influx, once it has been initiated, will further depolarize the cell and eventually cause voltage-gated K^{+} channels to open when their threshold potential is reached. The cell will then be repolarized by K^{+} efflux, $[Ca^{++}]_i$ will return toward resting values, and exocytosis stops. If plasma glucose continues to be high, the process repeats itself.

- Metabolic products resulting from digestion of proteins and fats can also generate ATP and thereby promote insulin release. They are stimuli of secondary importance.

— ***Nervous influences*** Both sympathetic and parasympathetic nerves supply the islets and the dominant autonomic receptors are α_2, β_2, M_1, and M_3. Insulin release can be expected to result from any nervous activity that increases intracellular Ca^{++} concentration, whereas decreases in $[Ca^{++}]_i$ inhibit insulin release.

- α_2-Adrenoceptor stimulation inhibits adenylate cyclase, inhibits cAMP, decreases phosphorylation of protein kinase A, and consequently decreases conductance of both leakage type and L type Ca^{++} channels. As a result $[Ca^{++}]_i$ decreases, and insulin secretion is inhibited by α_2-adrenoceptor agonists.
- β_2-Adrenoceptor activation increases insulin release by a pathway that involves activation of adenylate cyclase and increased cAMP.
- Muscarinic activation increases intracellular $[Ca^{++}]$ by activation of the phosphoinositide pathway and therefore increases insulin secretion.

— ***Ionic influences*** Changes in extracellular K^{+} affect insulin responses to glucose in a manner that can be predicted from the changes in resting membrane potential. Hypokalemia hyperpolarizes cells and will therefore decrease insulin secretion from B-cells. Conversely, hyperkalemia causes increased insulin secretion.

— ***Endocrine influences***

- Insulin secretion is directly stimulated by several GI hormones: gastric inhibitory polypeptide, gastrin, secretin, cholecystokinin, and glucagon.
- Insulin secretion is indirectly stimulated by the effects of other hormones on plasma glucose levels. Among these are cortisol, growth hormone, thyroxine, progesterone, estrogen, and testosterone.
- Insulin secretion is inhibited most importantly by somatostatin and by insulin itself.

● **Biologic actions of insulin** The biologic effects of insulin are mediated by a membrane receptor that is found predominantly in liver, muscle, and adipose tissue. The intracellular domain of the insulin receptor acquires tyrosine kinase activity when glucose binds to the extracellular domain. This leads to phosphorylation of the intracellular domain *(autophosphorylation),* a step that is necessary for the many biologic effects of insulin.

— ***Metabolic effects of insulin*** The net effect of insulin is intracellular storage of carbohydrate, protein, and fat. It expresses itself differently at different times after insulin release.

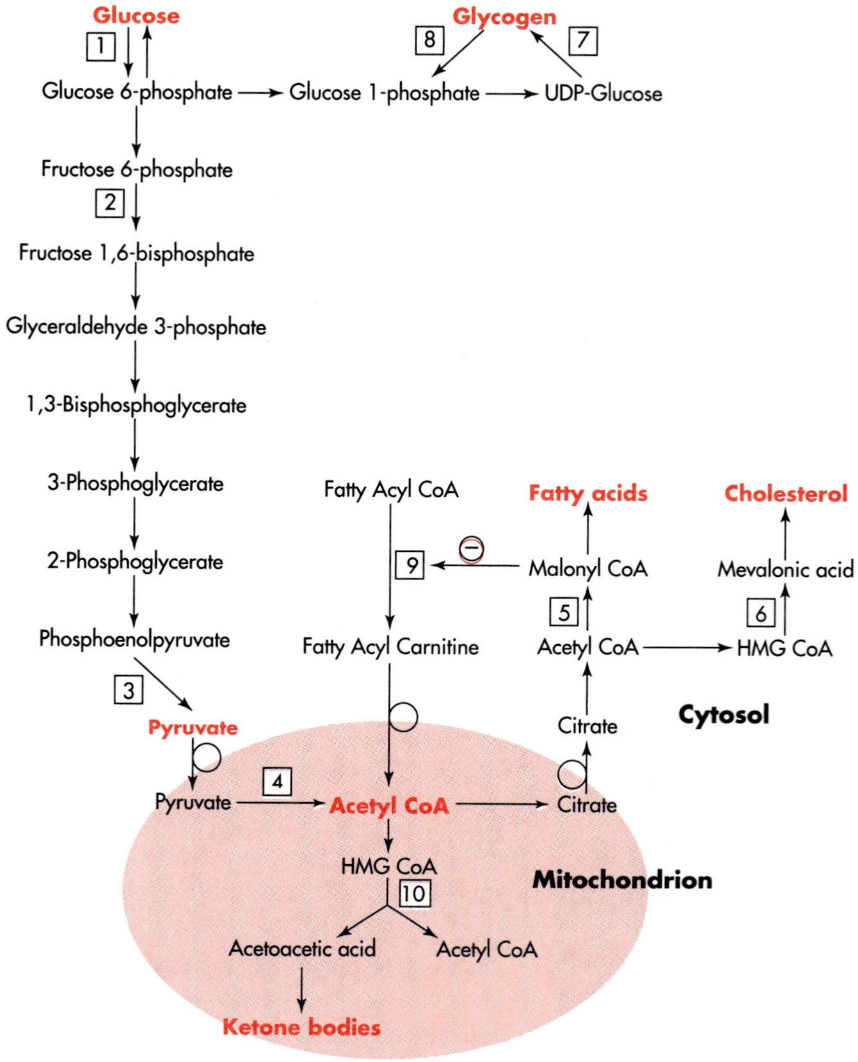

Fig. 9.11 Summary of pathways in the metabolism of glucose and fatty acids in liver cells. *1* refers to the enzyme glucokinase that is found only in liver parenchymal cells. *2* to *10* refer to enzymes that have significant influence on the metabolic effects of insulin.

Short-term effects (within seconds):

INCREASED TRANSPORT OF GLUCOSE, K^+, AND AMINO ACIDS INTO TARGET CELLS.

• In most peripheral tissues insulin effects on glucose transport are mediated by the insulin-sensitive GLUT 4 transporter. In these tissues a pool of GLUT 4 molecules exists in the cytoplasm. Insulin causes them to move toward and insert themselves into the plasma membrane.

• The liver has few if any insulin-sensitive glucose transporters. Liver parenchymal cells are unique in that they contain glucokinase instead of hexokinase. Glucokinase is an isoenzyme of hexokinase, the enzyme that converts glucose to glucose 6-phosphate (*1* in Fig. 9.11). Two significant features set glucokinase apart from hexokinase: (1) glucokinase is not inhibited by glucose 6-phosphate; (2) glucokinase is not saturated at physiologic levels of blood glucose. These features allow the liver to respond linearly to changes in blood glucose levels and to act as a buffer for blood glucose changes.

Intermediate-term effects (within minutes):

- Stimulation of both protein synthesis and glycogen synthesis
- Inhibition of both gluconeogenesis and protein breakdown

Long term effects (within hours):

- Increased transcription of lipogenic enzymes

The metabolic effects of insulin express themselves differently in different insulin-sensitive tissues. They are summarized in Table 9.3.

The metabolism of glucose and fatty acids are affected most prominently. Fig. 9.11 summarizes, for the liver, the products involved and the enzymes affected.

Glycogen synthesis. Insulin activates glycogen synthase (*7,* Fig. 9.11) and inhibits glycogen phosphorylase (*8,* Fig. 9.11). As a result formation of glycogen is promoted, and glycogenolysis is inhibited.

Fatty acid synthesis. Insulin promotes synthesis and storage of fatty acids in the liver. Two aspects are involved:

- Insulin promotes the glycolytic pathway of glucose metabolism by activating three enzymes in that path, 6-phosphofructo-1 kinase (*2,* Fig. 9.11), pyruvate kinase (*3,* Fig. 9.11) and pyruvate dehydrogenase (*4,* Fig. 9.11).
- Insulin activates acetyl CoA carboxylase (*5,* Fig. 9.11). This activation causes hepatic acetyl CoA to be metabolized to fatty acids rather than to cholesterol.

Ketone body synthesis. Insulin promotes formation of malonyl CoA, an intermediary in the synthesis of fatty acids in liver cell (see *5,* Fig. 9.11). This product inhibits the enzyme carnitine palmitoyl transferase 1 (*9,* Fig. 9.11), thereby inhibiting hepatic breakdown of fatty acids and keeping mitochondrial acetyl CoA at normal levels. Ketone bodies (β-hydroxybutyric acid and acetone) are formed in mitochondria when the intramitochondrial level of acetyl CoA is high. Under such conditions HMG CoA is formed within the mitochondria (see Fig. 9.11) and is broken down by HMG CoA lyase (*10,* Fig. 9.11) to acetoacetic acid and acetyl CoA. HMG CoA lyase is present in high concentration in liver mitochondria but not elsewhere. As a result:

- The liver is the primary producer of ketones.
- Cytosolic HMG CoA is used for biosynthesis of cholesterol instead of ketone bodies (see Fig. 9.11).

— ***Ion-transport effects of insulin*** Insulin stimulates entry of K^+, PO_4^{3-}, and Mg^{++} into muscle and liver cells. The magnitude of this effect is such that: insulin is considered to be a significant regulator of serum K^+ levels (see Chapter 7).

● **Insulin excess** Insulin excess causes hypoglycemia and leads to symptoms that arise from the effects of hypoglycemia on central nervous function:*

*Glucose is the major fuel used by the brain. It is only after prolonged starvation that the brain is able to use ketones as fuel (see Chapter 11 for the discussion of the control of fuel supply).

Table 9.3 *Summary of Metabolic Effects of Insulin in its Three Primary Target Organs*

Metabolic Target	Liver	Skeletal Muscle	Adipose Tissue
Carbohydrate Metabolism			
Insulin has a dual effect on carbohydrate metabolism: • it stimulates glucose utilization and glucose storage and • it inhibits glucose formation.	When plasma glucose levels are high, insulin converts the liver from a glucose-producing organ into a glucose-storing organ. These actions include: • facilitation of glucose uptake • promotion of glycogen formation • inhibition of glycogen breakdown • inhibition of amino acid uptake • inhibition of gluconeogenesis When plasma glucose levels are not high but insulin levels are (inappropriately) high, the liver escapes from the glucose-storing effects of insulin because: 1. Low glucose levels in the hypothalamus promote sympathetic nervous activity, which, in turn, increases hepatic glucose output. 2. There is increased pancreatic secretion of the insulin antagonist glucagon.	Insulin stimulates glucose transport into muscle cells. There it is converted chiefly to glycogen.	Insulin stimulates glucose transport into adipocytes. There it is converted chiefly to glycerol.
Fat Metabolism			
In blood, insulin activates insulin-dependent lipoprotein lipase. This promotes breakdown of dietary triglycerides to free fatty acids (see Chapter 8). In liver, skeletal muscle and adipose tissue insulin promotes fat storage and inhibits its mobilization and oxidation of fatty acids.	Insulin stimulates breakdown of dietary glucose by the glycolytic pathway (Fig. 9.11). This increases pyruvate levels and hence production of acetyl CoA in hepatic mitochondria. Intramitochondrial acetyl CoA is converted to citrate, and citrate is exported to the cytosol in exchange for malate. Cytosolic citrate is cleaved to oxaloacetate and acetyl CoA. Insulin promotes formation of malonyl CoA from acetyl CoA and thereby stimulates hepatic formation of fatty acids rather than cholesterol. Malonyl CoA inhibits carnitine palmitoyltransferase 1 (CPT 1), a key enzyme in the chain by which the liver uses fatty acids to form ketone bodies (Fig. 9.11).		The intracellular pool of free fatty acids (FFA) is increased. This results: • from increased extracellular FFA concentration and • from insulin-mediated activation of enzymes involved in the conversion of glucose to FFA. The intracellular glycerol pool is increased as a result of enhanced glucose entry. Triglyceride formation is increased as a result of: • increased substrate levels (FFA and glycerol) and • decreased breakdown of triglycerides into FFA or triglycerol.
Protein Metabolism			
Insulin promotes protein synthesis and sequestration and inhibits protein breakdown.	Insulin decreases uptake of gluconeogenic amino acids (alanine, glycine).	Insulin increases preferential uptake of branched-chain amino acids, increases RNA synthesis, and therefore increases amino acid incorporation into proteins.	

Table 9.4 ***Factors Controlling Synthesis and Secretion of Glucagon***

Synthesis and Secretion	
Increased by	Decreased by
hypoglycemia	**somatostatin**
β_2-adrenoceptor agonists	insulin
cortisol	α_2-adrenoceptor agonists

- Increased autonomic discharge leading to palpitations, sweating, and apprehension
- Mental confusion
- Lethargy, convulsions, and coma

Glucagon

- **Synthesis and secretion** Glucagon is a small peptide. Its major source is the A-cells of the pancreas. The gut can also produce some glucagon, but it is not a significant source in humans. Table 9.4 summarizes the significant factors influencing glucagon synthesis and secretion.

— ***Metabolic influences*** The most important metabolic stimulus for the secretion of glucagon is a decrease in plasma glucose levels.

— ***Nervous influences***

- The sympathetic nervous system acts by means of β_2-adrenoceptors to increase glucagon secretion. This can lead to significant increases in blood glucose levels after trauma, toxemia, or infection.
- Activation of α_2-adrenoceptors inhibits glucagon secretion, but the stimulatory effect of β_2-adrenoceptor activation normally dominates.

— ***Endocrine influences***

Epinephrine. Epinephrine reaches sufficiently high plasma levels during periods of severe stress to promote glucagon secretion by a β_2-adrenoceptor-mediated mechanism.

Somatostatin. Somatostatin is the strongest inhibitor of glucagon secretion.

Insulin. Insulin is a weak inhibitor, perhaps acting by its influence on glucose utilization by the A-cells.

Cortisol. Cortisol increases glucagon secretion.

- **Biologic actions of glucagon** The liver is the major target organ for glucagon. There the hormone acts at many of the same biochemical control points as insulin does, but its effects are the opposite.

  - Glucagon acts by means of a membrane receptor and uses cAMP as a second messenger and phosphorylation of protein kinase A as an effector mechanism.
  - Phosphorylated protein kinase A acts:
    - At several points in the biochemical pathways that link glycogen, glucose, and pyruvate. The net effects of these actions are to:

- promote conversion of glycogen to glucose
- inhibit breakdown of glucose to pyruvate
- inhibit conversion of glucose to glycogen

- To increase formation of glucose from available amino acids
- To direct incoming free fatty acids away from triglyceride synthesis and toward ketone body formation.*

Somatostatin

Synthesis and secretion Somatostatin, the hypothalamic inhibitor of growth hormone release, is also secreted by pancreatic D-cells and occurs throughout the GI tract.

Somatostatin synthesis:

- is promoted by glucose, some amino acids (arginine, luceine), β-adrenergic agonists, cholecystokinin (CCK), and secretin;
- is inhibited by α-adrenergic agonists and the parasympathetic neurotransmitter acetylcholine.

Biologic actions Somatostatin concentration is highest within the pancreatic islets. As a result, its major effect is the local, paracrine inhibition of insulin and glucagon secretion.

- Somatostatin also inhibits digestive function by inhibitory effects on:
  - gut motility
  - gastric secretion and pancreatic exocrine function
  - mucosal absorption

Overall it looks as if one of the physiologic roles of somatostatin is the pacing of foodstuff conversion. This is achieved by coordinated effects on gross digestive functions such as bulk transport, digestion, or absorption and on cellular disposition of nutrients.

Pancreatic Polypeptide

This 36–amino acid compound is produced by F-cells in the islets of Langerhans.

Synthesis and secretion

- Secretion is increased by protein-containing foods, hypoglycemia, and cholinergic nervous activity.
- Secretion is decreased by somatostatin and hyperglycemia.

Biologic actions Pancreatic polypeptide slows mucosal absorption of digestion products. Its function may be to smooth fluctuations in the absorption of nutrients.

*Ketones can be used as energy substrates in some circumstances.

Multiple Choice Review Questions

1. Which one of the following is true with respect to the interaction of thyroid hormone with cells?
   a. Thyroid hormones bind primarily to specific cell-surface receptors on target cells.
   b. Thyroid hormone interacts with the thyroid hormone receptors to activate adenylate cyclase, producing cyclic AMP as a secondary messenger.
   c. Modulation of the calcium channels and intracellular calcium concentrations is vital.
   d. Thyroid hormones alter DNA transcription by binding to specific nuclear receptors.
   e. Stimulatory and inhibitory subunits of cell-surface receptors are intricately involved with the effects of thyroid hormones on cells.
2. Which one of the following statements about thyroid function is true?
   a. In hyperthyroidism, increased production of T_3 and T_4 results in increased body weight.
   b. Cretinism results as a consequence of hypothyroidism after puberty.
   c. A hypothyroid goiter results from compensatory thyroid growth in the face of impaired hormone synthesis.
   d. The main effects of T_3 and T_4 on the functioning endocrine system are to increase the requirements for other systemic hormones.
   e. T_3 and T_4 act to decrease the rate of lipolysis, thereby inducing atherosclerosis.
3. Which one of the following hormones would be elevated after pituitary stalk section?
   a. Adrenocorticotropic hormone (ACTH)
   b. Growth hormone
   c. Luteinizing hormone
   d. Insulin
   e. Prolactin
4. Which one of the following statements regarding glucocorticoids is correct?
   a. They decrease plasma levels of free fatty acids.
   b. They are synthesized from tyrosine.
   c. They are synthesized in the adrenal medulla.
   d. They enhance the responses of the immune system.
   e. They increase hepatic gluconeogenesis from amino acids.
5. Indicate which of the following is associated with insulin's metabolic effects:
   a. Increased protein synthesis by elevating plasma amino acid concentrations
   b. Stimulation of lipolysis and increase in circulating free fatty acids as energy supply
   c. Elevation of plasma glucose concentration to ensure glucose supply to the brain
   d. Stimulation of glycogen synthesis in the liver
   e. Increased rate of ketone body production
6. Indicate one correct characteristic of glucagon:
   a. Glucagon is secreted by pancreatic G cells.
   b. Its secretion is stimulated by an increase in blood glucose.
   c. Glucagon promotes glycogenolysis and gluconeogenesis.
   d. Glucagon inhibits synthesis of ketone bodies by increasing blood glucose levels.
   e. Glucagon stimulates hepatic glycogen synthesis by increasing blood glucose levels.

Directions: Each of the numbered items in the following section is negatively phrased, as indicated by an italic word such as *except.* Select the *one* lettered answer or completion that is *best* in each case.

7. A woman has a deficiency of thyroid-stimulating hormone. Which one of the following steps in the production of the thyroid hormones would *not* be affected by this deficiency?
   a. Active transport of I^- into the follicular epithelial cells.
   b. Iodination of thyroglobulin to form diiodotyrosine.
   c. Coupling of two diiodotyrosines to form tetraiodothyronine coupled to thyroglobulin.
   d. Pinocytosis of colloid and fusion with lysosomes.
   e. Permeability of the follicular cell membrane to T_3 and T_4.
8. Removal of the hypophysis (pituitary) would result in each of the following *except:*
   a. Atrophy of the adrenal cortex
   b. Decreased blood glucose levels
   c. Decreased release of insulin-like growth factors from the liver
   d. Sterility
   e. Decreased corticotropin-releasing hormone (CRH) secretion

9. Each of the following may increase growth hormone release *except* one. Which one?
   a. Physical exercise
   b. Psychologic stress
   c. Being asleep
   d. Insulin-like growth factors
   e. Hypoglycemia

10. Which one of the following statements concerning vasopressin is *not correct?*
    a. Its secretion is stimulated by increases in blood volume and pressure.
    b. It is also known as antidiuretic hormone (ADH).
    c. It acts on the kidney to increase water reabsorption.
    d. It is synthesized as a prohormone with a neurophysin, which may serve as its own carrier protein in the blood.
    e. Its deficiency leads to decreased collecting tubule permeability to water.

Chapter 10

Cartilage and Bone

Cartilage

Two features distinguish cartilage from bone; one relates to structure, the other to sites at which growth can occur.

• Both cartilage and bone are built upon a collagen fiber matrix. However, cartilage is characterized by deposits of proteoglycans in the collagen matrix, whereas bone is characterized by abundant mineral deposits.

• Unlike bone, which can grow only on its surface, cartilage can grow from within.

Cartilage Structure The basic structural element of all connective tissues is formed by collagen fibers, 20 to 60 nm in diameter and arranged in a fine net. The interfiber spaces of the net contain a variety of noncollagenous proteins and mineral precipitates. Cartilage contains a high proportion of glycosaminoglycan chains and their associated proteins.

• Three types of cartilage can be recognized on the basis of fiber content and consequent visual appearance:

- *Hyaline* cartilage is the most abundant of the three types. The proportions of collagen and proteoglycan are such that no fibers can be seen macroscopically. The tissue has a glassy, translucent appearance. Hyaline cartilage is found, for example, at the ventral ends of ribs and on the joint surfaces of bones.
- *Elastic* cartilage differs structurally from hyaline cartilage by the presence of a large number of branching elastin fibers. This feature makes the flexibility of elastic cartilage greater than that of hyaline cartilage. Elastic cartilage is found, for example, in the external ear.
- *Fibrous* cartilage is a transitional form between cartilage and dense connective tissue. It has a high proportion of collagen and a very low proteoglycan content. As a result the fibers are visible to the naked eye. Fibrous cartilage contributes tensile strength and flexibility. It is found, for example, in intervertebral disks and at the attachment sites of some tendons to bone.

• Cartilage has no blood supply and no nerves.

• Cartilage is bounded by *perichondrium* or a calcified surface or a smooth articular margin. Perichondrium consists of two layers; the outer one is fibrous connective tissue and the inner one, called the *chondrogenic layer,* is rich in cells that are capable of differentiating into *chondrocytes.*

Cartilage cells In addition to the fiber matrix, cartilage contains cells, called *chondroblasts* and *chondrocytes.* The latter are isolated in small cavities *(lacunae)* within the matrix.

— ***Chondroblasts*** Chondroblasts are dividing, synthesizing cells. As they enlarge and differentiate, they secrete around themselves the components that quickly form the intercellular matrix.

- Type II collagen makes up the bulk of chondroblast secretion.
- Growth hormone is a major stimulus for chondroblast activity.

The increasing amount of matrix pushes apart the cells that are secreting it. Eventually each cell becomes isolated in its own lacuna and acquires the structural and functional characteristics of a *chondrocyte.*

— ***Chondrocytes*** Chondrocytes are mature cartilage cells. They continue to mitose, forming nests of daughter cells, each surrounding itself with matrix material and thereby causing cartilage to grow from within itself *(interstitial growth).*

Interstitial growth is the main mechanism of growth in mature cartilage. However, the chondrogenic layer of cells does retain the potential for *appositional growth* whereby the cells differentiate into chondrocytes, mitose, secrete matrix material, and add a new layer of cartilage.

Cartilage Function Most of the cartilage in the human body is present only in prenatal life where it serves as a precursor for bone. In the adult, cartilage is found at different sites, performing three different functions: flexible support, low-friction contact, or impact resilience.

- Cartilage is present at sites such as the trachea, the external ear, the junction between the sternum and the anterior end of ribs, or inside joints. At these sites cartilage provides flexible support so that the surrounding structures either do not collapse (trachea and external ear) or are capable of transient deformation (external ear, rib cage, and joint cartilage).
    - Bending deformation is possible because of the elastic fibers.
    - Compressive deformation and the ability to absorb compressive mechanical loads is possible because of the high water content associated particularly with articular cartilage (the cartilage "cushion" that occupies the space between the two bones of some joints). Under a compressive load the water can be slowly expressed through the cartilage surface and then be reabsorbed after the load has been removed.
- At sites such as the ends of articulating bones where they meet in a joint, cartilage provides a low-friction surface and helps absorb mechanical loads.

Bone

Bone Structure Bone is a specialized, harder form of connective tissue. It differs from cartilage in two respects:

1. The collagen matrix of bone holds precipitates of calcium and phosphate.
2. It has a blood supply.

The mature skeleton contains *trabecular bone* and *cortical bone.*

- **Trabecular bone (spongy bone)** Trabecular bone, also called spongy bone, is a network of interwoven partitions (the trabeculae, or trabeculas) surrounding cavities that contain red bone marrow. It is found in a small fraction of the skeleton, most prominently in ribs, vertebrae, and the ends of long bones.

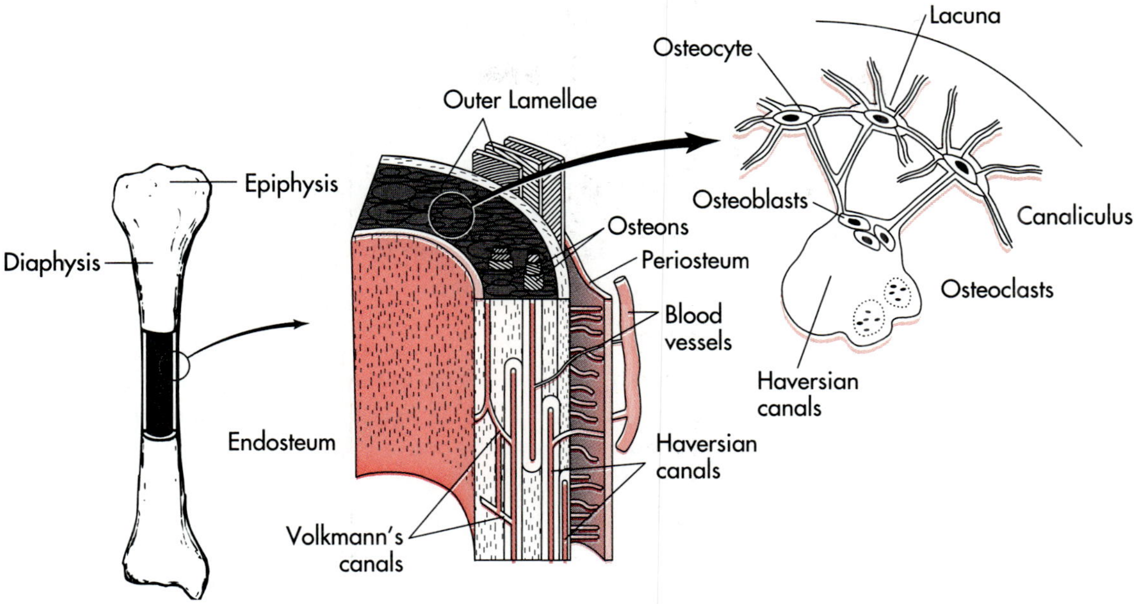

Fig. 10.1 Structure of cortical bone. Cortical bone is arranged in concentric layers around a central haversian canal. Each canal holds a capillary and also defines the core of an osteon, a unit element of bone. Larger canals (Volkmann's canals) allow blood-vessel interconnections among neighboring haversian canals. Minute channels, called *canaliculi,* traverse the bone. They originate from lacunae that are filled with one osteocyte each.

- **Cortical bone (dense bone)** The majority of skeletal bone is made up of hard, compact cortical bone.

  As shown in Fig. 10.1, cortical bone has a central cavity, filled with yellow marrow. Its mineralized portion is arranged in concentric layers that surround canals (*haversian canals*) through which the blood vessels run.

  A canal with its concentric lamellae forms an *osteon,* a unit element of bone. Within each osteon small, fluid-filled chambers, called *lacunae,* house one *osteocyte* each. Small channels, called *canaliculi* interconnect the lacunae. The canaliculae allow contact among bone cells and also provide a very large surface area for mineral exchange with extracellular fluid.

- **Bone cells** The internal and external surfaces of mineralized bone are covered with a layer that contains two types of specialized cells, *osteoblasts* and *osteoclasts.*

  In young, growing bone the osteoblast-osteoclast layer is about three cells thick. In older bone it is only one cell thick.

  A third type of bone cell, the *osteocyte,* is found imprisoned in the lacunae.

Osteoblasts

- Osteoblasts originate from a distinct line of pluripotent stem cells of the bone marrow. These common progenitors are called fibroblast colony–forming units (CFU_f) and give rise to fibroblasts, muscle cells, adipocytes, chondrocytes, or osteoblasts.
- Osteoblasts have two important functions. They
  - secrete substances that initiate formation of new bone (see the discussion of bone remodeling, p. 214) and
  - respond to parathyroid hormone (PTH), the 1,25-$(OH)_2D_3$

form of vitamin D, a variety of colony-stimulating factors, and cytokines by expressing, on their plasma membrane, one or more factors that influence the differentiation and subsequent activity of *osteoclasts.* Transcription of the osteoclast differentiation factor or factors is

- promoted equally by 1,25-$(OH)_2D_3$ (acting on a nuclear receptor) and cAMP (generally produced in this setting by membrane receptors for PTH or PGE_2) and
- inhibited by estrogen.

— ***Osteoclasts***

• Osteoclasts derive from the granulocyte-macrophage line of colony-forming units (CFU_{gm} in Fig. 2-3). Their differentiation and development are promoted by one or more osteoclast-differentiating factors produced by osteoblasts.

• Osteoclasts are rich in *tartrate-resistant acid phosphatase* (TRAP), which is widely used as their specific histochemical marker.

• Osteoclasts are highly mobile. Where they attach to a bone surface their secretory activity loosens, fragments, and resorbs bone. Collagenase, phosphatase, and lysosomal enzymes and creation of a highly acidic microenvironment are involved.

— ***Osteocytes*** Osteocytes are former osteoblasts, buried by freshly formed matrix.

• Osteocytes are covered by fine branching projections that extend through the canaliculi to communicate, through gap junctions, with neighboring osteocyte processes and with surface osteoblasts. Such junctions allow the flow of ions between neighboring cells.

• The osteocyte network is closely adjacent to the bone capillary network and communicates directly with the interfibrillar spaces of the bone matrix. This morphologic feature and the observation that osteocytes are stimulated by parathyroid hormone to lyse surrounding bone matrix indicate that they may form a link in the transfer of calcium from bone to blood.

■ **Bone Development and Growth** Bone begins as cartilage in the embryo. Those regions of embryonic cartilage that contain blood vessels will then become bone by a process of *ossification.* Once formed, bone grows in length, circumference, and thickness. When it has reached adult dimensions, bone undergoes continuous *remodeling;* the processes involved are identical to those involved in thickening or circumferential growth.

● **Bone lengthening** Bone grows longitudinally until adulthood is reached. Such growth occurs at the ends of long bones where chondroblasts, stimulated mainly by growth hormone (Fig. 10.2), deposit cartilage in the *epiphyseal disk.* Cartilage is subsequently transformed into bone by *endochondral ossification.*

• Epiphyseal disks separate the epiphyses from the diaphysis. During growth each disk deposits cartilage toward the epiphyseal surface and mineralizes into bone on the diaphyseal surface.

• When adult height is reached, growth stops and the cartilage of the epiphyseal disks is resorbed and replaced by bone trabeculae.

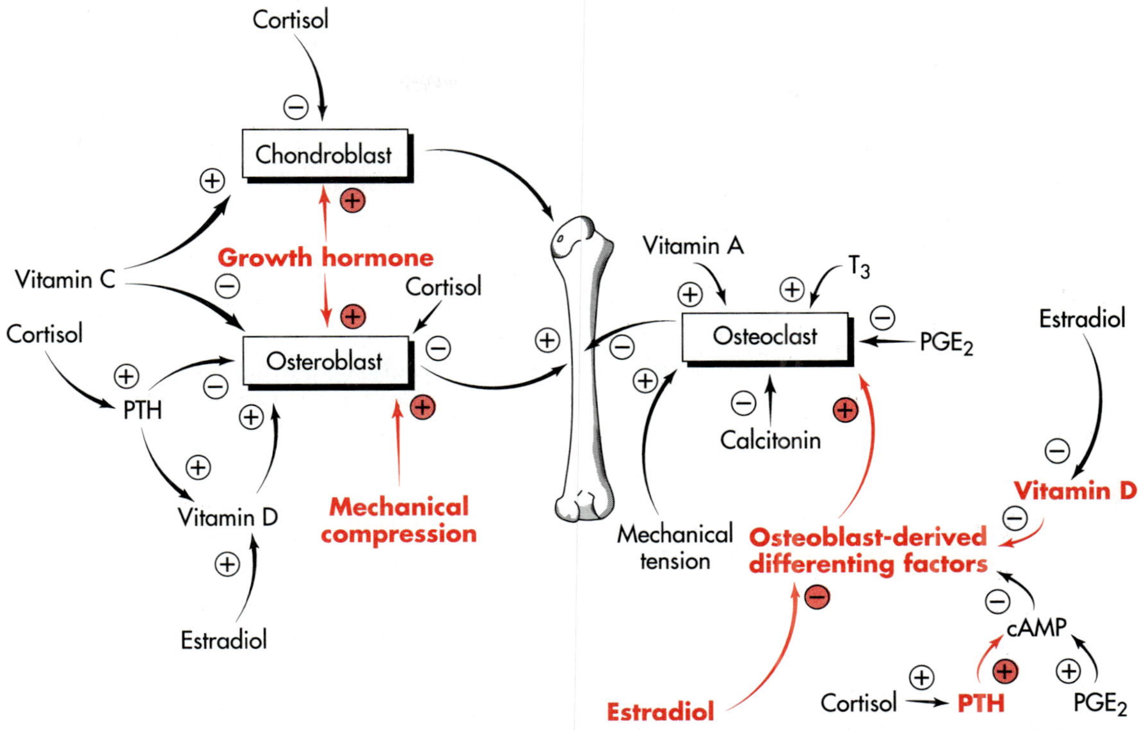

Fig. 10.2 Bone growth, remodeling, and skeletal homeostasis are determined by the activities of chondroblasts, osteoblasts, and osteoclasts. Mechanical stresses and a variety of blood-borne factors govern the activity of the three cell types. Of particular importance are growth hormone, PTH, vitamin D, and osteoblast-derived differentiating and growth factors. Although vitamin D stimulates activity in both osteoblasts and osteoclasts, its net effect on bone, in most physiologic settings, is resorption. Estrogen is significantly involved mostly because it inhibits release of the osteoblast-derived factors that stimulate osteoclast formation and action. Vitamin C is required for hydroxylation of collagen, which stabilizes the matrix. *cAMP,* Cyclic adenosine monophosphate; *PGE_2,* prostaglandin E_2; *PTH,* parathyroid hormone; *T_3,* thyroid hormone; *vitamin D,* 1,25-$(OH)_2D_3$.

— ***Endochondral ossification*** Most of the skeleton in the adult is formed by endochondral ossification. The process begins at about the middle of the cartilage cap. Three steps are involved:

- The first step is proliferation of blood vessels into the perichondrium. This alters the microenvironment of the chondrogenic layer, causes undifferentiated cells within it to differentiate into osteoblasts rather than chondroblasts, and leads to the deposition of a bone sleeve around the cartilage under it and under the surrounding perichondrium.
- The second step is hypertrophy of chondrocytes with secretion of *alkaline phosphatase.* This causes precipitation of calcium phosphate within the matrix, impairment of diffusion, and death of chondrocytes. Calcium phosphate is then slowly transformed into hydroxyapatite, an insoluble, crystalline mineral. As a result of these events thin trabeculae of calcified matrix are formed.
- The third step is invasion of the central area by macrophages from the newly proliferated blood vessels (step one, above). They remove calcified matrix and cellular debris, allowing *marrow* (capillaries and undifferentiated tissue) to proliferate into the space to form the

primary ossification center. Osteoblasts will quickly deposit a thin layer of bone upon any remaining trabeculae of calcified cartilage. Thus,

- all newly formed bone is trabecular. It is subsequently transformed into cortical bone by the processes of remodeling.

Bone thickening Living bone is never at rest. Both its matrix and mineral stores are constantly being remodeled along lines of mechanical stress. If bone resorption is not exactly balanced by bone rebuilding, there is a change in bone mass.

- Bone thickening or bone expansion (in the case of the flat bones of the skull) occur by removal of bone along the inner surface and concurrent deposition along the outer surface. Osteoclasts remove existing bone, whereas osteoblasts deposit new bone. The activity of both is regulated by mechanical factors and a variety of hormonal factors (Fig. 10.2).
  - The major long-term regulators are growth hormone and mechanical forces.
  - The major short-term regulators are parathyroid hormone (PTH), vitamin D, and calcitonin.
    - The major effect of PTH on bone is osteoclast stimulation and bone resorption. However, this effect is indirect via osteoblast-derived factors and therefore requires the presence of osteoblasts. Elevation of serum [PTH] causes hypercalcemia.
    - The major effect of vitamin D on bone is genomic upregulation of a family of matrix proteins, including osteopontin and osteocalcin, that are deposited in osteoid. They bind Ca^{++}, and by that mechanism they increase bone formation and mineralization. Thus elevation of vitamin D tends to cause hypocalcemia.
    - Calcitonin acts directly on membrane receptors in osteoclasts. As a result of the interaction, adenylate cyclase is activated, cytosolic cAMP is elevated, and bone resorption is inhibited. It is not yet known how elevated cAMP depresses osteoclast function.
- Estrogen plays a vital role in the local coupling of bone resorption and formation. Estrogen and to some extent testosterone act on osteoblasts and peripheral monocytes to inhibit the formation of osteoclast differentiation factors. As a result, the presence of estrogen ensures a low rate of bone resorption and precise coupling between osteoblast and osteoclast activities. Lack of estrogen promotes osteoclast formation and activity and leads to reduction of bone mass *(osteoporosis).*

Bone remodeling In the adult skeleton bone resorption and formation are balanced so that there is no net change in bone mass but a continuous turnover of substance. This process is called *remodeling.* It has been suggested that remodeling is initiated when a local osteocyte effect alters adjacent bone in such a way that it can be "recognized" by circulating mononuclear phagocytes. Phagocytes accumulate on the bone surface, fuse, and begin resorption.

- Osteoclast activity continues for about 3 weeks and results in a small tunnel (up to 1 mm in diameter).
- The tunnel is then invaded by osteoblasts, and their activity over the next several months fills the tunnel with new bone.

- Osteoblasts secrete collagen (mostly type 2) and "ground substance" (composed mostly of chondroitin sulfate and hyaluronic acid). The collagen polymerizes rapidly to form collagen fibers and *osteoid,* a cartilage-like material but differing from cartilage in that calcium salts precipitate in it.
- Within a few days after the osteoid is formed, calcium salts begin to precipitate on the surfaces of the collagen fibers. The precipitates appear at regularly placed intervals along each collagen fiber, forming minute nidi that gradually, over a period of days and weeks, grow into the finished product, *hydroxyapatite.*

Bone Function Bone has two principal functions. It exists to provide

- mechanical support for the soft tissues of the body and
- the major body pool of the minerals Ca^{++} and PO_4^{3-}.

Mechanical support Four factors determine the mechanical support provided by bone: its length, circumference, thickness, and composition.

- Changes in body size during childhood require both longitudinal and circumferential bone growth.
- Longitudinal bone growth stops when adult height is reached and the epiphyseal plates have been sealed.
- Circumferential adaptation to mechanical stress continues throughout life.

Bone as a reservoir for minerals Bone is the major reservoir for body calcium and phosphate. The minerals are deposited there by the action of osteoblasts and are released to the extracellular fluid by the action of osteocytes or osteoclasts.

- Osteocyte enzymes access the most recently formed bone crystals, a small, but rapidly exchanging pool for Ca^{++} and PO_4^{3-}. This process is called *osteolysis.*
- Osteoclast activity accesses the large but slowly exchanging compartment represented by mineralized bone.

Hormones in Mineral Metabolism

The physiologic mechanisms that regulate mineral metabolism have two concurrent objectives:

- To maintain appropriate plasma concentrations of Ca^{++} and PO_4^{3-}
- To maintain appropriate bone mass

Three hormones are involved: parathyroid hormone (PTH), calcitonin, and vitamin D. They act on three target organs: intestinal mucosa, nephron, and bone.

Parathyroid Hormone (PTH) Parathyroid hormone (parathormone, PTH) is secreted from parathyroid chief cells. Its most important stimulus is hypocalcemia. Its release is also promoted by agents that elevate cAMP in the cytosol of chief cells. Such agents are, for example, catecholamines and histamine. The major actions of PTH are:

- Promotion of bone resorption (by promoting formation of osteoclasts)
- Promotion of $1,25\text{-}(OH)_2D_3$ formation in the kidney (by activation of the rate-limiting enzyme 1α-hydroxylase)

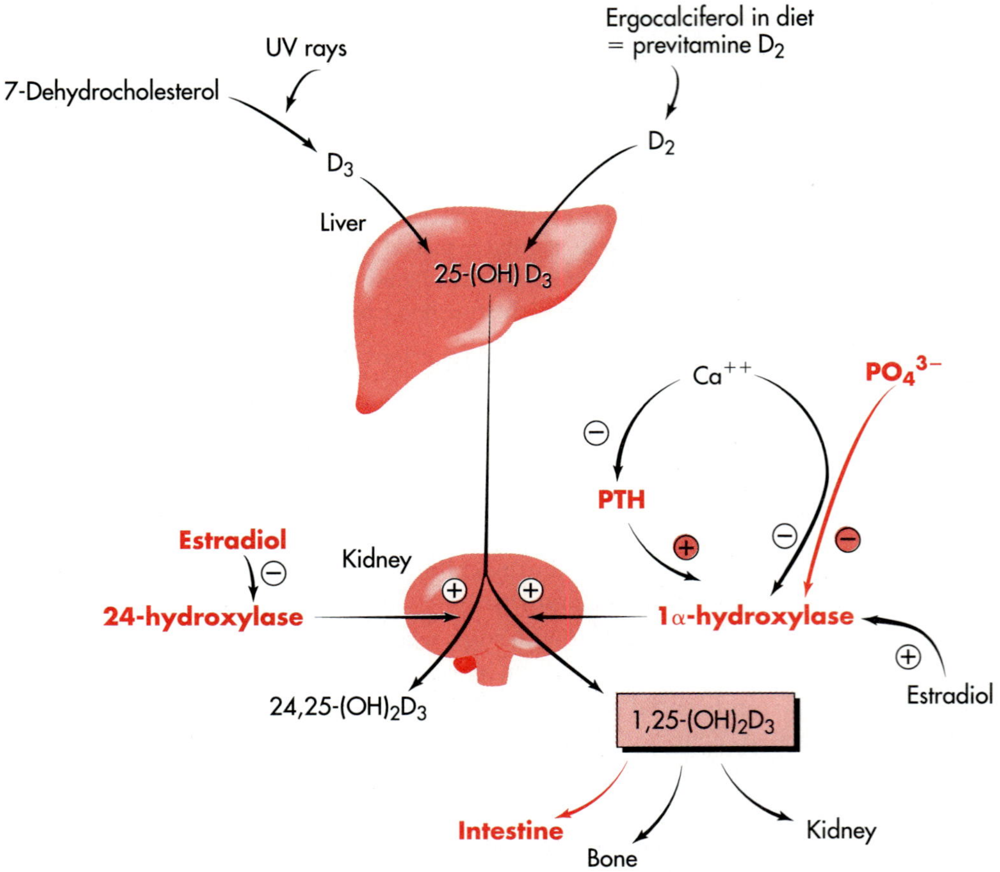

Fig. 10.3 Formation of vitamin D. Previtamins are concentrated in the liver and converted to 25-$(OH)D_3$, which is a biologically inactive precursor. The active hormone, 1,25-$(OH)_2D_3$, is produced in mitochondria of renal tubular cells and its production is regulated by feedback of serum [Ca^{++}] and PTH: hypocalcemia stimulates secretion of PTH, which, in turn, promotes conversion of 25-$(OH)D_3$to 1,25-$(OH)_2D_3$. The kidney also produces other forms of vitamin D. These forms have little biologic effect, but they use the same substrate from which active vitamin D could have been produced. The active hormone is transported to target cells in the gastrointestinal tract, bone, and other sites in the kidneys. Any excess is stored in adipose tissue. *D_2*, previtamin D_2; *D_3*, previtamin D_3.

- Enhancement of renal Ca^{++} reabsorption
- Inhibition of renal PO_4^{3-} reabsorption

- **Calcitonin** Calcitonin is synthesized in the C-cells of the thyroid (see Chapter 9). It depresses bone resorption and nephron reabsorption of both Ca^{++} and PO_4^{3-}. Therefore, it tends to decrease the extracellular concentrations of both Ca^{++} and PO_4^{3-}. However, its effects on plasma concentrations of Ca^{++} and PO_4^{3-} are small under physiologic circumstances.
- **Vitamin D** Vitamin D is a vital factor in Ca^{++} and PO_4^{3-} homeostasis. It is a steroid hormone that is not secreted by an endocrine gland. It exists in several biochemical forms (Fig. 10.3) that vary in biologic potency.
  - **Synthesis of vitamin D** Fig. 10.3 summarizes the formation of biologically active hormone from previtamins.
    - Vitamin D originates from the inactive previtamins D_2 or D_3. Previtamin D_3 is produced in the skin by ultraviolet irradiation of 7-dehydro-

cholesterol, whereas previtamin D_2 is *ergocalciferol,* a substance that is present in plants and introduced into humans by means of foodstuffs. The previtamins are transported by a specific carrier protein to the liver.

- The endoplasmic reticulum in hepatocytes concentrates and converts the previtamins to 25-(OH)D_3, which is the major circulating form of vitamin D. Although it is slightly more potent than its precursors, it has no significant biologic activity.

- The biologically active hormone 1,25-$(OH)_2D_3$ is produced mostly in mitochondria of epithelial cells of the renal proximal convoluted tubule. The enzyme 1α-hydroxylase is absolutely required (Fig. 10.3). It is under the control of several factors:

  — ***Short-term control*** Short-term control of vitamin D production is chiefly attributable to the effects of PTH, Ca^{++}, and PO_4^{3-}. They act by their respective influence on the levels of the enzyme 1α-hydroxylase, which is responsible for producing the active hormone 1,25-$(OH)_2D_3$.

  — ***Long-term control*** Estradiol is the major long-term controlling factor. It influences the level of 24-hydroxylase (see Fig. 10.3) and thereby governs the proportion of biologically active vitamin D [1,25-$(OH)_2D_3$] versus inactive vitamin D [24,25-$(OH)_2D_3$] produced by the kidney.

● **Functions of vitamin D** The major target organ of vitamin D is the intestinal mucosa. It interacts with a nuclear receptor and enhances the transcription of the *calbindin-D* family of Ca^{++}-binding proteins. An increase in their level is associated with increased uptake of both $Ca^{++}$and PO_4^{3-} (Figs. 10.4 and 10.5). The kidney and bone are secondary target organs.

- In the kidney 1,25-$(OH)_2D_3$ directly promotes Ca^{++} reabsorption from the nephron and increases renal reabsorption of phosphate.

  - The mechanism of enhanced Ca^{++} reabsorption is also by means of a receptor-coupled increase in calbindin-D in distal convoluted tubular cells.
  - The mechanism of enhanced PO_4^{3-} reabsorption is believed to depend on nongenomic actions of vitamin D and to involve changes in membrane lipid composition.

- In bone 1,25-$(OH)_2D_3$ increases production of osteocalcin, osteopontin, and alkaline phosphatase. In this tissue vitamin D acts paradoxically:

  - It stimulates young osteoclasts to increase bone resorption to increase serum [Ca^{++}]. However, mature osteoclasts lack nuclear receptors for vitamin D.
  - It interacts with nuclear receptors in osteoblasts to stimulate matrix deposition and thereby transfer Ca^{++} from extracellular fluid to bone.

- In addition to the genomic actions described above and requiring hours or days for maximal effectiveness, vitamin D also has nongenomic effects on all its target cells. They occur within seconds to minutes and include:

  - increasing Ca^{++} influx through voltage-regulated channels

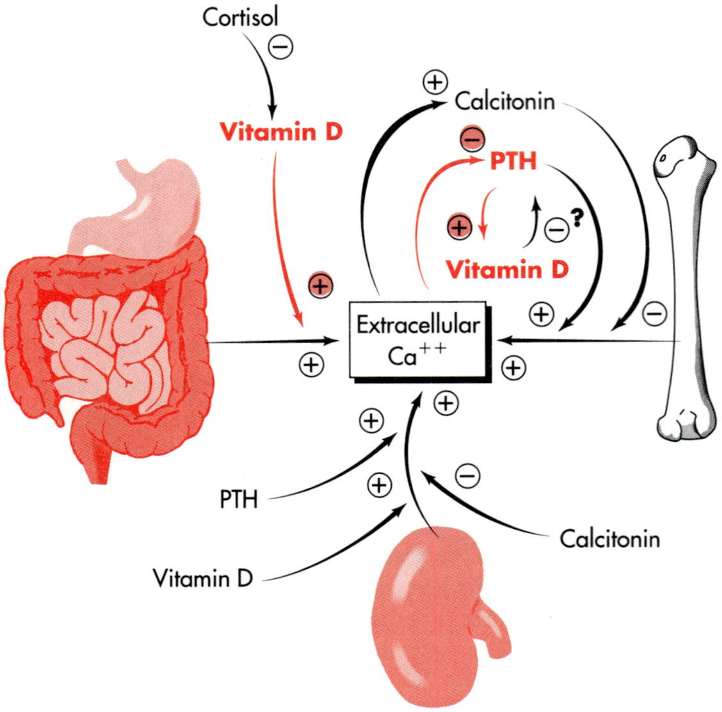

Fig. 10.4 Vitamin D and PTH are the major hormonal influences on Ca^{++} homeostasis. They act on target cells in the intestinal mucosa, bone, and kidney to elevate plasma Ca^{++} to levels that support normal mineralization of bone and normal function of the neuromuscular system.

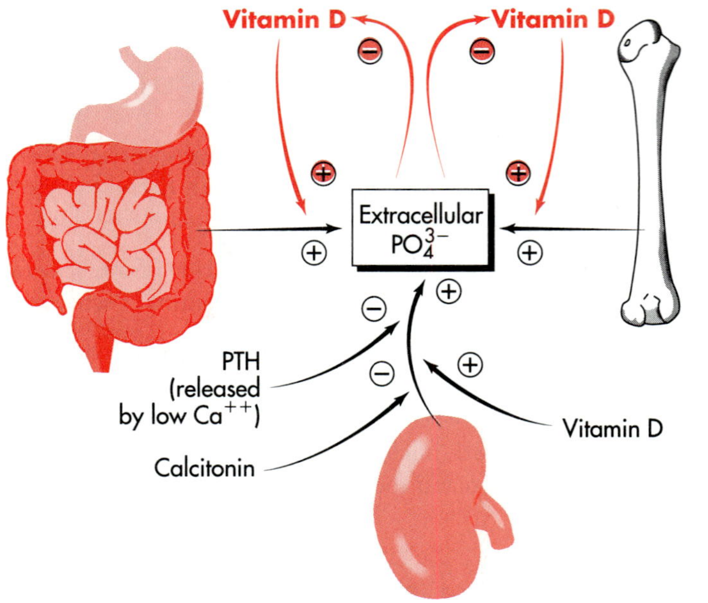

Fig. 10.5 Vitamin D levels are inversely related to extracellular $[PO_4^{3-}]$ and are the major hormonal influences on $[PO_4^{3-}]$ homeostasis. Vitamin D acts on target cells in the intestinal mucosa and bone to modulate PO_4^{3-} absorption.

- stimulation of IP_3 production and subsequent stimulation of Ca^{++} release from intracellular stores
- activation of phosphorylation processes

■ **Regulation of Extracellular Calcium** PTH, vitamin D, and calcitonin are the major hormones involved in the regulation of ionized calcium in extracellular fluid (Fig. 10.4).

● **PTH and vitamin D** Regulation of extracellular Ca^{++} concentration is accomplished by an inverse relationship between plasma $[Ca^{++}]$ and the secretion of PTH. The molecular relationship between the two is unusual in that elevated $[Ca^{++}]$ suppresses PTH synthesis in parathyroid chief cells. As a result of this direct effect of Ca^{++} on PTH:

- A fall in plasma $[Ca^{++}]$ will increase PTH levels (Fig. 10.4). PTH, in turn, will
  - increase bone resorption by promoting osteoclast formation and activity (see Fig. 10.2),
  - decrease bone formation by suppressing osteoblast activity (see Fig. 10.2),
  - promote renal reabsorption of Ca^{++}, and
  - increase production of physiologically active vitamin D in the kidney (see Fig. 10.3).
- Vitamin D stimulates absorption of Ca^{++} from the intestine and the nephron. Its usual net effect on bone is increased resorption with an accompanying increase in extracellular $[Ca^{++}]$.

● **Calcitonin** Secretion of the peptide calcitonin from C-cells in the thyroid is stimulated by hypercalcemia.

- Its principal action is one of short-term lowering of extracellular $[Ca^{++}]$ and $[PO_4^{3-}]$. This is accomplished by receptor-coupled mechanisms that operate to decrease osteolytic activity in osteocytes and thereby promote Ca^{++} deposition in the rapidly exchanging bone mineral pool.
- Calcitonin inhibits osteoclast activity and renal Ca^{++} reabsorption (see Fig. 10.4).
- In summary, calcitonin acts as an antagonist to PTH and vitamin D in calcium homeostasis. However, its long-term effect on Ca^{++} handling is small because any tendency toward hypocalcemia is quickly compensated by increases in PTH and vitamin D.

■ **Regulation of Extracellular Phosphate** The major influence on PO_4^{3-} regulation is exerted by vitamin D. PTH is involved only secondarily.

● **Vitamin D** Extracellular phosphate concentration inversely affects renal production of biologically active vitamin D (Fig. 10.5).

- When extracellular $[PO_4^{3-}]$ increases, vitamin D production decreases.
- When vitamin D levels decrease, absorption of PO_4^{3-} from intestinal mucosa decreases and so does its resorption from bone.

● **PTH** When hyperphosphatemia depresses vitamin D and leads to a compensatory decrease in PO_4^{3-} absorption, Ca^{++} absorption from intestine and bone will also be decreased. The resulting decrease in serum $[Ca^{++}]$ will stimulate PTH secretion.

- PTH acts on proximal tubular cells to depress renal PO_4^{3-} reabsorption.
- PTH also promotes vitamin D production (see Fig. 10.3), which is in opposition to the vitamin D–suppressing action of hyperphosphatemia.

In the long term, therefore, the opposing effects of similarly directed changes in extracellular $[PO_4^{3-}]$ and $[Ca^{++}]$ tend to normalize plasma levels of vitamin D and with that normalize mineral absorption from intestine and bone.

Multiple Choice Review Questions

1. Which one of the following correctly identifies the primary site of longitudinal growth in long bones?
   a. Epiphysis
   b. Diaphysis
   c. Epiphyseal disk
   d. Medullary cavity
   e. Primary ossification center

2. Which one of the following statements about the hormones that affect bone growth is correct?
   a. Decreased estrogen in menopause results in increased osteoid formation by osteoblasts.
   b. Vitamin D increases both bone formation and bone resorption.
   c. Glucocorticoid excess increases protein synthesis and therefore increases bone growth.
   d. Elevated T_3 levels in hyperthyroidism cause increased osteoblast activity.

3. The major stimulus for the release of parathyroid hormone (PTH) is:
   a. hypocalcemia.
   b. low levels of $1,25\text{-}(OH)_2D_3$.
   c. hypercalcemia.
   d. high levels of $1,25\text{-}(OH)_2D_3$.
   e. hyperphosphatemia.

4. Hyperparathyroidism may be associated with which one of the following?
   a. Decreased plasma levels of 1,25-dihydroxyvitamin D_3
   b. Renal calcium stones caused by increased plasma calcium levels above the precipitation constant
   c. Increased plasma phosphate levels
   d. Hypocalcemia

5. Active vitamin D and PTH have many similar effects. Which of the following physiologic effects is specific only for calcitriol?
   a. Increased renal phosphate reabsorption
   b. Increased renal Ca^{++}reabsorption
   c. Decreased intestinal Ca^{++} absorption
   d. Elevation of plasma $[Ca^{++}]$

6. A 7-year-old girl is admitted to the hospital with low plasma calcium levels and high parathyroid hormone (PTH) levels. Which one of the following is the most likely diagnosis?
   a. Calcitonin deficiency
   b. Hyperparathyroidism
   c. 1,25-Dihydroxyvitamin D_3 deficiency
   d. Hypoparathyroidism

7. Which one of the following statements regarding calcium and phosphate homeostasis is true?
   a. Plasma phosphate levels are tightly regulated to prevent crystallization in the blood.
   b. Alkaline phosphatase increases local phosphate concentrations causing the formation of hydroxyapatite.
   c. Osteoclast activity is exactly balanced by osteoblast activity before menopause.
   d. 100% of the calcium received in the diet is absorbed by the gut.
   e. Excessive clearance of phosphates by the kidney leads to formation of kidney stones.

8. Which one of the following statements about the hormones involved in calcium and phosphate homeostasis is correct?
   a. Parathyroid hormone stimulates conversion of 25-hydroxyvitamin D_3 to 1,25-dihydroxyvitamin D_3 in the kidney.
   b. Calcitonin increases plasma levels of Ca^{++} and PO_4^{3-} through inhibition of osteoblast activity.
   c. 1,25-Dihydroxyvitamin D_3 is required for reabsorption of Ca^{++} from kidney tubules.
   d. Parathyroid hormone secretion is stimulated by increases in plasma PO_4^{3-} levels.
   e. Hypoparathyroidism is associated with increased absorption of Ca^{++} from the gut.

9. Each of the following statements relating to compact bone is true, *except:*
   a. It is made up of parallel bony columns.
   b. It contains neurovascular channels.
   c. It is composed of a network of trabeculae.
   d. It contains haversian canals.
   e. It contains osteocytes that communicate with one another by means of gap junctions.

10. Which one of the following statements regarding the synthesis of 1,25-dihydroxyvitamin D_3 is *incorrect?*
    a. Kidney failure leads to a greatly reduced ability to convert 25-hydroxyvitamin D_3 to 1,25-dihydroxyvitamin D_3.
    b. Calcium-stimulated ATPase is required for absorption of 1,25-dihydroxyvitamin D_3 from the gut.
    c. Excess vitamin D_3 is stored in body fat.
    d. PTH stimulates conversion of 25-hydroxyvitamin D_3 to 1,25-dihydroxyvitamin D_3.

Chapter 11

Integrative Physiology

Control of Fuel Supply; Starvation

The Fed State

- **Daily energy balance** North American adults have an average daily energy turnover of 2500 kcal (10×10^6 J). Food intake and liver function drive the energy supply, whereas basal metabolism (1200 kcal), voluntary muscle activity (1100 kcal), and processes of food absorption are the major consumers of energy.

  Glucose and fat are the major sources of energy and tissues differ in their preference for energy substrate:

  - Most of the glucose is used by the central nervous system.
  - Most of the fat is used by resting skeletal muscle.

- **Energy sources** Dietary intake of carbohydrates and fats form the major energy sources in the fed state, and the liver makes additional contributions of glucose and ketone bodies.

  — ***Glucose*** The daily usage is about 670 kcal. Most of it derives from intestinal absorption (after dietary intake). Some glucose is contributed by the liver by *glycogenolysis* or *gluconeogenesis* (Fig. 11.1).

  — ***Free fatty acids*** Dietary intake forms the major source of free fatty acids

  — ***Ketones*** Ketones (mostly in the forms of β-hydroxybutyric acid and acetone) are formed by the liver (see Chapter 9 on the discussion of insulin) and can be used as an energy substrate by, for example, striated muscle. Under circumstances of food deprivation, the brain also is able to adjust its preference for metabolic substrate from glucose to ketones.

- **Body weight** Excess caloric intake is accumulated in body energy stores. These amount to approximately 2600 kcal/kg of body weight and are composed of fat (76%), protein (23%), and glycogen (1%).

  • Physical activity increases the need for energy substrate. Therefore it is normally accompanied by increased caloric intake to ensure stable body weight.

  • The mechanisms that maintain body weight through the regulation of appetite are critically dependent on hypothalamic areas that respond to a variety of stimuli such as plasma glucose level, glucose consumption rate, rate of heat production, and brain concentration of gastrointestinal hormones.

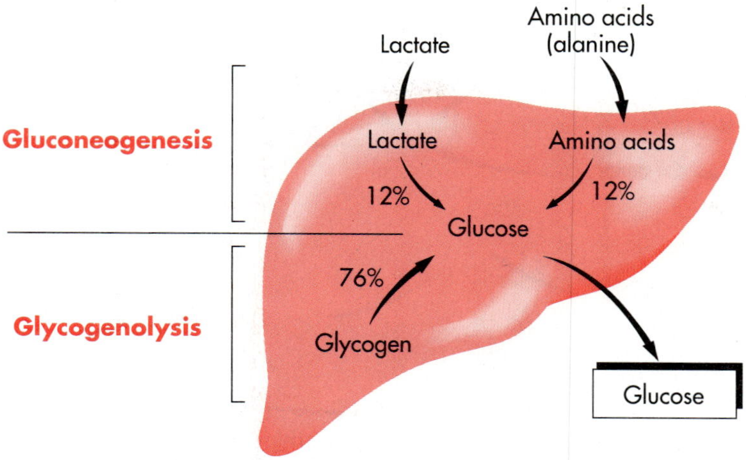

Fig. 11.1 The liver as glucose producer. Most glucose produced by the liver derives from breakdown of glycogen. Lactate, derived from glycolysis in muscle, erythrocytes, and leukocytes as well as amino acids from muscle proteolysis, provide the precursors for gluconeogenesis. Gluconeogenesis normally accounts for only one fourth of hepatic glucose production.

Energy Sources During Starvation When there is no dietary intake, plasma glucose tends to fall, glucagon becomes the dominant hormonal influence on metabolism, and the liver is the only source of energy.

- **The liver as an energy source** Circulating glucose stores are only as large as the output of the liver in 1 hour. Therefore hepatic glucose production is of vital importance in the fasted state. It supplies both glucose and ketones (Fig. 11.2). Hepatic glucose production can be increased by glucagon, epinephrine, and sympathetic neural stimulation.
  - Early increases in glucose production are derived from glycogenolysis.
  - Gluconeogenesis assumes greater prominence as glycogenolysis falls when hepatic glycogen levels are depleted by prolonged fasting.
- **The first days of starvation** During the first few days of fasting:
  - Glycogenolysis provides the major and immediate source of glucose.
  - The total glycogen stored in the body is only about 450 g (1800 kcal) so that it can provide a fuel reservoir for only a short time.
- **The first weeks of starvation** During the first week or two of fasting:
  - Nervous tissue, erythrocytes and leukocytes continue to require glucose. It derives from hepatic gluconeogenesis, with amino acids being used as a substrate.
  - Adipose tissue provides ketones and free fatty acids for the other tissues.
- **Prolonged starvation** When starvation is prolonged:
  - Glucose requirements decline gradually because metabolic adaptations in the brain allow it to use ketones as a fuel.
  - As ketone utilization becomes more effective, muscle proteolysis diminishes and fat reserves are used to a greater extent until they are near depletion (after about 6 weeks of starvation).

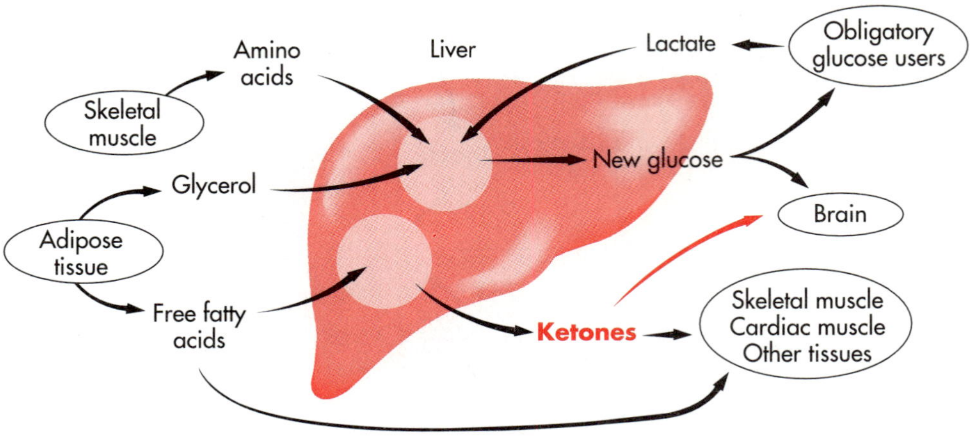

Fig. 11.2 During starvation the liver uses glycerol, amino acids, and lactate to form glucose. Ketone production assumes greater importance when insulin levels decline. Simultaneously there is an increase in the ability of the brain to utilize ketones as energy substrate.

- At that time protein stores are the only remaining energy substrate, and the rate of proteolysis increases again.
- Death occurs when metabolic requirements have depleted body protein stores to a level where protein-dependent cellular functions can no longer be maintained. In an adult this occurs after about 8 weeks of starvation.

Exercise

Cardiovascular Responses Exercise is the most severe physiologic stress that can be put on the cardiovascular system. Measurements performed in world-class cyclists show that their heart rates span the range 38 to 260 beats per minute (rest to peak activity) and their corresponding cardiac outputs range from 4.5 L/min at rest to 50 L/min at full tilt. The short-term cardiovascular adaptations to exercise can be divided into two phases: anticipation and execution.

- **The anticipation of exercise** This phase is initiated in the cerebral cortex. The resulting autonomic pattern that influences cardiovascular function is mild vagal inhibition and sympathetic activation. They cause the following:
  - A small rise in plasma levels of epinephrine and norepinephrine
  - A small increase in cardiac output
  - A slightly preferential distribution of cardiac output toward large skeletal muscle groups. Two causes have been postulated for this anticipatory increase in muscle blood flow:
    - General β_2-adrenergic relaxation of vascular smooth muscle in arterioles of skeletal muscle, mediated by a slight increase in plasma epinephrine
    - Specific sympathetic cholinergic vasodilatation of selected muscle vascular beds
- **The execution of exercise** The cardiovascular responses to exercise are summarized in Fig. 11.3. They are driven by two factors:
  - The effects of metabolites

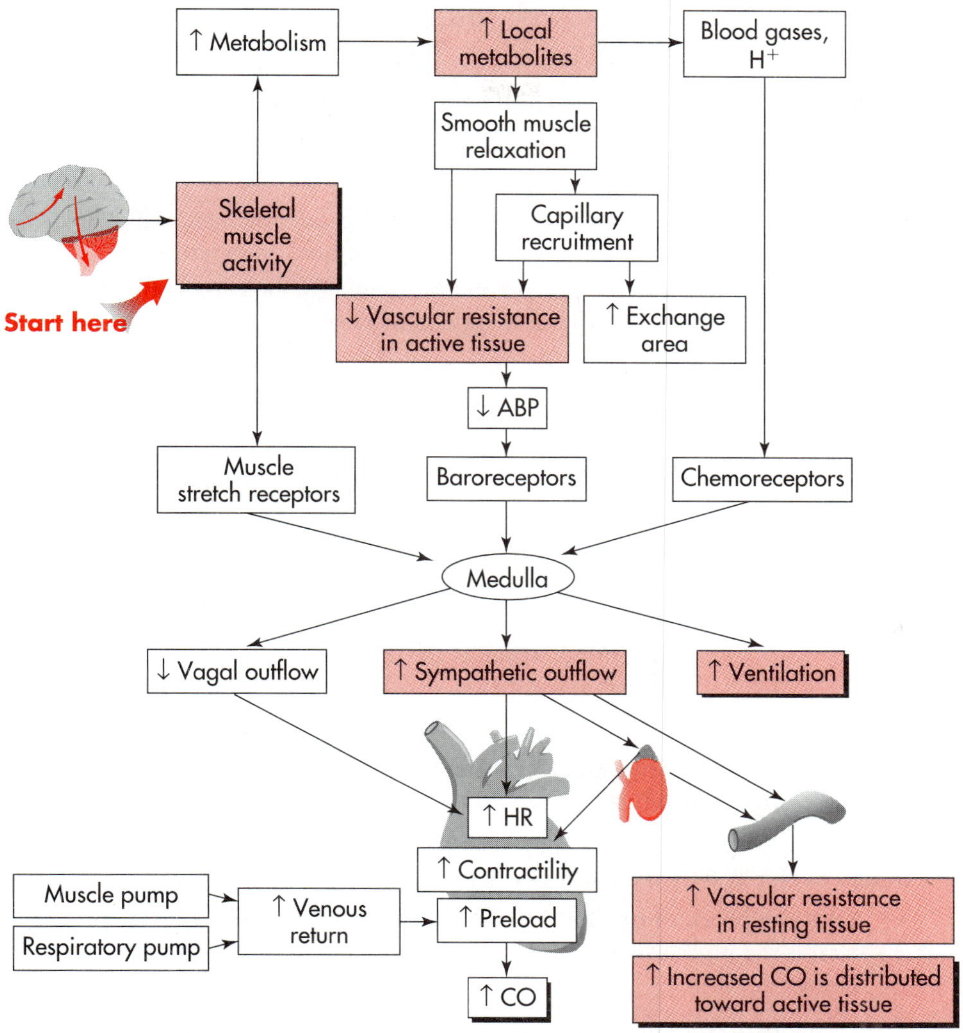

Fig. 11.3 Cardiovascular short-term adaptations to exercise. Cardiovascular responses to skeletal muscle activity are driven in part by chemical products of metabolism and in part by medullary responses to increased nervous activity in muscle stretch receptor fibers. The net effects of these adjustments are increased capillary exchange area, increased cardiac output, and preferential distribution of cardiac output to metabolically active muscle. *ABP,* arterial blood pressure; *HR,* heart rate.

- Neural input from muscle receptors (muscle spindle and Golgi tendon organ)

— ***Effects of metabolites*** Significant effects of metabolic activity are reflected in altered plasma P_{CO_2} and $[H^+]$.

These changes

- stimulate ventilation
- cause relaxation of vascular smooth muscle. This causes both capillary recruitment and vasodilatation, predominantly in active tissue where metabolite concentrations are highest.

— ***Medullary reflex patterns*** Skeletal muscle stretch receptors, arterial baroreceptors, and chemoreceptors influence medullary activity during exercise and initiate reflex responses in both cardiovascular function and respiration. There are three major net cardiovascular effects of these responses:

- A large increase in cardiac output results from the following:
  - Increased heart rate caused by autonomic nerves and elevated epinephrine
  - Increased cardiac contractility caused by increased sympathetic nerve activity and by elevated epinephrine
  - Increased cardiac preload caused by the effects of increased respiratory pumping and increased muscle pumping on venous return

- Distribution of cardiac output toward active muscle occurs because
  - inactive regions are constricted by both increased sympathetic nerve activity and the α-adrenergic effect of epinephrine, and
  - active muscle regions are dilated by the direct smooth muscle-relaxant effects of metabolites and the β_2 effect of epinephrine, which is the dominant epinephrine effect in skeletal muscle blood vessels.

- Specific vasodilatation in blood vessels of the skin is observed if the exercise continues long enough to threaten an increase in body core temperature.

Metabolic Responses During exercise increased utilization of glucose in muscle is matched by increased glucose production in the liver.

- **Glucose production** The primary stimulus for increased glucose production is the high level of sympathetic activity during exercise. Such an increase
  - depresses insulin release (because the α_2-adrenoceptor–mediated inhibition of insulin secretion dominates over any β_2-adrenoceptor-mediated stimulation of insulin secretion)
  - promotes glucagon release (because the β_2-adrenoceptor-mediated stimulation of glucagon secretion dominates over any α_2-adrenoceptor-mediated inhibition of glucagon secretion)

  The net outcome of these hormonal changes is increased hepatic glucose production and release.

- **Glucose utilization** Exercise is a state characterized by elevated plasma levels of catecholamines and depressed levels of insulin. Both tend to decrease fractional glucose uptake in skeletal muscle. Nevertheless, net glucose utilization increases during exercise. The following factors are responsible:
  - Muscle blood flow is increased greatly.
  - Muscle glycogenolysis is increased (a local, direct catecholamine effect).
  - Glucose delivery is increased greatly by the combined effects of increased blood flow and increased glucose production.
  - The affinity of muscle insulin receptors may be increased.

Body Temperature Regulation

Maintenance of constant body temperature in the face of changing ambient temperature requires regulation of both heat production and heat loss.

- **Heat Production** Heat is produced in tissues as a byproduct of metabolic activity. Total heat production is directly proportional to *basal metabolic rate* and *activity-induced metabolic rate.*

  - **Basal metabolic rate** Basal metabolic rate (BMR) is the rate of metabolism that exists in the fully awake, 12-hour, fasted state in a thermoneutral environment.

  - **Activity-induced metabolic rate** Mammals can respond to a cool environment by producing extra heat to supplement basal metabolic heat production. This occurs by increased metabolic activity, primarily in muscle and brown fat, and involves three, generally progressive physiologic activities:

    - Nonshivering thermogenesis
    - Involuntary muscle activity, such as shivering
    - Voluntary muscle activity

    — ***Nonshivering thermogenesis*** Nonshivering thermogenesis is a cold-induced increase in oxygen consumption.

    • It is mediated by the sympathetic nervous system and is a common feature in neonates, but is found in adults chiefly as a long-term adaptive response to cold environments.

    • It is represented chiefly by metabolism of brown fat and therefore depends directly on the amount of brown fat that is present.

    Brown fat. Brown fat differs from white fat in several respects. These are summarized in Table 11.1.

    — ***Shivering*** Shivering occurs when nonshivering thermogenesis is incapable of generating sufficient heat to maintain body temperature. Shivering is quantitatively the most important short-term involuntary thermogenic mechanism though it is less efficient than nonshivering thermogenesis because the rhythmic body oscillations associated with shivering increase convective heat loss.

- **Heat Loss** Heat can be dissipated only after it has been transferred to an interface with cooler surroundings. In adults skin is the most significant such interface. Human neonates and fur-covered animals also use the respiratory tract.

  • Neonates show a sharp increase in respiration rate (panting) just before onset of sweating. Adult humans do not use panting as a significant means of heat loss.

  • In adult humans heat loss is regulated mostly by control of skin-blood flow, the major avenue of heat transport to the body surface. Once heat reaches the body surface four paths are involved in its elimination: conduction, convection, radiation, and water evaporation. The relative importance of each depends on ambient conditions such as temperature, wind velocity, and humidity.

  - **Conduction** Conduction of heat requires physical contact between skin and another medium and represents transfer of heat to the medium that is in physical contact with the skin.

  - **Convection** Convection of heat represents removal by mass transport, as by a stream of air.

Table 11.1 *Comparison of Brown and White Fat*

Brown Fat	White Fat
Reddish brown	White
Small cell; contains large number of small fat droplets; centrally located nucleus	Large cell; contains one large fat droplet; peripherally located nucleus
Many mitochondria	Few mitochondria
Mitochondrial metabolism can be uncoupled from ATP production to produce more heat	Mitochondrial metabolism produces mainly ATP
High oxygen consumption	Low oxygen consumption
Extensive sympathetic innervation • Nerve activity increases lipolysis (β_1-adrenoceptor effect) • Fatty acid metabolism in mitochondria produces heat	Cells are not innervated
Selective distribution over the torso	General subcutaneous distribution
Total amount decreases with age but deposition can be increased at any time in response to prolonged cold exposure	Total amount tends to sneak up with age
Extensive blood supply and local blood flow increases with decrease in temperature	Large blood supply and local blood flow decreases with decrease in temperature
Main physiologic function is to be metabolically active and thereby to produce heat	Main physiologic function is to store fat

- **Radiation** Radiation of heat is loss by means of electromagnetic waves in the infrared region.
- **Evaporation** Evaporative heat loss occurs when water is evaporated, and the vapor is removed from the body surface.
  - In humans, the source of the water is sweat; in animals it is mostly saliva.
  - In humans, heat of evaporation is removed mostly by convective air currents, created by fans if necessary; in animals it is removed mostly by panting.

Temperature Homeostasis Different regions of the body contain sensory structures that apprise the hypothalamus of the temperature in those regions.

- Skin is rich in sensors that respond to cold.
- The hypothalamus, spinal cord, airway, and abdomen contain sensors that respond to heat.
- Specific hypothalamic nuclei use sensory inputs from the periphery to derive an integrated body temperature and then regulate that temperature by balancing heat production and heat loss.
  - Defenses against cold are dominated by cutaneous cold receptors, whereas
  - Defenses against heat are dominated by internal warm receptors.
- The regulation set point can be altered by agents such as leukocyte py-

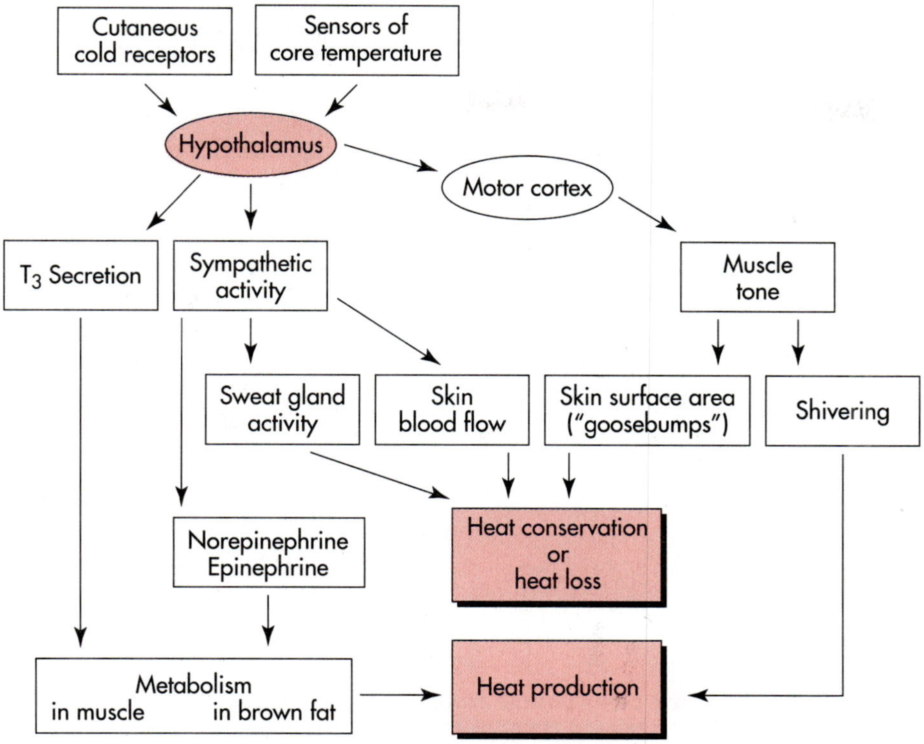

Fig. 11.4 Reflex mechanisms of body temperature regulation are coordinated responses to sensory information from temperature sensors in the skin and in the body core. Efferent mechanisms involve the thyroid gland, the autonomic nervous system, and the systems controlling skeletal muscle. T_3, triiodothyronine.

rogens. However, they can cause fever only if they also activate heat-conserving mechanisms such as cutaneous vasoconstriction.

Most humans live in environments that are colder than normal body temperature. As a result, temperature homeostasis is dominated by production and conservation of heat. Fig. 11.4 summarizes the mechanisms that are involved in body temperature regulation.

Acid-Base Regulation

Sources of H^+ Hydrogen ions originate from the metabolism of dietary carbohydrate, fat, and protein. Five specific sources of dietary H^+ can be identified:

1. Metabolism of carbohydrate and fat produces CO_2 in the Krebs cycle. CO_2 combines with water to generate 13,000 to 20,000 mM of H^+ each day.
2. Incomplete oxidation of carbohydrates produces lactic acid and pyruvic acid.
3. Incomplete oxidation of fats produces ketoacids.
4. Oxidation of dibasic amino acids like lysine or arginine produces HCl.
5. Oxidation of sulfur-containing amino acids like methionine or cystine produces H_2SO_4.

• On a typical North American diet sources 2 to 5 produce, each day, 40 to 50 mM of H^+ in the form of nonvolatile acid.

• If the volatile acid that is produced each day by metabolism of foodstuffs were dissolved in the body water space, the resulting H^+ concentration would be 325,000 to 500,000 nM/L.

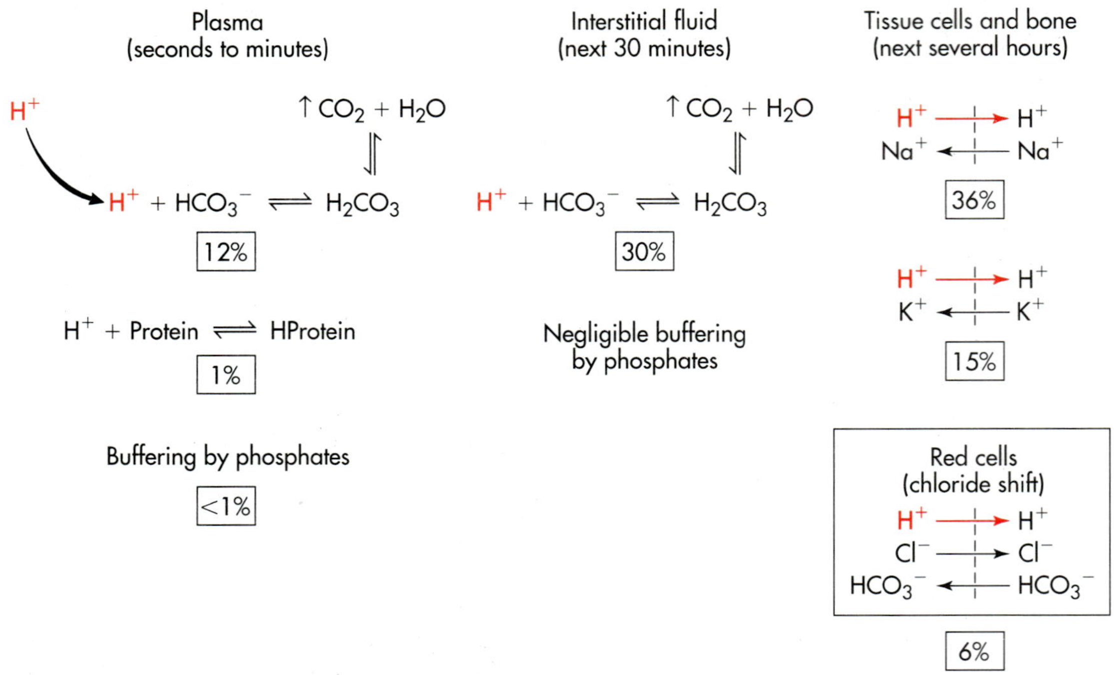

Fig. 11.5 Summary of mechanisms and time scale of nonvolatile acid-buffering. For clarity, the anion accompanying H^+ is not shown. Percentage figures refer to fractions of total H^+ load. Approximately 42% of an acid load is buffered in the extracellular space over a period of about 30 minutes. The remaining acid is buffered by intracellular mechanisms over the next several hours. If the H^+ is not accompanied by a freely permeant anion, intracellular buffering involves transmembrane exchange of Na^+, K^+, Cl^-, or HCO_3^- to maintain intracellular charge neutrality.

- If only the nonvolatile acid that is produced each day were dissolved in the body water space, the resulting H^+ concentration would be 1000 to 1500 nM/L.
- Extracellular H^+ concentration is normally maintained in the range 38 to 43 nM/L (pH of 7.42 to 7.37).
- Values outside of the range 16 to 160 nM/L (pH 7.8 to 6.8) are *not* compatible with life.

Disposal of the Daily Acid Load

- **Volatile acid** Volatile acids are those whose dissociation produces a gaseous component. Carbonic acid (H_2CO_3) is a volatile acid because it partially dissociates into H_2O and CO_2. CO_2 is then excreted by the lungs, and the chemoreceptor-ventilation feedback mechanism maintains extracellular P_{CO_2} near 40 mm Hg.
- **Nonvolatile acids** Nonvolatile acids are also called *fixed acids.* They do not produce a gaseous component that can be eliminated by the lungs. The immediate defense against nonvolatile acids is *buffering* by the body buffer systems. In subsequent hours and days the buffered acids are excreted by the kidneys and new buffer, chiefly in the form of HCO_3^-, is made. The renal involvement in acid-base regulation is described in Chapter 7.
  - *Buffering by the body buffering systems* The total capacity of the body buffers is 200 mM of nonvolatile acids per day. Fig. 11.5 summarizes the buffers and the extent of their involvement in the defense against an acid load.

Buffering of H^+ in erythrocytes involves the phenomenon known as the *chloride shift.* It is characterized by a transfer of HCO_3^- out of cells in exchange for Cl^- moving into cells. The chloride shift occurs mainly in erythrocytes because they are rich in both CO_2 (picked up in tissues) and carbonic anhydrase. This enzyme causes rapid formation of HCO_3^- within erythrocytes. When HCO_3^- leaves, driven by a concentration gradient, Cl^- is attracted into the cell by electrical forces.

Physiologic Compensations During Acid-Base Imbalance

Respiratory acidosis

— ***Definition*** A state of respiratory acidosis exists when the arterial P_{CO_2} is higher than appropriate.

— ***Cause*** Respiratory acidosis is caused by impaired respiration because such impairment increases arterial P_{CO_2} and raises extracellular $[H^+]$ as a result of the following reaction:

$$CO_2 + H_2O \leftrightharpoons H_2CO_3 \leftrightharpoons H^+ + HCO_3^-$$

Notice that this reaction is concerned with concentration changes in the nM/L range and therefore makes negligible direct contributions to the changes in extracellular $[HCO_3^-]$, which is normally near 25 mM/L.

— ***Disposal of H^+ in respiratory acidosis*** Buffering by combination with extracellular HCO_3^- is counterproductive because it would simply form more CO_2.

The short-term response, observed in states of acute respiratory acidosis, is intracellular buffering. As a result of the chloride shift in red cells, extracellular $[HCO_3^-]$ rises slightly.

Long-term compensation in states of chronic respiratory acidosis is by increased renal NH_4^+ excretion. As a result, new HCO_3^- is generated by the kidney and extracellular $[HCO_3^-]$ rises more.

— ***Acute respiratory acidosis***

Definition. Acute respiratory acidosis exists if $\Delta[H^+] = 0.77\ \Delta P_{CO_2}$.*

Compensation. If the metabolic compensation to acute respiratory acidosis is appropriate, then every ΔP_{CO_2} of + 10 mm Hg results in a $\Delta[HCO_3^-]$ of + 1 mM/L.

Otherwise a mixed acid-base disturbance is present.

— ***Chronic respiratory acidosis***

Definition. Chronic respiratory acidosis exists if $\Delta[H^+] = 0.32\ \Delta P_{CO_2}$.

Compensation. If the metabolic compensation to chronic respiratory acidosis is appropriate, then every ΔP_{CO_2} of + 10 mm Hg results in a $\Delta[HCO_3^-]$ of + 3.5 mM/L.

Otherwise a mixed acid-base disturbance is present.

Metabolic acidosis

— ***Definition*** A state of metabolic acidosis exists if $[HCO_3^-]$ is lower than appropriate.

---

*Δ in this and all subsequent equations represents change from normal value. The following are the usual normal values for humans: $[H^+] = 40$ nM/L; $P_{CO_2} = 40$ mmHg; $[HCO_3^-] = 25$ mM/L; $[Na^+] = 140$ mM/L; $[K^+] = 5$ mM/L.

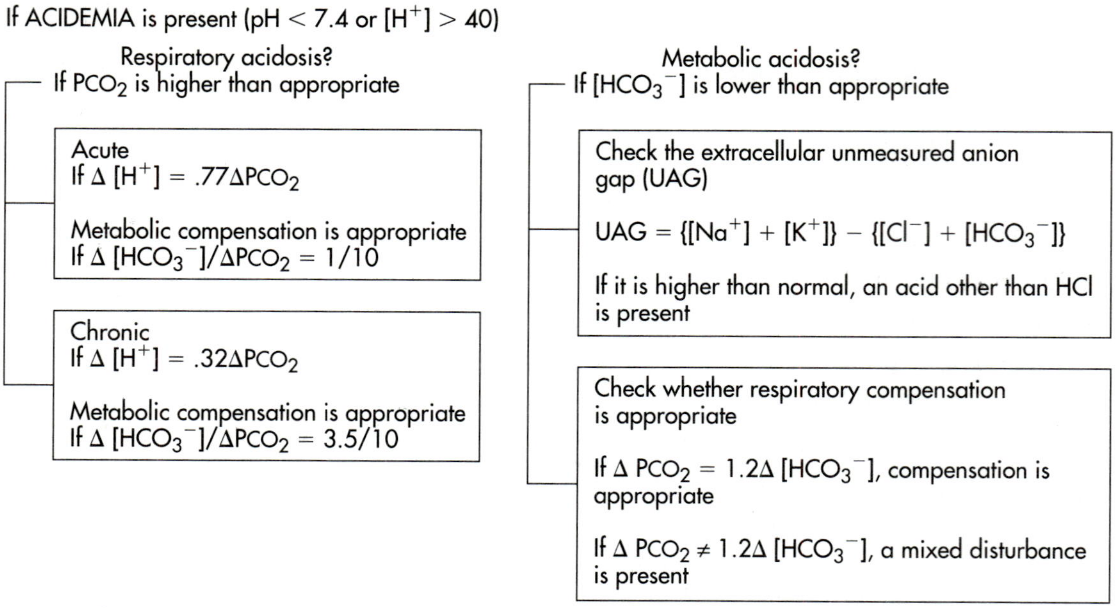

Fig. 11.6 Summary of analysis to diagnose the cause or causes of acidemia. Determination of whether compensation is appropriate reveals whether a mixed acid-base disturbance is present.

— ***Cause*** The possible causes of metabolic acidosis are:
1. Loss of HCO_3^- (from the GI tract or the kidneys) accompanied by elevation of extracellular $[Cl^-]$.
2. Gain of a metabolic acid (diabetic ketoacidosis; lactic acidosis in ischemia; retention of organic, sulfuric, or phosphoric acids in renal failure) accompanied by elevation of the *unmeasured anion gap.*

— ***Role of the kidney***
- To excrete H^+ (buffered and free)
- To generate new HCO_3^- for the body buffer stores

— ***Compensation*** If respiratory compensation is appropriate in this disturbance, then

$$\Delta P_{CO_2} = 1.2\ \Delta[HCO_3^-]$$

Fig. 11.6 summarizes the steps that will permit diagnosing the causes of a state of acidemia.

Metabolic alkalosis

— ***Definition*** A state of metabolic alkalosis exists if $[HCO_3^-]$ is higher than appropriate.

— ***Cause*** Metabolic alkalosis is caused by an initial event that causes HCO_3^- gain. Such an event might be, for example, prolonged vomiting. This event must then be followed by a set of circumstances preventing the kidney from excreting the excess HCO_3^-. Typically, such a circumstance exists when there is a drive for renal Na^+ reabsorption that is so intense that it requires accompanying reabsorption of anions in excess of available Cl^-.

— ***Compensation*** If respiratory compensation is appropriate, then $\Delta P_{CO_2} = 0.6\ \Delta[HCO_3^-]$

If ALKALEMIA is present (pH > 7.4 or $[H^+] < 40$)

Respiratory alkalosis?
If PCO_2 is lower than appropriate

- Large decreases in PCO_2 produce minimal changes in pH
- Acute

  Metabolic compensation is appropriate
  If $\Delta HCO_3^-/\Delta PCO_2 = 2/10$
- Chronic

  Metabolic compensation is appropriate
  If $\Delta HCO_3^-/\Delta PCO_2 = 5/10$

Metabolic acidosis?
If $[HCO_3^-]$ is higher than appropriate

- Check whether respiratory compensation is appropriate

  If $\Delta PCO_2 = .6\Delta [HCO_3^-]$, compensation is appropriate

  If $\Delta PCO_2 \neq .6\Delta [HCO_3^-]$, a mixed disturbance is present

Fig. 11.7 Summary of analysis to diagnose the cause or causes of alkalemia. Determination of whether compensation is appropriate reveals whether a mixed acid-base disturbance is present.

Fig. 11.7 summarizes the steps that will permit diagnosing the causes of a state of alkalemia.

Ovarian Function

■ **Anatomy** The dominant anatomic feature of the ovaries is their cortex. It is composed of three areas: interstitium, follicles, and the corpus luteum.

● **Follicles** Follicles are the sites of oocyte nurture.

- Almost all follicles are *primary follicles.* They are protective spherical structures in which an oocyte is surrounded by a layer of granulosa cells, a basement membrane, and a layer of theca cells (Fig. 11.8).
- Some follicles grow, for a brief period each month, into concentrated hormone sources that profoundly influence the emotional, physical, and reproductive potential of a woman. These are the *graafian follicles.*

— ***Primary follicles*** This type of follicle dominates in the mature woman.

- These follicles were present at birth, and at maturity they have attained only double or triple their original size.
- They have a centrally placed oocyte (see Fig. 11.8), and the most advanced among them show an oocyte in a small pool of antral fluid that separates the oocyte from a layer of granulosa cells.
- There are also follicles that are clearly undergoing atresia, a process of deterioration that steadily reduces the number of oocytes during the woman's lifetime.

— ***Graafian follicles*** Although most of the 2 million oocytes that were present at birth are lost to attrition over the next 50 years,

- each month during a woman's lifetime before menopause, 6 to 12 follicles develop beyond the primary stage. Within the month these follicles grow to be up to 30 times larger.
- Their granulosa cells and theca cells grow so rapidly that the

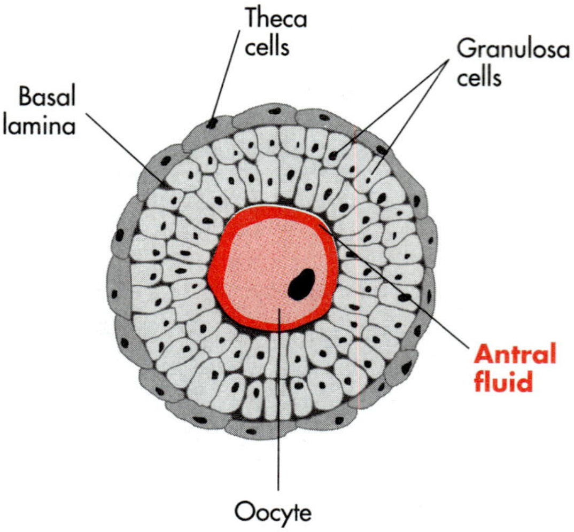

Fig. 11.8 Anatomy of a primary follicle showing the oocyte and associated layers of cell types.

interior structure becomes asymmetrical. Such follicles are then called *graafian follicles.*

Regional Membrane Receptor Activity and Hormone Synthesis

While the follicle is in the primary stage:

- Theca interna cells (the layer of theca cells closest to the basement membrane, see Fig. 11.8) have few if any receptors.
- Granulosa cells contain only receptors for follicle-stimulating hormone (FSH) but have many of them. Their activation promotes growth of granulosa cells and stimulates aromatase, a group of enzymes that convert testosterone to estradiol (Fig. 11.9).

As the follicle grows to the preovulatory stage:

- Theca interna cells develop a large number of luteinizing hormone (LH) receptors. LH receptors are membrane receptors. Their activation increases intracellular cAMP, causes enhanced activity of cholesterol side-chain cleavage enzyme (P-450_{SCC}), and thereby promotes conversion of cholesterol to pregnenolone (see Fig. 11.9). As a result, these cells, which contain the enzymes required for the full synthetic pathway from acetate or cholesterol to testosterone, produce ample androstenedione and testosterone.
- Granulosa cells use androstenedione and testosterone as substrates for the production of estradiol (see Fig. 11.9).
    - Activation of FSH receptors stimulates estradiol formation and cell growth.
    - Mature granulosa cells also develop LH receptors, and their activation further promotes estradiol synthesis.
    - Granulosa cells also produce polypeptides, named *activins* and *inhibins,* by cleavage from larger precursors. One of their major functions is related to FSH release from the anterior pituitary:
        - Activins stimulate FSH secretion.
        - Inhibins depress FSH secretion.

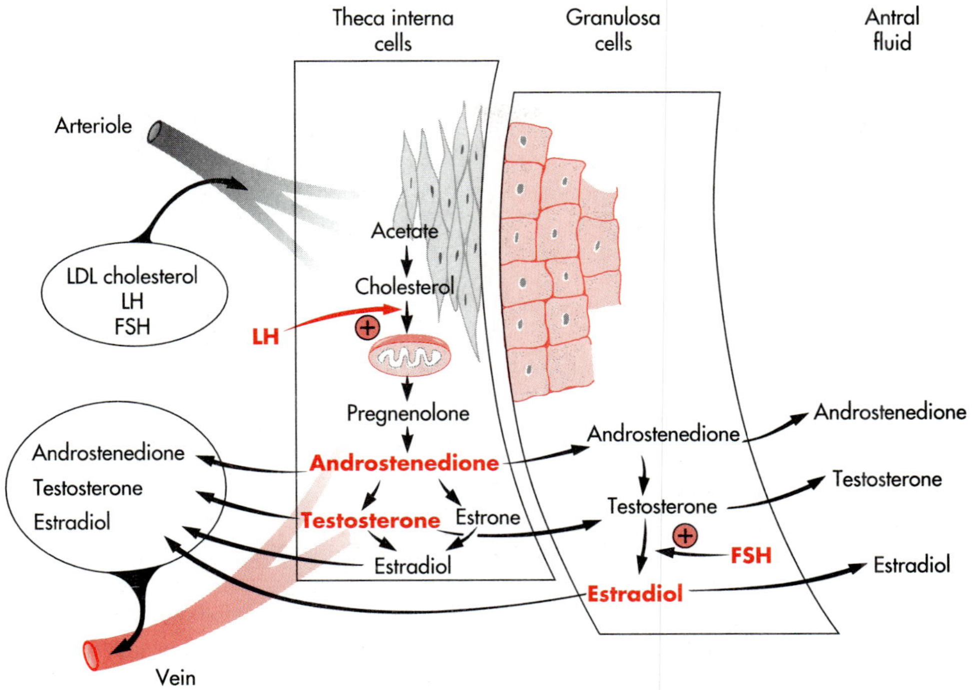

Fig. 11.9 Physiology of theca and granulosa cells in a maturing ovarian follicle. Theca cells can synthesize cholesterol (from acetate) and can also use cholesterol from circulating low-density lipoproteins (LDL) for steroid synthesis. Their principal function is to convert cholesterol primarily to androstenedione and testosterone. Granulosa cells are rich in the aromatase that converts androgens to estrogen. They use locally produced androgens as substrate for estradiol synthesis. Blood-borne luteinizing hormone (LH) and follicle-stimulating hormone (FSH) stimulate both follicle growth and steroid synthesis. The steroids are partly used to influence the oocyte and partly exported into the general circulation.

- **Corpus luteum** The corpus luteum is the final stage of the follicle from which the oocyte was expelled during that month's ovulation. It is richly supplied with LH receptors. Their stimulation increases synthesis of progesterone, the major steroid product of the corpus luteum.

Women's Monthly Rhythm

The monthly cycle of women exists for two purposes:

- To create at some point during each cycle the best possible conditions for reproduction and nurture.
- To ensure that normally only a single oocyte is fertilized at one time.

The hormones estradiol, progesterone, and inhibin are of central importance for achieving the purposes of the monthly rhythm.

Hormonal Changes and Their Effects on Ovarian Function The cycle is driven by gonadotropin-releasing hormone (GnRH), released in periodic bursts from hypothalamic cells. The amplitude and frequency of these bursts are vital features in generating the other hormonal changes responsible for the monthly cycle.

- **Day 24 of the previous cycle** At about the twenty-fourth day of each cycle FSH output from the pituitary begins to rise (Fig. 11.10). As a result of this rise in FSH:

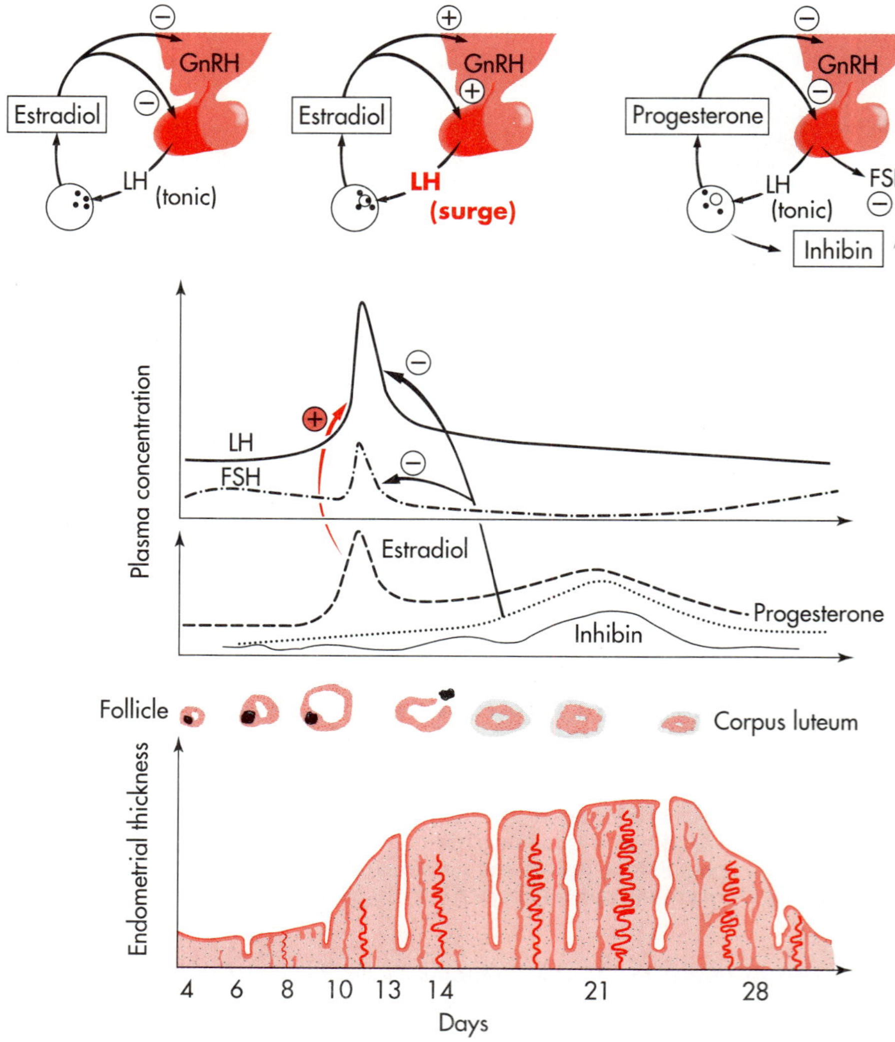

Fig. 11.10 Hormonal and endometrial changes during the course of a normal menstrual cycle. At normal plasma levels (50 to 60 pg/mL), estradiol inhibits LH synthesis by negative feedback on pituitary gonadotrophs as well as on hypothalamic GnRH-producing neurons. However, when plasma estrogen levels have been maintained at a level above 150 pg/mL for more than 36 hours, estrogen exerts a transient positive feedback and produces a surge in the release of LH from the anterior pituitary. This surge causes ovulation from one mature follicle. *FSH,* follicle-stimulating hormone; *LH,* luteinizing hormone.

- Growth is induced in 3 to 6 primary follicles in each ovary. They will reach a peak size of about 5 mm in diameter near day 10 of the next cycle, and a single dominant follicle will be selected from among them.

- Estradiol synthesis is stimulated in granulosa cells. Not enough estradiol is produced to increase its plasma levels at this time or to bring about systemic effects such as inhibition of FSH secretion, but its local concentration rises enough to induce more FSH receptors in granulosa cells. This promotes even more estradiol production and causes a gradual increase in estradiol toward day 10 of the present cycle (see Fig. 11.10). The rising plasma estrogen level is an inhibitory influence on FSH and LH secretion.

- **Day 10 of the present cycle** At days 10 to 12 of the present cycle there is a steep rise in plasma estradiol concentration.

  - At 36 to 48 hours before ovulation the increased estradiol briefly exerts positive feedback on pituitary LH release and an LH surge triggers ovulation from the dominant follicle (see Fig. 11.10).
  - A surge in FSH release occurs as well. This induces growth in a further group of follicles. The function of this latter group is to contribute to estradiol and inhibin synthesis in the luteal phase of the cycle (days 12 to 28).
  - After the LH and FSH surges there is a rapid downregulation of LH and FSH receptors. This causes decreased output of androgens from theca interna cells and estradiol from granulosa cells.

  The released oocyte is not under any immediate further hormonal influence. On the other hand, the remaining granulosa cells, set free from an apparent inhibition exerted either by the oocyte or by the follicular fluid, proliferate over the next 3 days and develop into the *corpus luteum.*

The Corpus Luteum The formation, growth, and secretory behavior of the corpus are under hormonal control.

- It is initially driven by LH.
- Its major secretory product is progesterone.
  - Progesterone inhibits the hypothalamus and the anterior pituitary, and in this way luteal secretion of progesterone gradually and progressively inhibits secretion of LH and FSH.
  - Progesterone causes a rise in body temperature and is probably responsible for the slight increase in body temperature at the time of ovulation.

- Rising levels of inhibin also inhibit FSH secretion. Progressive decrease in LH and FSH levels leads to atresia in the postovulatory follicles, causes regression in the corpus luteum (provided that fertilization has not occurred),* and diminishes plasma levels of estradiol, progesterone, and inhibin.
- By the twenty-sixth day of the cycle the levels of estradiol, progesterone, and inhibin will be so low that (1) the pituitary can begin the FSH rise that starts the next cycle and (2) the endometrium is without adequate steroid support and disintegrates to be sloughed into the menstrual flow.

Steroid Support of the Endometrium Estradiol and progesterone contribute vitally to the purposes of the monthly rhythm by creating optimum conditions for reproduction and nurture.

- **Estradiol**
  - Promotes endometrial growth and vascularization.
  - Increases vaginal secretions, which provide lubrication as well as a capacitating medium for sperm.
  - Stimulates growth and development of breasts.

*If fertilization has occurred, the corpus luteum is maintained in the later luteal phase by human chorionic gonadotropin (hCG), an LH-equivalent of placental origin.

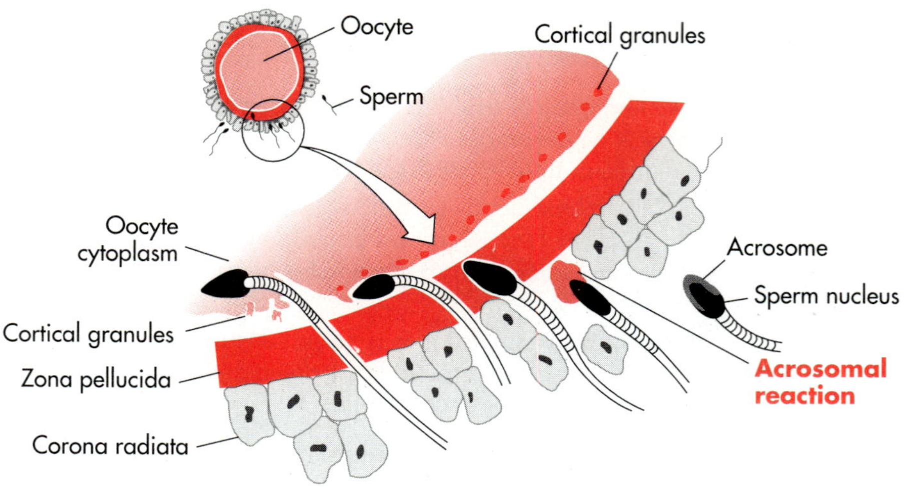

Fig. 11.11 Steps in penetration of an oocyte by a sperm. The sperm nucleus is surrounded by the acrosome, an organelle that resembles a lysosome and contains several enzymes involved in penetrating the defenses of an oocyte. The interaction of these enzymes with the cells of the *corona radiata* and the *zona pellucida* is called an *acrosomal reaction.* This reaction prepares a path that may be followed by other sperm. One of them will eventually reach the cytoplasm of the oocyte and generate within it a wave of Ca^{++} that sets in motion all subsequent events.

- **Progesterone**
  - Facilitates sperm transport by decreasing viscosity of vaginal secretions.
  - Slows endometrial growth but induces supportive secretory activity instead.
  - Stimulates mammary growth and development so that lactation may occur.
  - Inhibits motility of the uterus.

Fertilization and Pregnancy

Fertilization Fertilization occurs when the genetic material of a sperm combines with that of an oocyte.

- **Sperm transport** Of the several million sperm that are deposited during ejaculation near the cervix:
  - many leak out through the vagina;
  - many are incapacitated by the ambient acidity;
  - many are immobilized by the viscous cervical mucus;
  - many are phagotized by intrauterine leukocytes;
  - only a few experience full capacitation (that is, they undergo sufficient membrane deterioration to permit release of hydrolytic enzymes from their acrosomes).
- **Penetration of the ovum** Hydrolytic enzymes released from the acrosomes of many sperm clear a path by dissolving the ground substance of the *corona radiata* and *zona pellucida* (Fig. 11.11). This is called an *acrosomal reaction.* The path is then available for subsequent sperm.

- As soon as one sperm reaches the oocyte cytoplasm, entry of further sperm is prevented (1) by depolarization of the oocyte membrane (inward Na^+ current) and (2) by an alteration of the space between the oocyte membrane and the zona pellucida. This involves exteriorization of cortical granules from the oocyte cytoplasm subsequent to a sperm-mediated increase in intracellular [Ca^{++}].
- The sperm-mediated surge in intracellular [Ca^{++}] also initiates the first cell division.

Hormone Synthesis

Placenta Over a period of 6 to 8 weeks after fertilization and implantation, synthesis of estrogens and progesterone is shifted from the corpus luteum to the placenta and fetus. Thus after fertilization:

- estrogens and progesterone continue to provide important hormonal support for the endometrium.
- two new trophic hormones appear: human chorionic gonadotropin (hCG) and human chorionic somatomammotropin (hCS).

Estrogens During pregnancy the estrogens (chiefly estradiol):

- foster continuous growth of the myometrium;
- foster growth of the mammary ducts;
- act to soften and reshape abdominal and genital structures so as to accommodate the expanding uterus.

Progesterone

- Ensures a quiescent uterus by blocking spread of contractile activity
- Promotes nutritive secretions from fallopian tubes and endometrium

Human chorionic gonadotropin (hCG) As the levels of estrogens and progesterone increase during early pregnancy, secretion of LH and FSH are progressively suppressed, and hCG emerges as the controlling hormone.

- It appears early after fertilization.
- It is synthesized by a layer of cells (trophoblasts) whose major function is attachment of the blastocyst to the endometrium.
- It operates through the LH receptor and therefore has primarily LH activity and little FSH activity.
- It acts to maintain corpus luteum function beyond its usual life-span and promotes steroid synthesis in the placental-fetal unit.

Human chorionic somatomammotropin (hCS) hCS resembles human growth hormone (GH) in structure and shares many GH actions.

- It is a peptide hormone that is unique to pregnancy.
- It appears by the fifth week, and its plasma concentration rises continuously until birth, in direct proportion to the size of the placenta.
- It adapts maternal metabolism to the needs of the growing fetus and diverts glucose to the fetus by promoting lypolysis and decreased glucose utilization in the mother.

The Role of Steroids in Limiting Fertilization to one Oocyte at a Time

- When an oocyte has been fertilized, the pregnant uterus maintains a functioning corpus luteum and develops a placenta.
- The corpus luteum secretes progesterone.

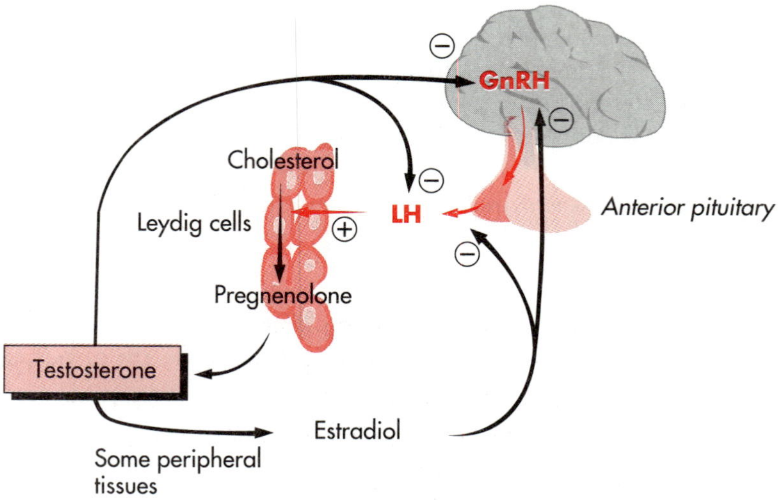

Fig. 11.12 Testosterone is synthesized in Leydig cells under LH stimulation. The process is feedback-regulated partly by testosterone itself and partly by estradiol, the major metabolite of testosterone. Estradiol is produced in those tissues that contain the enzyme P-$450_{AROMATASE}$. Adipose tissue is the most important peripheral source of estradiol.

- Progesterone blocks the positive-feedback effect of estradiol on the pituitary.
- The LH surge is prevented.
- There is no further ovulation.

Testicular Function

The testes have two major functions. These occur in different cells:

- Androgen synthesis occurs in Leydig cells.
- Sperm formation takes place in seminiferous tubules.

Anatomy

- **Leydig cells (interstitial cells)** These cells are few in number and are located in the connective tissue that lies between the seminiferous tubules.
- **Seminiferous tubules** These coiled structures make up the bulk of the testes.
- **Rete testis** This is a sperm-collecting chamber that is connected to the epididymis by efferent ductules.
- **Epididymis** The epididymis is a coiled tube that conveys sperm from the rete testis to the vas deferens and makes them potent in the process.

Androgens

Androgens Testosterone is the major androgen of the testes. It is carried in blood by a specific testosterone-binding globulin as well as by albumin and other plasma proteins.

- **Synthesis of androgens** Testosterone is synthesized in Leydig cells under the stimulation and control of LH (Fig. 11.12). These cells have three relevant features:
  - Their mitochondria contain the enzyme systems that allow steroid synthesis from cholesterol.

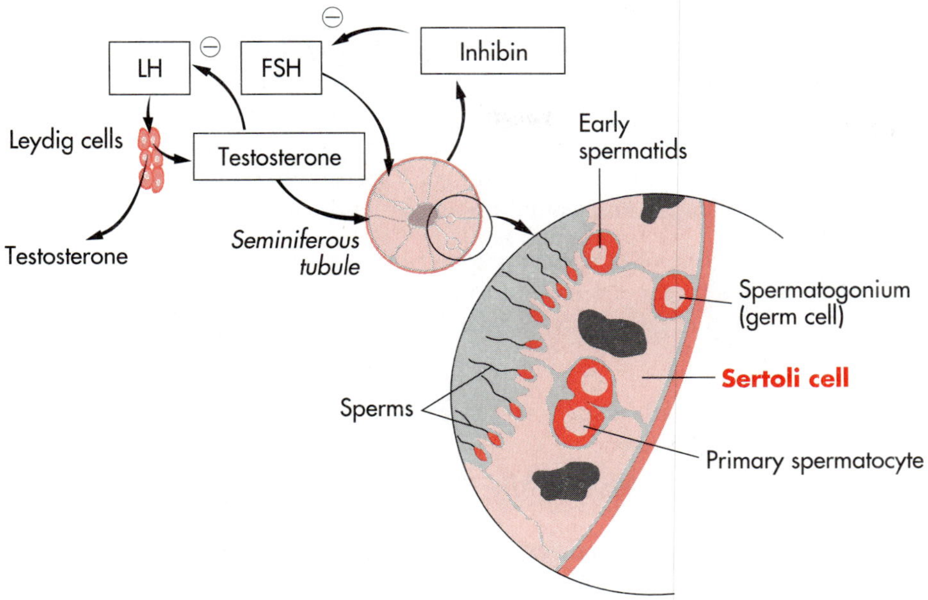

Fig. 11.13 The epithelium of seminiferous tubules is formed by Sertoli cells. Their special function is promotion of sperm formation and secretion of several peptides, including inhibin. Sperm formation begins near the basement membrane, with maturation of spermatogonia and ends on the luminal surface of Sertoli cells when sperms have been formed.

- They contain LH receptors and have therefore a high rate of pregnenolone formation when LH receptor ligands are present. Testosterone is synthesized in the fetus, while human chorionic gonadotropin (hCG) is present and after puberty when LH levels are sufficient, but it is not synthesized in childhood.
- Their cytoplasm contains the enzyme, *17β-hydroxysteroid oxidoreductase,* which is necessary for conversion of androstenedione to testosterone.

Actions of androgens Testosterone has six major functions:

- Regulation of gonadotropins (GnRH and LH in adults or hCG in the fetus)
- Formation of male phenotype during sexual differentiation
- Induction of sexual maturation and of secondary sexual characteristics at puberty
- Modulation of sexual desire
- Modulation of bone formation
- Promotion of protein transcription, especially contractile proteins in skeletal muscle (this is called *anabolism*)

Sperm Production Spermatogenesis takes place in seminiferous tubules, within the highly regulated environment of the intercellular spaces between neighbouring Sertoli cells. Over a period of 60 to 80 days undifferentiated germ cells (spermatogonia) become highly differentiated spermatozoa (Fig. 11.13). They

- are independently mobile
- carry in their heads 23 unpaired chromosomes and the enzymes required for penetration of the ovum

The process of spermatogenesis begins with spermatogonia near the basement membrane (see Fig. 11.13). They mature, form *primary spermatocytes,* and then undergo meiotic division twice. The first division yields *secondary spermatocytes,* the second yields *early spermatids.* Early spermatids have the haploid number of 23 chromosomes and mature into sperm. As maturation of each cell type proceeds, the cells move closer and closer to the lumen of the seminiferous tubule. The last stage, maturation of the early spermatids into mature sperms (spermatozoa), occurs in deep infoldings of the luminal surface of Sertoli cells.

- **Hormonal control of sperm production** Although no androgens appear to be needed for the maturation and divisions involved in making early spermatids from germ cells, spermatogenesis does require LH, testosterone, and FSH.

  • LH exerts all its spermatogenic effects by means of high intratesticular levels of testosterone.

  • Testosterone is required for the final stages that make sperms from early spermatids.

  • FSH acts mainly on Sertoli cells and is required for maturation of newly formed spermatozoa. It also stimulates Sertoli cells to secrete several factors, including *androgen-binding protein* and *inhibin.*

- **The role of Sertoli cells in sperm production** Sertoli cells are specialized endothelial cells lining the inside of seminiferous tubules. They

  - create around themselves an appropriate chemical milieu for sperm formation,
  - nurture spermatogonia, spermatocytes, and spermatids by several means:
    - Maintaining a constant physical contact,
    - Supplying substrates
    - Absorbing redundant cytoplasm from the transforming spermatids
  - synthesize and secrete a variety of proteins:
    - Transport proteins for Cu^{++}, Fe^{3+}, and androgens
    - Proteins that have hormone-like activity (including inhibin)
    - Proteins that have enzymatic activities (such as plasminogen activator)
    - Proteins that help maintain the basement membrane

- **The role of the epididymis in sperm maturation** The end product of Sertoli cell nurture is nonmotile (and therefore impotent) spermatozoa that are washed out of the tubule into the rete testis. They must pass through the epididymis to undergo extensive structural and functional changes that will make them motile and fertile. The ability of the epididymis to provide the appropriate environment for these changes depends on the presence of androgens.

The Human Sexual Response

This area of human physiology is dominated by misinformation, psychologic characteristics, and cultural mythology.

Misconceptions and Realities

Misconceptions

- Individuals with stereotypically attractive physique are capable of better sexual performance and greater sexual enjoyment.
- Large breasts are more satisfying to a male sexual partner.
- A large penis is more satisfying to a sexual partner.
- Penis size is correlated with body build, race, and frequency of sexual experience.
- Only vaginal penetration can lead to a full orgasm in the female sexual partner.

Reality

- Body configuration and sexual function are not anatomically or physiologically correlated. They are, however, linked by the psychologic effects of a perceived mismatch between an individual's body configuration and the culturally fostered stereotype.
- A large penis may cause a partner to be apprehensive about inability to accommodate.
- There is great variation in the size of the flaccid penis, but when it is erect, it is 15 ± 2.5 cm (5 to 7 inches) in length and 3 to 4 cm (1 to 1.5 inches) in diameter. Its size is not influenced by sexual experience.
- Sensory nerve endings are highly concentrated in the clitoris. In addition, there may be particularly responsive areas along the vaginal walls (such as the Grafenberg [G] spot). As a result, a woman experiences different kinds of orgasm, varying in intensity of feeling and depending on the degree of stimulation of each of her centers of sexual sensory perception.

The Sexual Response Cycle

Phases of the sexual response cycle Most often we are in a nonsexual state, characterized by attention to other matters and by a receptiveness to sexual stimuli. When one of these stimuli draws our attention and a full sexual response is allowed to develop, four phases can be recognized in the response: excitement, plateau, orgasm, and resolution.

— ***Excitement*** The excitement phase is initiated by physical (visual, olfactory, or tactile) or psychogenic factors. Its continuation and development depend on situational reinforcement.

Excitement begins with the onset of erotic feelings. It is characterized by

- progressive increase in autonomic nervous activity, leading to
  - regionally specific changes in blood flow,
  - increased ventilation,
  - increased heart rate, and
  - increased arterial blood pressure;
- progressive increase in vasocongestion and muscular tension in the erogenous areas. This causes
  - skin mottling,
  - erection of nipples or penis,

- clitoral enlargement,
- vaginal lubrication.

— ***Plateau*** The plateau phase is an advanced state of arousal in which vasocongestion is at its maximum. As a result:

- The *orgasmic platform* forms in the woman. This is an advanced state of engorgement of the vaginal wall at the introitus.
- The penis attains its maximum size in the man.

Additional physical changes include:

- Lifting and tilting of the uterus to form a seminal receptacle
- Raising of the testes to the perineum
- Discharge of a few mucoid drops from the penis

The duration of the plateau phase varies with the effectiveness of erotic stimuli, the effectiveness of situational reinforcement, and the desire, training, and age of the individuals involved. The plateau phase can lead to orgasm.

— ***Orgasm*** This phase is brief and is characterized by rapid release of the vasocongestion and muscular tension. It involves the whole body:

- Reactions in the genitalia include rhythmic contractions, emission and ejaculation of semen, or increased liberation of vaginal secretions.
- Reactions in skeletal muscles include voluntary and involuntary contractions including nearly spastic contractions of facial, abdominal, and intercostal muscles.
- Reactions in whole-body sympathetic activity induce changes in respiration, heart rate, and blood pressure.
- Central nervous effects operate to exclude all other sensory perceptions.

Orgasm is followed by a refractory period in the man. During this period he will not be able to respond physically to further sexual stimuli. The length of the refractory period increases with age.

Although the woman may not wish to be further aroused after orgasm, she does not have a refractory period during which she cannot be aroused.

- Older women have an increasing ability to reach orgasm more quickly and to have multiple orgasms.
- Men show the opposite age-related pattern. Contrary to the cultural Western stereotype, their ability decreases with age.

— ***Resolution*** During resolution all physiologic variables return to their resting state within 30 minutes after orgasm, and there frequently is a desire to sleep.

Often a desire to urinate is felt as well. This is most likely caused by activation, during orgasm, of sympathetic efferents that have branches innervating the bladder.

● **The role of nerves in the human sexual response** As shown in Fig. 11.14, both sympathetic and parasympathetic nerves are involved.

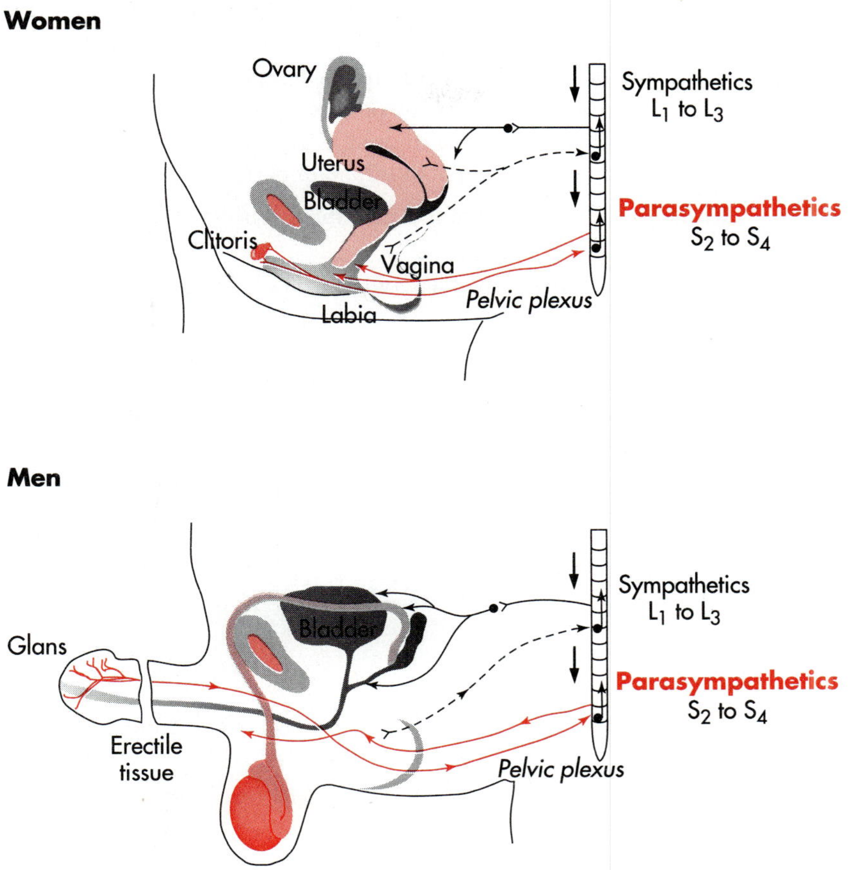

Fig. 11.14 Innervation of the genitalia in women and men. Sympathetic and parasympathetic nerves are involved as well as a separate system of vasoactive intestinal peptide (VIP)-secreting nerves (not shown). The control of sexual function is partly by local, segmental mechanisms and partly by centrally coordinated mechanisms.

- Sympathetic fibers come from L1, L2, and L3 (lumbar levels) and travel through the inferior mesenteric ganglion.
- Parasympathetic fibers come from S2, S3, and S4 (sacral levels). They travel in the pelvic and pudendal nerves and distribute to the clitoris or glans penis, the perineal region, and the rectal area. As a result of this distribution, sensations received from the clitoris, penis, labia, scrotum, or anus may be perceived as similarly pleasurable.
- Since nerves play a more essential role in the physical component of sexual performance in men than they do in women, only male physiology is described in this section.

— ***Erection*** Erection is a complex autonomic activity. Two hemodynamic events are of key importance:

  - There must be an initial dilatation of the inflow arteries.
  - There must then be a subsequent opening of shunt vessels to divert flow into the cavernous spaces. The extent of this dilatation is such that the hydrostatic pressure in the corpora cavernosum increases to near 100 mm Hg from its resting value near 20 mm Hg.

    - A set of nerves, separate from the cholinergic fibers and using vasoactive intestinal peptide (VIP) as neurotransmitter seems to be of critical importance in the erectile process.

- The process is largely under local control and does not require higher nervous centers. If the local spinal cord area (S2 and below) is injured, erection is possible by psychogenic factors in about 25% of patients.

— ***Ejaculation*** Ejaculation is a two-stage process involving *emission* and *expulsion.*

Emission. Emission involves contraction of internal structures such as the vas deferens, prostate, seminal vesicles, and the internal urethra. It is a reflex that is centrally coordinated. Patients with cord injuries in the sacral area will have emission if erection is possible.

Expulsion. Expulsion involves contraction of bulbar muscles to eject semen and seminal fluid. It involves motor nerves to the bulbar muscles. They are under strong local control. Therefore "expulsion" is possible in patients with cord injuries above S2. Since such patients cannot have emission, they experience "dry ejaculation."

Multiple Choice Review Questions

1. After an earthquake, Martin was isolated in his basement for 3 days before he was rescued. During this period, he had nothing to eat, though drinking water was available. Which one of the following hormonal and metabolic changes would you *not* expect to be taking place in him?
   a. Increased cortisol secretion
   b. Increased glucagon secretion to stimulate hepatic glucose release
   c. Decreased circulating free fatty acid levels
   d. Increased aldosterone secretion
   e. Increased plasma amino acid levels

2. A middle-aged man is exercising at 75% VO_2 max. Suppose that while he is exercising his parasympathetic nervous system suddenly failed completely to function. Which one of the following statements about possible consequences of this event is *true?*
   a. He would collapse because there would be no action potentials going to any skeletal muscle fibers.
   b. He would experience a large decrease in heart rate.
   c. There would be no significant cardiovascular effect on him, and he would just continue to exercise.
   d. He would suddenly produce a great deal of saliva and get an erection. The resulting stretch on his track suit would pull him into a crouch position, compress his chest, and impair his respiration.

3. A middle-aged man is exercising at 75% VO_2 max. Suppose that while he is exercising the sympathetic nervous system suddenly failed completely to function. Each one of the following statements about possible consequences of such an event is true *except:*
   a. He would lose a major part of the ability to direct blood flow preferentially to exercising muscle.
   b. His cardiac output would not be affected despite a decrease in his total peripheral resistance.
   c. His cardiac output would decrease because his heart rate would decrease.
   d. His plasma level of norepinephrine would decrease.

Directions: For each of questions 4 to 6 a group of plasma biochemistry values is given. Assume the following normal values:

$[Na^+] = 140$ mM/L; $[Cl^-] = 100$ mM/L; $[K^+] = 5$ mM/L

$[HCO_3^-] = 24$ mM/L; $[H^+] = 40$ nM/L; $P_{CO_2} = 40$ mm Hg

4. If the following plasma values were found: $[Na^+] = 140$, $[Cl^-] = 100$, $[HCO_3^-] = 14$, $[H^+] = 70$, $P_{CO_2} = 40$, what is the simplest acid-base diagnosis compatible with the data?
   a. Acute respiratory acidosis
   b. Chronic respiratory acidosis
   c. Metabolic acidosis
   d. Metabolic alkalosis
   e. Combined metabolic acidosis and respiratory acidosis

5. If the following plasma values were found: $[Na^+] = 150$, $[Cl^-] = 100$, $[K^+] = 8$, $[HCO_3^-] = 19$, $[H^+] = 40$, $P_{CO_2} = 32$, what would be the best explanation for the elevated $[K^+]$?
   a. High dietary intake of K^+
   b. Evidence of the operation of a reflex mechanism for the regulation of the unmeasured anion gap
   c. Evidence that a solution of KCl has been ingested
   d. Evidence of intracellular buffering of an acid load
   e. Evidence of excessive nervous activity

6. If the following values were found: $[Na^+] = 147$, $[Cl^-] = 107$, $[HCO_3^-] = 32$, $[H^+] = 47$, $P_{CO_2} = 63$. What is the simplest acid-base diagnosis compatible with the data?
   a. Acute respiratory acidosis
   b. Chronic respiratory acidosis
   c. Metabolic acidosis
   d. Metabolic alkalosis
   e. Combined metabolic acidosis and respiratory acidosis

7. Which one of the following statements regarding the menstrual cycle is *incorrect?*
   a. Increased estradiol and progesterone during the luteal phase feed back on the anterior pituitary to inhibit luteinizing hormone (LH) and follicle-stimulating hormone (FSH) release.

b. The corpus luteum is a short-lived endocrine tissue that secretes estradiol and progesterone.
c. Follicle-stimulating hormone (FSH) release during the early part of the cycle induces maturation of a primary follicle into a graafian follicle.
d. In humans, fertilization of an oocyte leads to secretion of chorionic gonadotropin (hCG) by the trophoblast cells, thus maintaining the corpus luteum.
e. Secretion of estradiol by the maturing graafian follicle prevents ovulation.

8. Your 10-year-old brother is a champion grade school track runner. His coach has illicitly given him testosterone. Which one of the following will *not* occur as a result of the testosterone administration?

a. His basal metabolic rate will be increased.
b. He will lose weight because of increased protein catabolism.
c. Spermatogenesis will be stimulated
d. Luteinizing hormone (LH) secretion will be inhibited
e. Development of secondary sex characteristics will be stimulated.

9. Which one of the following cells of the male reproductive system contains a half set of doubled chromosomes:

a. Spermatogonium
b. Secondary spermatocyte
c. Primary spermatocyte
d. Spermatid
e. Leydig cell

10. Which of the following is *not* a function of Sertoli cells?

a. To provide binding sites for LH
b. To secrete seminiferous tubule fluid
c. To secrete inhibin
d. To provide nourishment for developing sperm
e. To secrete androgen-binding protein

11. Each of the following statements relating to excitement, the first phase of the human sexual response cycle, is true *except*:

a. Changes in regional blood flow are one of the features of this early stage.
b. Excitement does not necessarily lead to the next more intense phase of feeling.
c. Excitement involves changes in cardiovascular and respiratory function.
d. Excitement can be initiated only by a physical stimulus.

Answers and Explanations to Multiple Choice Review Questions

Chapter 1: Cell Physiology

1. **Answer a:**

   Oxidative Phoshorylation occurs in mitochondria.

   b: NADH is one of the end products of the citric acid cycle (Krebs cycle) and represents an intermediate product, not the end product, of cell metabolism. NADH is one of the *reducing equivalents* that enters the *electron transport chain* where it is used to liberate free energy that is subsequently used to generate ATP by the process of *oxidative phosphorylation.*

   c: Acetyl CoA is not a product of the citric acid cycle. It is produced by different metabolic processes from each of carbohydrates, fats, or proteins. Each of them eventually forms acetyl CoA and uses it as the one entry into the citric acid cycle (Krebs cycle).

   d: Animals cannot convert fatty acids into glucose because they are unable to convert acetyl CoA into pyruvate.

   e: In the absence of oxygen (anaerobic metabolism) pyruvate is converted to lactate rather than acetyl CoA.

2. **Answer a:**

   The equilibrium potential of an ion is that electrical force that will exactly counteract the concentration gradient-force for transmembrane transport of that ion. An equilibrium potential of positive polarity will oppose the entry of positive charges into the cell. Hence, the concentration gradient for Ca^{++} must be from extracellular (high concentration) to intracellular space.

   b: The question of whether active processes are likely to be involved in the net transport of an ion across cell membranes can be addressed only by comparing the equilibrium potential of the ion with the resting membrane potential of the cell.

   c: In mammalian cells action potentials are electrical events by which the membrane potential moves transiently to a more positive value. The instantaneous electrical driving force for the entry (or exit) of any ion is the difference between the ion equilibrium potential and the membrane potential existing at that instant. With respect to the Ca^{++} equilibrium potential this difference decreases progressively during the upstroke of the action potential.

   d: Ca^{++} is in equilibrium across the cell membrane at its resting potential. If the resting membrane potential were made more positive by 80 mV, then voltage-gated Ca^{++} channels would open and Ca^{++} would move down its electrochemical gradient until the membrane potential reaches the Ca^{++} equilibrium potential.

3. **Answer b:**

   In view of the fact that the equilibrium potential for Sp^+ is different from the resting membrane potential, active transport mechanisms are likely to be involved in maintaining the transmembrane concentration difference for Sp^+. Its direction of pumping is derived from a comparison of the ion equilibrium potential for Sp^+ and the resting membrane potential of the cell. The existing concentration gradient tends to move Sp^+ into the cell and this movement would stop if the inside of the cell were at +350 mV. The cell interior is at only +280 mV, not sufficient to maintain the existing gradient. Hence, the electrochemical gradient would cause Sp^+ to enter the cell. An active pumping mechanism must remove Sp^+ from the cell to maintain the observed concentration gradient.

a: The active transport mechanism pumps Sp^+ out of the cell, as explained above, not into the cell.
c: This answer states electrophysiologic nonsense.
d: The equilibrium potential for any one ion species is determined only by the steady-state transmembrane concentration difference and is not affected by the presence of other ion species. Note, however, that the *resting membrane potential* of a cell is very much influenced by all of the ion species that are present on the two sides of the cell surface membrane.

4. **Answer a:**
Reduction of the concentration difference for Fd^{++} in this setting would most likely result from the transfer of Fd^{++} from the intracellular space to the extracellular space. In order to maintain steady-state charge neutrality within each region, such a change in Fd^{++} would lead to translocation of other positive ions (namely Sp^+) into the cell and of negative ions (namely Rh^-) out of the cell. These changes in the concentration gradient for Sp^+ and Rh^- will change their respective ion equilibrium potentials.

b: If Sp^+ moves into the cell, then the concentration gradient for this ion will be decreased, and its equilibrium potential will also decrease in magnitude.

5. **Answer d:**
Membrane cholesterol imparts mechanical rigidity to the phospholipid bilayer of the plasma membrane. It has no signal transduction properties.

a: Ion channels do participate directly in signal transduction across the plasma membrane. Two mechanisms are used. In some cases the change in membrane potential resulting from the ion current provides the mechanism for further events. In other cases the transported ion acts directly as a messenger and precipitates further events.
b: G-proteins are involved in signal transduction in a large number of cell surface receptors. When they are activated by ligand-receptor interaction they split off one of their subunits and use it to interact with an intracellular effector for the purpose of a designated biologic action.
c: Guanylate cyclase participates in transmembrane signal transduction as an enzyme in the formation of the second messenger, cGMP.
e: Adrenoceptors participate in signal transduction by specific binding of catecholamines, permitting subsequent activation of G-proteins and formation of second messengers.

6. **Answer a:**
Proteoglycans is a component of the extracellular matrix. Its functions relate to cell-matrix interactions.

b: Fibronectin fragments promote cell adhesion to the extracellular matrix and thereby play a significant role in wound healing.
c: In the absence of vitamin C proline hydroxylation is prevented and as a result the collagen pro-α chains that are synthesized are defective in that they fail to form a stable triple helix, the hallmark of collagen structure. Such defective chains are immediately degraded within the synthesizing cell, and no new collagen is formed.
d: Laminin is one of the first extracellular matrix proteins to be synthesized in a developing embryo. It is a strong stimulus for the formation of new capillary branches from existing capillaries.

7. **Answer c:**
Phospholipids do not permit free diffusion of ions and water-soluble molecules. They do permit relatively free diffusion of water molecules.

a: Proteins may penetrate the bilayer.
b: Phospholipids are amphipathic. Amphipathic means consisting of one part that is water-insoluble and another part that is water-soluble.
d: Proteins permit selective transport of ions.

8. **Answer d:**

An increase in extracellular potassium concentration will decrease the transmembrane concentration gradient for K^+, make the potassium equilibrium potential less negative, and make the resting membrane potential less negative because the $[K^+]$ gradient has a profound influence on resting membrane potential (by virtue of the high membrane permeability to K^+). Moving the resting membrane potential to a less negative value without changing anything else will have these effects:

1. In excitable cells the resting membrane potential will be closer to the threshold potential, and this will cause such cells to generate action potentials in response to weaker stimuli. The cells will be more excitable.
2. The amplitude of an action potential is the difference between its peak value (near the Na^+ equilibrium potential) and the resting membrane potential. If the resting membrane potential moves to a less negative value and nothing else changes, then the action potential amplitude will be decreased.

9. **Answer c:**

Inositol trisphosphate (IP_3) cannot be enzymatically cleaved to produce arachidonic acid.

a: Protein kinase C requires Ca^{++} for full activation.
b: Protein kinase C is activated in part by diacylglycerol (DAG).
d: Protein kinase C requires membrane lipids for full activation.
e: The biologic effects of inositol trisphosphate (IP_3) require the presence of intracellular Ca^{++} receptors.

Chapter 2: Blood

1. **Answer a:**

Antibodies are produced by lymphocytes.

b: Basophils constitute no more than 1% of all leukocytes.
c: Neutrophils have an average half-life of about 6 hours, requiring that more than 100 billion be produced per day in order to maintain steady state.
d: The major site of erythropoietin production is cells near the proximal convoluted tubules in the kidney.

2. **Answer d:**

Activation of the clotting cascade via the extrinsic pathway differs from the intrinsic pathway in that it requires the presence of factor VII.

a: In spite of its name, the extrinsic path is triggered in response to blood contact with collagen in the blood vessel adventitia.
b: No matter which path initiates the clotting process, prothrombinase formation requires membrane phospholipid (and Ca^{++}) to form the Xa/Va complex.
c: The enzymatic cleavage of fibrinogen to form fibrin is performed by thrombin, a protease whose structure is not influenced by the triggering event in the clotting cascade. Not even the structure of prothrombinase (the enzyme that produces thrombin from prothrombin) is affected by the nature of the antecedent enzymatic paths.

3. **Answer b:**

Warfarin competes with Vitamin K in the hepatic synthesis of vitamin K-dependent clotting factors.

a: Heparin is a cofactor for antithrombin III and therefore assists in blocking the action of thrombin on fibrinogen.
c: The site of action of aspirin is the cyclooxygenase pathway of phospholipid metabolism in platelet membranes. Aspirin inhibits the formation of the endoperoxides and therefore retards clot formation. Two things should be noted:

1. Aspirin-mediated inactivation of cyclooxygenase is irreversible during the lifetime of a given platelet.
2. Aspirin treatment prolongs bleeding time but does not normally cause frank bleeding. This suggests that other mechanisms must compensate for the aspirin-mediated defect in thromboxane A_2 synthesis.

d: The substrate for t-PA is plasminogen, not thrombin.

4. **Answer e:**

The platelets are activated, and the release action is initiated after vascular trauma.

a: Tissue factor participates in the clotting cascade without being chemically changed. That is, it does not need to be activated. (This may seem like semantics, but remember that examinations are often composed by people who specialize in counting beans or measuring margins.) When blood comes in contact with tissue factor and its tightly associated membrane phospholipid, then a complex is formed with factor VII-activated, provided Ca^{++} is present.

b: The anticoagulant, heparin, would prevent clot formation and would therefore not be good for the health after vascular trauma.

c: The statement does not correctly describe the role of collagen in the events that follow vascular injury, and it ascribes an incorrect function to thrombin.

d: PGI_2 is not produced in collagen fibrils, and PGI_2 counteracts platelet aggregation; it does not promote it.

5. **Answer b:**

The function of vitamin K is to insert residues of gamma-carboxyglutamic acid (GLA) into four clotting factors: prothrombin and factors VII, IX, and X. GLA residues form a negatively charged region that binds Ca^{++}. These bound Ca^{++} ions form bridges between the clotting factors and adjacent phospholipids.

a: Thrombin requires no cofactors for its enzymatic cleavages of fibrinogen.

c: Vitamin K is not necessary for normal platelet function.

d: Vitamin K is not necessary for normal differentiation of platelets in the bone marrow.

e: Vitamin K is found in green leafy plants like alfalfa.

6. **Answer a:**

The objective of blood coagulation is formation of fibrin from fibrinogen by the action of thrombin which must be generated from prothrombin by the enzymatic activity of prothrombinase.

b: This states that fibrinogen is an enzyme that transforms prothrombin to thrombin, and blood clots are formed directly from thrombin.

c: This shows prothrombinase as an intermediate step in the conversion of prothrombin to thrombin. Prothrombinase is a complex that permits factor Xa to perform two sequential proteolytic cleavages in prothrombin to yield thrombin, the final active protease of coagulation.

d: Collagen is not produced from platelets, and collagen does not induce collagen fibers to activate platelets.

7. **Answer c:**

The hematocrit is defined as the fraction of total blood volume that is occupied by erythrocytes. The total volume of leukocytes and platelets is insignificantly small in this regard and may be neglected. Therefore the erythrocyte volume is the difference between blood volume and plasma volume. That is,

$$\text{hematocrit} = (5 - 3)/5.$$

a: This represents the fraction of blood volume that is occupied by plasma; namely 3/5.

b: This represents the volume of erythrocytes as a fraction of plasma volume; namely $(5 - 3)/3$. This is not generally a useful number. However, it might be of use in some forms of data analysis.

d: This represents the ratio of blood volume to plasma volume, namely 5/3. This is not generally a useful number. However, it might be of use in some forms of data analysis.

8. Answer d:

Antithrombin III does not act in the manner that is described. Instead, it (with heparin as a cofactor) forms a stable complex with serine residues to block several serine proteases (IX_a, X_a, XI_a, and XII_a) completely and irreversibly so that they cannot act as clotting factors. The results is that thrombin cannot be formed.

a: Thrombin, like many of the clotting enzymes, is a serine protease.
b: Vitamin K is required for the formation and insertion of Ca^{++}-binding, paired residues of gamma-carboxy glutamic acid (GLA) into four essential clotting factors.
c: Conversion of soluble fibrinogen to insoluble fibrin requires the proteolytic action of thrombin.

9. Answer c:

This does not describe correctly the action of warfarin. Warfarin resembles vitamin K and competes with it in liver cells that produce vitamin K- dependent clotting factors. Warfarin is not an effective anticoagulant when it is added to blood in a test tube because such blood already contains the correctly configured vitamin K-dependent proteases.

a: Aspirin inhibits the formation of the endoperoxides and therefore interferes with normal phospholipid metabolism in platelet membranes and retards clot formation. These actions of aspirin are mediated by inactivation of cyclooxygenase and are irreversible during the lifetime of a given platelet.
b: Heparin acts as a cofactor for antithrombin III, allowing it to block the active serine sites on the activated forms of factors IX, X, XI, and XII completely and irreversibly. This inhibits the formation of prothrombinase and is therefore common to all coagulation paths whether they are triggered by contact with a test tube surface or by blood vessel trauma.
d: Antithrombin III blocks the formation of prothrombinase and thereby retards the formation of clots. Lack of antithrombin III tends to favour clotting.

Chapter 3: The Autonomic Nervous System

1. Answer d:

The sympathetic division of the ANS is characterized by thoracolumbar outflow from the spinal cord.

a: The postsynaptic receptors of sympathetic neuroeffector junctions are of the adrenergic type and are not affected by muscarinic antagonists.
b: All preganglionic fibers in the autonomic nervous system are cholinergic.
c: It is the parasympathetic postganglionic fibers that are short, not the sympathetic postganglionic fibers.
e: The vagus nerves are a major outflow path in the parasympathetic nervous system, not in the sympathetic system.

2. Answer b:

Unloading of cardiovascular mechanoreceptors is an afferent signal whose normal reflex response includes sympathetically mediated vasoconstriction and sympathetically mediated release of epinephrine from the adrenal medulla. Neither response is possible when the spinal roots are cut.

a: The arterioles would still be capable of constricting in response to mechanical stretch or blood-borne vasoconstrictor substances like vasopressin, angiotensin II, or endothelin.
c: Sympathetic excitation is only one of the ways by which heart rate can be increased. Heart rate can also be increased by inhibition of cardiac parasympathetic neurons in the vagus nerve. The parasympathetic neurons are left intact after this operation.
d: Sympathetic excitation is only one of the ways by which cardiac performance can be increased. The other ways are increased preload (Starling's Law), increased afterload (Anrep Effect), and increased heart rate (Bowditch Effect).

3. Answer d:

Parasympathetic neurons mediate the contraction of the detrusor muscle (smooth muscle) of the urinary bladder.

a: Small doses of atropine do not interfere with neural stimulation of skeletal muscles like the diaphragm because their cholinoreceptors are nicotinic.
b: Acetylcholine modifies the intrinsic activity of the heart but does not initiate contractions. Therefore blocking acetylcholine action will not stop the heart from contracting.
c: Acetylcholine causes pupillary constriction. Blocking its action would cause pupillary dilation.
e: Postganglionic vasoconstrictor neurons are sympathetic fibers that release norepinephrine. Atropine does not block the action of norepinephrine, nor does it block the action of acetylcholine on postganglionic neurons because this is mediated by nicotinic receptors.

4. Answer c:

If X blocks all β adrenoceptors without affecting the vasoconstrictor action of epinephrine, then such action cannot be caused by activation of β adrenoceptors. It must be the result of an α activation effect of epinephrine.

a: A nonspecific beta antagonist will block all β adrenoceptors, including β_2.
b and e: The net effect of α adrenoceptor activation is increased cytosolic $[Ca^{++}]$ and vasoconstriction, not vasodilatation.
d: The stem of the question states that epinephrine has cardioaccelerator effects.

5. Answer c:

The detrusor muscle of the urinary bladder is smooth muscle, whereas the diaphragm is striated muscle. The action of smooth muscle is often initiated by the hypothalamus; the action of striated muscle is not influenced by it.

a: As it is usually defined, the autonomic nervous system consists of efferent nerves only. They are conveyors of action potentials rather than initiators of biologic activity.
b: The ganglia act as relay stations and modulators of autonomic nervous activity. They do not initiate it.
d: The cholinoreceptors of the urinary bladder are muscarinic, whereas those of the diaphragm are nicotinic.
e: The adrenergic system exerts little influence on bladder function.

6. Answer c:

The hypothalamus acts as a reflex center for autonomic function.

7. Answer d:

Preganglionic fibers in both the sympathetic and parasympathetic divisions are nicotinic fibers.

Chapter 4: Muscle

1. Answer c:

Smooth muscle does not show a striated pattern because it has a less regular organization of thin and thick filaments than skeletal muscle.

a: The statement describes a functional aspect of smooth muscle, not a structural aspect.
b: There are no T-tubules in smooth muscle cells. However, their presence in striated muscle is not the reason for the striated appearance of the tissue, and their absence from smooth muscle is not the reason for the lack of striations in this tissue.
d: This is a description of the relative numbers of mitochondria in smooth muscle versus striated muscle. The presence of mitochondria is not responsible for the striated appearance.

2. Answer d:

Energy released from the hydrolysis of ATP is utilized in the rotation of myosin crossbridges.

a: ATP-consuming Ca^{++} transport across the membranes of the sarcoplasmic reticulum (SR) moves Ca^{++} into the SR, not out of it.
b: Acetylcholine activates muscarinic membrane receptors allowing Na^+ and K^+ to follow their electrochemical gradients more easily. They move by passive mechanisms to depolarize the membrane and generate an endplate potential.
c: Crossbridge release does not require energy. It does, however, require the presence of molecular ATP.

3. **Answer a:**

In skeletal muscle the relationships among action potential duration, stimulus-contraction latency, and duration of twitch tension are such that electrical excitation and its attendant twitch are possible before the tension from the preceding excitation has returned to baseline. As a result, rapidly applied electrical stimuli will lead to temporal summation of "new" twitch tension upon the remainder of the preceding twitch tension.

b: The statement implies that a higher frequency of action potentials always leads to a shorter duration of action potentials. It also implies that muscle excitation and development of tension are related to the release of bound extracellular calcium.

c: The statement implies that a higher frequency of action potentials always leads to a shorter duration of action potentials. It also implies that the number of actomyosin crossbridges is related to extracellular potassium concentration.

d: Action potentials cannot fuse. The reason for this is the inability of membrane Na^+ channels to be activated again until the membrane potential has repolarized to a value near threshold potential. This imposes a refractory period on excitable cells.

4. **Answer c:**

When at least 3 of the 4 Ca^{++} binding sites on the calmodulin molecule are occupied, the molecule undergoes a conformational change, allowing it to bind to myosin light-chain kinase (MLCK) and thereby activate it. Activated MLCK is then able to catalyze ATP hydrolysis. ATP hydrolysis leads to crossbridge formation and conformational changes in the myosin head so that the power stroke can follow.

a: Thick filaments do not shorten in the presence of Ca^{++}, and shortening of these filaments is not the basis of muscle contraction.

b: This answer is nonsense and should be instantly recognized as such.

d: Smooth muscle does not contain troponin. It uses the related compound, calmodulin.

5. **Answer b:**

The plateau of the cardiac muscle action potential is maintained by Ca^{++} currents. $I_{Ca,L}$ moving into the cell drives E_m towards positivity. Net current moving out of the cell as a result of Na^+-Ca^{++} exchange operating in the Ca^{++}-in mode (3 Na^+ out per Ca^{++} in) provides the balancing drive towards negativity. In the absence of extracellular Ca^{++}, K^+ currents would be unopposed, the cell would repolarize quickly, its action potential duration would be reduced, and tetanization may be possible.

a: It is true that of the three muscle types, smooth muscle is most dependent on extracellular Ca^{++} and would be the first to stop contracting in its absence. The second part of the statement, however, is not true. Smooth muscle would not be the last to stop generating action potentials in response to electrical stimulation. In most smooth muscle, action potentials result from influx of Ca^{++}, whereas action potentials in striated muscle result from Na^+ influx. As a result smooth muscle would be the first of the three types to stop generating action potentials. Note that smooth muscle contraction can occur in the absence of action potentials because in some smooth muscle the amount of Ca^{++} required for contraction is less than that required for generating an action potential.

c: See a.

d: The active Ca^{++} pump in the cardiac sarcolemma transports Ca^{++} out of the cell, not into it as is required for actomyosin crossbridge formation.

6. **Answer d:**

The strength of a muscle contraction is directly proportional to the total number of motor units activated simultaneously.

a: During rest and activity levels that are at most moderate and include rest periods, fatty acids are favoured as metabolic fuel. Fatty acids have high energy storage capacity (typically 138 mol of ATP per mol of fatty acid).

b: Anaerobic metabolism requires up to 18 times more fuel to produce the same amount of ATP.

c: Extraction of energy from fats is a relatively slow process and cannot keep pace with the demands of heavy exercise. For such exercise, glucose liberated from glycogen is the dominant energy substrate.

7. **Answer c:**

Inhibition of glycogenolysis would eliminate a process that can rapidly supply ATP, even though in relatively limited amounts. Consequently, types of muscle activity that require rapid replenishment of ATP would be compromised. Levels of activity that can be sustained by the slower processes of fat metabolism can be performed without compromise.

a and b: Neither is related to the availability of energy substrates.

d: Action potential propagation depends on the presence of voltage-gated Na^+ and K^+ channels and the maintenance of appropriate concentration differences for Na^+ and K^+ across the cell membrane. Such differences can be maintained with amounts of ATP derived from fatty acids.

8. **Answer d:**

Smooth muscle, including vascular smooth muscle, is dependent on Ca^{++} entry from extracellular sources. Inhibition of channels that convey Ca^{++} flux will therefore inhibit smooth muscle excitation and inhibit blood vessel constriction.

a: Ca^{++} uptake into the sarcoplasmic reticulum is an active process driven by a variety of Ca^{++} ATPases. It cannot be passive because the electrochemical gradient for Ca^{++} favours exit from the sarcoplasmic reticulum.

b: There are no physiologic mechanisms by which an increase in cardiac performance would cause a decrease in coronary artery constriction. Note, however, that an increase in the workload of the heart will lead to an increase in coronary flow. Inhibition of L-type Ca^{++} channels would, indeed, reduce the duration of the action potential plateau, but it would decrease cardiac performance because diminished Ca^{++} entry would have long-term deleterious effects on sarcoplasmic Ca^{++} stores.

c: Inhibition of Ca^{++} entry into vascular smooth muscle would prevent muscle constriction, not promote it.

9. **Answer a:**

The I bands are regions composed of thin filaments only; the H band is composed of thick filaments only, and the sections of A band on each side of the H band are regions where thin and thick filaments overlap. The total width of the A band is set by the length of the thick filaments. During contraction, the thin filaments on each side of the H band "walk" along the thick filaments towards the H band.

b: The width of the H band decreases.

c: The width of the I band decreases.

d: The Z lines move closer to the A band.

10. **Answer d:**

Release of Ca^{++} from the sarcoplasmic reticulum is as essential a source of Ca^{++} in skeletal muscle as it is in cardiac muscle. However, the mechanisms by which this release is triggered differ between the two tissues.

a: A motor nerve fiber that exits from the spinal cord normally innervates several muscle fibers.

b: All the muscle fibers innervated by one spinal efferent nerve fiber and its branches are collectively described as a motor unit.

c: Small muscles that react rapidly and require fine control have a large number of muscle fibers per spinal efferent.

Chapter 5: Respiratory Physiology

1. **Answer e:**

Since most of the O_2 carried in blood is carried in association with hemoglobin, the hemoglobin dissociation curve is the focus for answering this question.

Increasing the fraction of red cells in blood will not alter hemoglobin saturation but will increase the number of hemoglobin molecules per 100 mL of blood.

a: Breathing pure O_2, although it would raise $Pa{O_2}$ to a value near 700 mm Hg, would not increase hemoglobin saturation above 100%, which is the level reached at $P{O_2}$ near 100 mm Hg.

b and c: An increase in body temperature as well as an increase in $[H^+]$ would shift the dissociation curve to the right. The meaning of such a shift is that O_2 carrying capacity decreases at any level of $Pa{O_2}$.

d: Increasing alveolar ventilation could not raise hemoglobin saturation above 100%.

2. **Answer d:**

At the end of expiration the lungs and chest wall assume an equilibrium volume (residual volume) at which the intrapleural pressure prevents the thoracic cavity from expanding and the lungs from collapsing further.

a: The pressure inside the lungs is either alveolar pressure or pulmonary interstitial pressure. Intrapleural pressure is the pressure that exists outside the lungs, in the virtual space between the visceral and parietal pleura.

b: The difference between alveolar and atmospheric pressures is the pressure gradient that drives airflow between mouth and alveoli. It is not intrapleural pressure.

c: If the residual volume (RV) is increased, then the lumgs must be contracted less than normal and less elastic force must be exerted on the inside of the chest wall. In this setting the intrapleural pressure cannot be more negative than normal. (Most often an increase in RV is caused by premature closure of the smaller airways and is observed in diseases in which there is an increased airway resistance.)

e: An intrapleural pressure more positive than alveolar pressure would compress the lungs, and it would not be possible to draw air into them.

3. **Answer c:**

In the capillaries CO_2 follows its partial pressure gradient and diffuses into red cells. In the presence of carbonic anhydrase it is combined with water, forms H_2CO_3, and subsequently dissociates into H^+ and HCO_3^-. HCO_3^- diffuses out of the red cells, down its concentration gradient. Cl^- enters the red cell in response to the electrical gradient created by the diffusion of HCO_3^-. This is called the *chloride shift.*

a: The concentration of Cl^- in venous blood is not lower than in arterial blood.

b: Red cells would gain water only if there were an osmotic gradient for water movement. This would require transport of osmotically active particles into red cells. No such transport occurs. Gas diffusion takes place, but gases are not osmotically active. The addition of Cl^- ions does create an increase in osmotic force. However, the increase is balanced by the departure of an equal number of HCO_3^- ions.

d: The concentration of Cl^- in venous bloood is higher than in arterial blood because Cl^- is gained in capillary transit.

4. **Answer c:**

When the intrapleural space is permitted to equilibrate with the atmosphere then it is not possible to develop, by physiologic means, a pulmonary transmural pressure gradient that will expand the lung. Accordingly, the affected lung will collapse to a very small residual volume.

a and d: The residual volume in the affected lung would be very small.

b: RV can never be greater than FRC.

5. **Answer b:**

Alveolar ventilation = (tidal volume − deadspace) × Respiration rate = (300 − 100) × 20 = 4,000 mL/min = 4 L/min.

6. **Answer a:**

The right (affected) lung will have lttle or no ventilation, but its perfusion will be closer to normal than its ventilation. If overall alveolar ventilation were normal, then it could be that only if ventilation of the left (unaffected) lung were almost doubled from the normal level. As a result the overall ventilation/perfusion ratio must be elevated.

c: Air does not magically cross the visceral pleural membrane from the interpleural space. Even if it did cross in that manner it could not increase ventilation above that accomplished by drawing air into the alveoli via the trachea.

7. **Answer d:**

The collapsed lung should contribute blood of low Po_2 and high CO_2 and influence pulmonary venous blood accordingly.

8. **Answer c:**

Pao_2 = (atmospheric pressure-water vapour pressure) × fraction of O_2 in alveolar air $= (500 - 47) \times 0.14 = 63$ mm Hg. Hemoglobin saturation is near 90%, still on the upper, flat part of the dissociation curve at a Po_2 of 63.

a: Po_2 depends on prevailing atmospheric pressure as well as on the fractional composition of air.

b: The vapour pressure of water depends only on the temperature. It is a measure of the energy with which water molecules attempt to escape to the gaseous phase from the surface of the liquid phase. It is not influenced by atmospheric pressure.

d: The prevailing Pao_2 will be 63 mm Hg. At that Pao_2 hemoglobin saturation will be near 90%.

9. **Answer b:**

FVC will be decreased. The reason is thought to be premature closure of airways towards the end of a full expiration.

a and c: During an attack all indices of expiratory flow rate are markedly reduced. This includes FEV_1 (a) and FEV_1/FVC (c) even though FVC is reduced.

d: Total lung capacity (TLC) is generally increased during an asthma attack. The reason for this may be a loss in elastic recoil.

10. **Answer d:**

If the surface tension were reduced then less pressure would be required across the alveolar wall to keep them expanded. That is, the intrapleural pressure would have to be less negative at any one value of lung volume, and the static pressure-volume curve of the lungs would be shifted to the right.

a: As explained above, the static pressure-volume curve of the lungs would be shifted to the right.

b: A rightward shift of the pressure-volume curve would result in an elevated volume at zero pressure across the alveolar wall (i.e., the residual volume).

c: Alveolar pressure would, indeed, decrease because less pressure is required to distend alveoli when surface tension is reduced.

11. **Answer b:**

The ventilatory response to hypercapnia is linear over the full range of physiologic levels of Pco_2.

a: Peripheral chemoreceptors show a poor response to changes in O_2 when Pao_2 is above 50 mm Hg and when $Paco_2$ is normal. However, they are very responsive to changes in Pao_2 in any range below 50 mm Hg and moderately responsive to changes in Pao_2 when, simultaneously, $Paco_2$ is above normal (see Fig. 5.9 [B]).

c: The respiratory control centers do NOT stop functioning when Po_2 levels fall below 60 mm Hg.

d: As explained under (a), changes in Pao_2 do provide a stimulus for chemoreceptors, particularly the carotid chemoreceptors. The nature of the transduction process is not yet known, but it does occur within the neuronal ending and does not require a change in the hematocrit.

Chapter 6: Cardiovascular Physiology

1. **Answer a:**

Increased extracellular $[K^+]$ will make the resting membrane potential of myocytes less negative and thereby decrease the electrochemical force on Na^+ ions. (In phase 4 of the cardiac action potential this force is equal to the difference between the Na^+ equilibrium potential and the resting membrane potential). If the force driving Na^+ ions is reduced, then the rate of Na^+ entry in phase 0 cannot be higher than normal, and the action potential upstroke cannot be steeper than normal.

b, c, and e: These are true at normal extracellular $[K^+]$ and would not be affected by moving E_m slightly towards the positive direction.

d: Resting membrane potential would be reached a little sooner during repolarization if it is not as negative as it normally is. The difference, if it exists, would be very small.

2. **Answer d:**

Compared to all other pacemaker cells, SA nodal pacemaker cells normally show a steeper rate of diastolic membrane potential rise. Hence one of them will be the first to reach the threshold for generation of an action potential.

d: SA nodal pacemaker cells are richly supplied with autonomic nerves and activity in these nerves does modulate the rate of pacemaker activity. However, it is only under extreme circumstances that activity in these nerves can decrease SA nodal activity to a rate that is so low that the next pacemaker in the hierarchy takes over.

e: It is true that pacemaker cells do not contain contractile proteins, but that has nothing to do with the fact that of all pacemakers, one of the SA nodal cells is the first to be spontaneously depolarized.

f: This region is not the first to show contractile activity. In fact it does not ever show contractile activity because the pacemaker cells do not have thin and thick filaments.

h: The AV node, not the SA node, could be considered, more correctly, to be the origin of the Purkinje conduction system.

3. **Answer e:**

Soon after birth, pressures on the left side of the circulation rise above those on the right side, and from then on aortic pressure is higher than pulmonary arterial pressure throughout the cardiac cycle. As a result blood would flow from aorta to pulmonary artery at all times in the cardiac cycle.

4. **Answer d:**

It is important to recognize that venous return will be increased on first assuming the head-down position, but only momentarily. Once steady state has been reached and, assuming that there is not a continuing state of nervousness, cardiac output and venous return will be normal. However, during the maintenance of the head-down position, the top of the head is several centimeters below the reference level for cardiovascular measurements. As a result hydrostatic pressures in the head will be increased by an amount ρGh above those created by cardiac and peripheral vascular factors.

5. **Answer d:**

Gravitational influences during the head-down position will cause the heart to move physically in such a way that its apex rotates towards leads I and III and away from lead II. This will result in more parallel alignment of the average QRS vector with lead I and closer to perpendicular alignments with both other leads. The QRS scalar amplitude will be increased in lead I and decreased in both other leads.

6. **Answer d:**

Stimulation of the carotid sinus nerve at a higher-than-normal frequency is interpreted centrally as increased activation of carotid sinus mechanoreceptors. Reflex responses that depress cardiovascular function are the normal sequel to excitatory stimuli.

a: Decreased blood pressure in the carotid sinus would elicit reflex responses capable of increasing blood pressure. These would include vasoconstriction and tachycardia, not vasodilatation and bradycardia.

b: Cutting the carotid sinus nerve would eliminate action potentials from that sensory nerve. This would be centrally perceived as a decrease in carotid sinus pressure, requiring systemic vasoconstriction and tachycardia.

c: Cutting the vagus nerves will interrupt both afferent fibers (which would be expected to lead to a pressor response) and efferent fibers to the cardiac atria. The vagi contain no fibers that innervate peripheral blood vessels. The normal, resting heart rate is lower than the intrinsic pacemaker rate as a result of vagal influences. As countless student experiments in cardiovascular physiology have shown, the dominant effect of vagotomy is an increase in heart rate.

e: Occlusion of the common carotid artery would erect a barrier between the carotid sinus mechanoreceptors and the left ventricle. This will decrease carotid sinus pressure and be centrally perceived as a condition requiring systemic vasoconstriction and tachycardia.

7. **Answer d:**

Of the items listed, leukocyte transport is the only one without influence on the forces that modify the rate of fluid exchange between blood and interstitial fluid across the capillary wall.

a: Fluid removal via lymphatic vessels tends to decrease interstitial hydrostatic pressure and thereby influence the gradient for hydrostatic pressure that tends to push fluid out of the capillary into the interstitial space.

b: Protein metabolism may influence plasma protein concentration and, with that, the plasma protein osmotic pressure. Plasma protein osmotic pressure is one of the factors affecting the rate at which interstitial fluid is being drawn into the intravascular space.

c: The pressure generated by the left ventricle can influence capillary hydrostatic pressure and thereby influence the transmural forces governing fluid exchange between intravascular and interstitial compartments.

8. **Answer e:**

Dilatation of arterioles will certainly affect capillary hydrostatic pressure. If the total number of dilated arterioles is small enough so that mean arterial blood pressure is not changed, then hydrostatic pressure in capillaries that lie beyond dilated arterioles will be increased because less energy is lost to precapillary friction. If the number of dilated arterioles is large enough to decrease mean arterial blood pressure, then capillary hydrostatic pressure will probably decrease. It is possible, but unlikely, that there may be no change in capillary hydrostatic pressure if there happens to be an exact balance between the decrease in systemic arterial pressure and the relatively increased pressure downstream from a decreased hydraulic resistance.

a: Recruitment (opening of previously closed capillaries) increases the area of endothelium that is available for blood/tissue exchange.

b: Recruitment is a result of relaxation of precapillary sphincters.

c: Recruitment results in decreased resistance to blood flow.

d: Relaxation of vascular smooth muscle surrounding arterioles results in decreased resistance to blood flow.

Chapter 7: Body fluids and Electrolytes

1. **Answer c:**

By definition an isotonic solution is one that causes no change in the volume of a cell that is placed into it.

a: If the osmolality of the external solution is derived from a solute to which the erythrocyte membrane is permeable, then the solute would move into the cell, down the concentration gradient, and drag water with it.

b: Water moves from a hypotonic solution to one that is hypertonic.

d: See a.

2. **Answer d:**

Saline solution contains no protein and must therefore decrease plasma protein concentration.

a: Hypertonic solutions will cause an increase in vasopressin secretion so that increased water reabsorption in the renal collecting tubule may restore extracellular hyperosmolality towards normal.

b: Hypertonic fluids will draw water out of the intracellular space.

c: NaC1, the solute constituent of saline, does not exert an osmotic effect across the capillary membrane. Hence in this setting interstitial fluid volume would increase by the same fraction as extracellular fluid volume.

3. **Answer c:**

Inulin has the property of being freely filtered and then neither secreted nor reabsorbed. As a result, any substance that has a clearance greater than C_{INULIN} is likely to be filtered and secreted.

a: If Diodrast were reabsorbed from the tubule after having been filtered at the glomerulus then its clearance would be less than that of inulin.

b: If a molecule is too large to be filtered at the glomerulus, then it is also too large to be secreted at more distal tubular sites.

d: A substance that is filtered and then neither secreted not reabsorbed would have a clearance identical to the clearance of inulin.

4. **Answer a:**

The solution must be capable of drawing water from the intracellular space and electrolyte solution from the interstitial space.

b: If b were correct then the infusate would expand intracellular, extracellular, and interstitial fluid volumes in the proportions they, respectively, represent of total body water.

c: If c were correct then intracellular volume would be increased more than extracellular volume, and interstitial volume would be increased by a greater fraction than extracellular fluid volume.

d: If d were correct then intracellular volume would be increased more than extracellular volume, and interstitial volume would be increased by a smaller fraction than extracellular fluid volume.

5. **Answer b:**

Increased NaC1 transport in the ascending limb of the Loop of Henle tends to increase medullary interstitial osmolality and thereby promote water reabsorption in the collecting duct. Such reabsorption will increase solute concentration in urine.

a: Such a condition would tend to increase plasma aldosterone levels and cause increased reabsorption of Na^+ and associated anions. This would tend to decrease urine solute concentration.

c: Increased extracellular fluid volume would lead to diminished plasma levels of vasopressin. Reduction in vasopressin levels will cause reduction in water reabsorption from the collecting duct and thereby decrease solute concentration in urine.

d: A decrease in afferent arteriolar resistance will increase GFR without changing filtration fraction. The result of increased GFR will be increased tubular flow rates and consequent reduction in reabsorption.

6. **Answer d:**

When a substance like PAH that is filtered and secreted but not reabsorbed reaches its Tm for tubular secretion, further increases in its plasma concentration do not yield further increases in tubular secretion. It behaves, above those plasma levels, like a substance that is filtered but not reabsorbed or secreted. Inulin behaves like that at physiologic concentrations.

a: If the plasma concentration increases then the clearance must have decreased.

b: Glucose is filtered and reabsorbed by a Tm-limited mechanism. The value of Tm is below normal plasma glucose levels. Therefore normal glucose clearance is zero.

c: If the plasma concentration increases then the clearance must have decreased.

7. **Answer b:**

Increased afferent arteriolar resistance would increase renal vascular resistance without changing the factors that determine intraglomerular flow distribution between filtrate and efferent arteriole.

a: Increased constriction will increase local resistance and is more likely to decrease flow than increase it.

c and d: Filtration fraction is the ratio of GFR to renal plasma flow. That ratio will not change when there is a greater degree of afferent arteriolar constriction.

8. **Answer d:**

The collecting tubule is the major site for vasopressin V_2 receptors and aggrephores, the vesicles that contain water channels.

a: The proximal convoluted tubule is a major site for reabsorption and secretion of solutes by both active and passive transport mechanisms. Water follows readily in the direction of osmotic gradients that result from solute transport. Vasopressin has no effect in this segment.

b: The thin descending limb of the Loop of Henle has no especially remarkable transport properties. In this area, water and electrolytes move across the tubule wall in the direction of concentration gradients. The segment is highly permeable to water and moderately permeable to most solutes.

c: The thick ascending Loop of Henle is characterized by active electrolyte reabsorption without accompanying water reabsorption. This section of the nephron is impermeable to water. As a result it is possible to create a highly concentrated interstitium.

9. **Answer a:**

The proximal convoluted tubule is a major site for reabsorption and secretion of solutes by both active and passive transport mechanisms. Water follows readily, in the direction of osmotic gradients that result from solute transport. Approximately 80% of glomerular filtrate is reabsorbed in the proximal convoluted tubule.

b: Reabsorption in the thin descending limb accounts for only about 5% of filtered load.

c: The thick ascending limb is impermeable to water.

d: Less than 5% of the filtered load is reabsorbed in the collecting tubule.

10. **Answer c:**

The thick ascending Loop of Henle is characterized by active electrolyte reabsorption without accompanying water reabsorption. This section of the nephron is impermeable to water. As a result it is possible to create a highly concentrated interstitium.

a: Reabsorption in the proximal convoluted tubule is isoosmotic.

b: Reabsorption in the thin descending limb is isoosmotic under most circumstances. It is true that the osmolality of the tubular fluid increases towards the bend in the Loop. However, this is not due to any special transport properties of the thin descending epithelium.

d: It is true that the tubular fluid can get concentrated as it flows along the collecting tubule, and water is reabsorbed in response to the high transmural osmotic gradient if vasopressin is present. Nevertheless, since the collecting tubule is not responsible for creating the highly concentrated interstitium, it is not called a concentrating segment.

Chapter 8: GI Physiology

1. **Answer d:**

Local circumferential contractions, called segmenting contractions, are the dominant motility pattern in the small intestine. Such contractions are not coordinated with activity above or below the contracting segment. As a result intestinal contents are pushed to regions both above and below the contracting segment. During the period of relaxation they flow back. This, as all kneaders of bread dough have discovered, causes mixing.

a and b: The primary effect of the infrequently occurring peristaltic waves and the even less frequent migrating motor complex is transport of intestinal contents towards the colon.

c: Haustrations are segmenting contractions of the colon. They differ from those seen in the small intestine in two ways: haustra are formed because the contracting and receiving segments in the colon each maintain their respective states for longer periods of time, and the sites of contraction repeat so regularly that they appear to be anatomically fixed.

2. **Answer b:**

Cells near the neck of the salivon secrete a fluid that is rich in K^+ and HCO_3^-. As a result, salivary K^+ and HCO_3^- increase in concentration as the acinar secretion flows towards the neck. Electroneutrality is maintained by decreasing $[Cl^-]$.

3. **Answer c:**

Gastric acid is secreted from the parietal cells of oxyntic glands. These cells have muscarinic receptors of the M_1 type, gastrin receptors, and histamine H_2 receptors. The first two increase HCl secretion by an IP_3-dependent mechanism. The predominant effector pathway of the H_2 receptor is via cAMP.

a: Histamine is not the only mechanism for secreting gastric acid.

b: Phospholipids do not have agonist properties.

d: Cimetidine and atropine are antagonists of receptors that promote HCl secretion. Cimetidine is an antagonist of H_2 receptors. Atropine is an antagonist of muscarinic receptors.

e: Angiotension II is a potent vasoconstrictor agent and does not enhance HCl secretion.

4. **Answer b:**

The source of HCO_3^- is the ductule cells in the neck region of the secreting unit. Secretin and acetylcholine are the major direct stimuli for these cells.

a: Cholecystokinin (CCK) is the major stimulus for the enzyme-secreting acinar cells, and CCK is an agonist for the action of secretin on the HCO_3^--secreting ductule cells, but it is not the major stimulus for these cells.

c: Glucagon has no appreciable effect on secretory activity of the exocrine pancreas.

d: Insulin has no appreciable effect on secretory activity of the exocrine pancreas.

e: Histamine has no appreciable effect on the secretory activity of the exocrine pancreas.

5. **Answer b:**

Although the liver synthesizes bile acids, recirculation via the enterohepatic circulation of previously formed bile acid is a major source. Its loss cannot be completely overcome and results in diminished bile acid concentration in the small intestine. Since the major role of bile is formation of micelles so that fat can be absorbed more readily, lack of bile causes fat malabsorption.

a: Impaired ileal reabsorption of bile salts tends to increase their loss in the feces, not decrease it.

c: Reduced bile concentration will reduce absorption of dietary cholesterol. Even though dietary cholesterol generally forms a small part of total body cholesterol, a deficit in its absorption would certainly not increase serum cholesterol levels.

d: The liver would, in fact, increase synthesis of bile acids in order to compensate for loss of recirculated bile acids.

6. **Answer c:**

Proteins are digested, in part, by peptidases. Peptic cells in the oxyntic glands secrete pepsinogen, the precursor to the proteolytic enzyme, pepsin. Note, however, that pepsin digestion of proteins accounts for only 10% to 15% of total protein digestion.

a and b: Fats and carbohydrates require lipases and amylase, respectively. These are contained in saliva and begin digestive processes there. However, they are inactivated by the high acidity that exists in the stomach.

d: Vitamins are organic compounds that require no breakdown before they can be absorbed. They are either water soluble or fat soluble. Fat soluble vitamins are solubilized within bile salt micelles and then absorbed. Absorption occurs mostly in the small intestine.

e: Minerals are absorbed in the small intestine.

7. **Answer c:**

Amino acids and simple sugars like glucose and galactose require a Na^+-dependent cotransport for absorption.

a and d: Chylomicra and mixed micelles are required for the absorption of dietary lipids and are not used by carbohydrate and protein end-products.

b: Only absorbed lipids are carried by lymphatic channels to the bloodstream after absorption from the intestine.

8. **Answer a:**

Salivary flow is entirely dependent on the autonomic nervous system. Vagal stimulation produces a large volume of watery fluid, whereas sympathetic stimulation causes the secretion of proteins (mucus and some enzymes).

b-e: Secretion of HCl, pepsin, pancreatic juice, and bile is influenced by vagal stimulation but can occur without it.

9. **Answer d:**

Food has to pass through the pharynx in order to reach the esophagus.

a: One of the first steps in the swallowing sequence involves the tongue pushing the bolus to be swallowed into the pharynx.

b: A centrally coordinated reflex causes contraction of the upper pharyngeal constrictors just as distal portions of the tongue elevate and retract against the palate to push the food bolus into the pharynx. At the same time the soft palate moves downward. The combination of upper pharyngeal constriction and movement of the soft palate closes off the nasopharynx so that food does not move up into the nose.

c: Relaxation of the upper esophagcal sphincter permits the food bolus to enter the esophagus.

e: Once the food bolus has been placed into the upper esophagus a vagally mediated peristaltic wave propels it down the esophagus.

10. **Answer a:**

Vitamin E is a fat-soluble vitamin and is absorbed by passive diffusion along with other lipids in the small intestine.

b: Amino acids are absorbed by Na^+-dependent active transport systems.

c: Glucose is absorbed in the small intestine by Na^+-dependent transport systems.

d: Bile salts are reabsorbed by a Na^+-dependent active transport system located within the terminal ileum.

Chapter 9: Endocrinology

1. **Answer d:**

Thyroid hormones, like steroid hormones, interact with cells by binding to specific nuclear receptors, and this results in altered DNA transcription.

a and e: There are no cell surface membrane receptors for thyroid hormone.

b: Only G-protein coupled cell surface membrane receptors activate adenylate cyclase, an enzyme that is associated with the plasma membrane.

c: Thyroid hormone does not influence cytosolic $[Ca^{++}]$, and its efficacy does not depend on changes in cytosolic $[Ca^{++}]$.

2. **Answer c:**

Certain forms of hypothyroidism, called *goitrous hypothyroidism,* are associated with thyroid enlargement (goiter). Simple (nontoxic) goiter occurs when there is impairment of the capacity of the thyroid to secrete sufficient hormone to meet the need of the peripheral tissues. The resulting increase in thyroid mass (and usually in follicular activity) can overcome a mild impairment. If the impairment is severe, then there is both goiter and hypothyroidism.

a: Increased levels of thyroid hormone will stimulate body metabolism to such an extent that it is not possible to take in sufficient calories to maintain body weight.
b: Cretinism is the condition that results when there is extreme hypothyroidism during fetal life, infancy, and childhood.
d: T_3 does interact with several endocrine systems. Two of its more prominent interactions are with insulin and catecholamines. T_3, when it is present in normal physiologic amounts, will potentiate the actions of insulin in the promotion of glycogenesis and glucose utilization. With respect to catecholamines, T_3 will stimulate formation of β-adrenergic receptors and thereby increase the sensitivity of tissues to the metabolic or vascular effects of β-stimulation.
e: Hypothyroidism is associated with decreased lipolysis and increased serum triglycerides, increased cholesterol, and increased incidence of atherosclerosis.

3. **Answer e:**

Of all the hormones listed, only prolactin is tonically inhibited by hypothalamic factors that are transported through the portal circulation in the pituitary stalk. The factor that is responsible is prolactin inhibitory factor (PIF). It is mostly dopamine.

a, b, and c: Each of these factors will show decreased plasma levels after section of the pituitary stalk because such section would prevent the relevant releasing factors from reaching the anterior pituitary.
d: Insulin levels will be decreased mostly because low cortisol levels will decrease hepatic glucose output but increase tissue utilization of glucose. This will lead to decreased blood glucose levels and decreased insulin levels.

4. **Answer e:**

The major effect of glucocorticoids on metabolism of carbohydrate, fat, and protein is to help supply glucose to critical tissues. To that end they increase the breakdown of proteins so there is an increase in amino acid substrate for hepatic glucose production, and they increase the activity of several enzymes involved in hepatic gluconeogenesis.

a: Glucocorticoids normally increase plasma levels of free fatty acids and ketones, provided that these actions are not inhibited by elevated insulin levels.
b: They are synthesized from cholesterol.
c: They are synthesized by cells in the *zona fasciculata* and *zona reticularis* of the adrenal cortex.
d: Large doses of glucocorticoids will suppress cell-mediated immunity. This is one of the bases for the use of glucocorticoids in inhibition of inflammatory responses to tissue injury.

5. **Answer d:**

Glycogen is a readily converted storage form of glucose. The ability of the liver to store glycogen is very large. Insulin activates glycogen synthase and inhibits glycogen phosphorylase. This promotes formation of glycogen from UDP-glucose and inhibits breakdown of glycogen to glucose-1-phosphate and the subsequent formation of glucose.

a: Insulin does promote protein synthesis in liver and muscle cells and increases amino acid uptake in muscle cells. However, the effect does not arise from elevated levels of amino acids in the blood. On the contrary, insulin decreases protein breakdown.
b: Insulin does promote an increase in free fatty acids but not because of lipolysis. On the contrary, insulin promotes lipid synthesis by the liver. Two aspects are involved: insulin promotes formation of pyruvate by activating 2 enzymes involved in the glycolytic pathway, 6-phosphofructo-1 kinase and pyruvate kinase. Thereafter, insulin activates pyruvate dehydrogenase and acetyl CoA carboxylase. The former promotes formation of acetyl CoA and the latter promotes carboxylation of acetyl CoA to malonyl CoA. This last step is the metabolic process that commits acetyl CoA to synthesis of fatty acids rather than cholesterol (see Fig. 9.10).

c: Insulin lowers plasma glucose concentration by stimulating glucose uptake into tissue cells, glucose utilization by these cells, and storage of the products of glucose metabolism within those cells.

e: Insulin inhibits formation of ketone bodies by inhibiting hepatic breakdown of fatty acids. This maintains normal levels of mitochondrial acetyl CoA and production of ketone bodies.

6. **Answer c:**

Glucagon activates glycogen phosphorylase and thereby promotes breakdown of glycogen and formation of glucose 6-phosphate (Fig. 9.10). Two actions of glucagon then favor the formation of glucose:
1) Glucagon inhibits pyruvate kinase and thereby inhibits the glycolytic pathway by which glucose 6-phosphate is eventually converted to pyruvate (Fig. 9.10).
2) Glucagon promotes conversion of fructose 1,6-bisphosphate to fructose 6-phosphate (Fig. 9.10) with a corresponding increase in the rate of gluconeogenesis.
Finally, glucagon inhibits glycogen synthase, thereby inhibiting the formation of glycogen from glucose 6-phosphate (Fig. 9.10).

a: Glucagon is secreted by pancreatic A cells. G cells are found in the mucosa of the gastric antrum, the duodenum, and jejunum. They secrete the gastrointestinal hormone, gastrin.

b: An increase in blood glucose level will inhibit glucagon secretion. The most important metabolic stimulus for glucagon secretion is a fall in plasma glucose levels.

d: Glucagon promotes formation of ketone bodies and does not do it by increasing blood glucose levels. Promotion of ketone body formation occurs in the presence of glucagon because, as explained in c, glucagon depresses the sequence glucose 6-phosphate to pyruvate to acetyl CoA to citrate to malonyl CoA (Fig. 9.10). Malonyl CoA inhibits a crucial step in fatty acid oxidation (Fig. 9.10). Therefore glucagon attenuates the inhibition of fatty acid oxidation, promotes formation of fatty acyl carnitine (Fig. 9.10), and subsequent formation of intramitochondrial acetyl CoA and ketone bodies (Fig. 9.10).

e: Glucagon does increase glycogen synthesis, but it does it by activating or inhibiting relevant enzymes (see explanation for choice c), not by elevating blood glucose levels.

7. **Answer e:**

TSH works via activation of adenylate cyclase and affects each of the listed steps in the synthesis and secretion of T_3 and T_4 except the final diffusion of T_3 and T_4 out of the follicular cell into the blood. T_3 and T_4 are soluble in lipid and can therefore easily cross the plasma membrane. The major influence on the rate of diffusion is the concentration gradient. Nothing regulates the solubility of T_3 and T_4 in the plasma membrane.

a: Active transport of I^- into the follicular epithelial cells would be affected.

b: Iodination of thyroglobulin to form diiodotyrosine would be affected.

c: Coupling of two diiodotyrosines to form tetraiodothyronine coupled to thyroglobulin would be affected.

d: Pinocytosis of colloid and fusion with lysosomes would be affected.

8. **Answer e:**

Corticotropin releasing hormone (CRH) is synthesized in and secreted from the hypothalamus. Its action on the pituitary yields ACTH, which, in turn, promotes cortisol secretion from the adrenal cortex. Cortisol is an important inhibitor of hypothalamic CRH release (Fig. 9.8). Hence removal of the pituitary would enhance CRH secretion, not decrease it.

a: Removal of the pituitary would eliminate the source for ACTH and would lead to atrophy of the adrenal cortex.

b: Removal of the pituitary would eliminate the source for ACTH and would therefore greatly diminish blood levels of glucocorticoids. As a result hepatic glucose output would decrease and tissue utilization of glucose would increase, leading to decreased blood glucose levels.

c: Removal of the pituitary would eliminate the source for growth hormone and thereby remove the stimulus for synthesis of IGFs.

d: Removal of the pituitary would eliminate the source for follicle stimulating hormone (FSH) and leuteinizing hormone (LH). In men LH is required for testosterone synthesis, and both testosterone and FSH are required for sperm production. In women LH is required for ovulation. Hence lack of FSH and LH will cause sterility.

9. Answer d:

Growth hormone stimulates IGF synthesis. IGFs inhibit GH secretion via two pathways: by a direct inhibitory effect on pituitary somatotrophs and by a stimulatory effect on hypothalamic release of somatostatin. Somatostatin, in turn, inhibits pituitary somatotrophs.

a and b: Both exercise and psychologic stress increase GH secretion. The mechanism may be via elevated levels of central nervous noradrenaline.

c: GH secretion does increase during normal sleeping hours. Details of the mechanism are not known.

e: Hypoglycemia and several other conditions in which there is an actual or threatened decrease in the substrates for cellular energy production will promote GH secretion.

10. Answer a:

Vasopressin acts on the renal collecting tubule and causes increased reabsorption of water from the tubular fluid to the body water space. It also acts on vascular smooth muscle to cause vasoconstriction. Its secretion is stimulated by decreases in blood pressure or blood volume so that its blood vessel action can compensate for decreased blood pressure and its renal action can compensate for both decreased blood volume and decreased blood pressure.

b: Vasopressin is also known as antidiuretic hormone (ADH).

c: Vasopressin acts on the kidney to increase water reabsorption.

d: Vasopressin is synthesized as a prohormone with a neurophysin, which may serve as its own carrier protein in the blood.

e: Vasopressin deficiency leads to decreased collecting tubule permeability to water.

Chapter 10: Cartilage and Bone

1. Answer c:

The epiphyseal disk is a narrow band of cartilage between epiphysis and diaphysis. It and a narrow region of diaphysis bordering it form the growing zone of a long bone. Once adult body proportions are reached, growth stops and the epiphyseal disks are replaced by trabecular bone.

a: The epiphysis is the bony region at the end of long bones, including the region that forms the joint. Longitudinal growth occurs only in the narrow cartilage disk, the epiphyseal disk that separates the epiphysis from the diaphysis.

b: The diaphysis is the area of hollow, calcified bone. This area can change in thickness or in composition if there are appropriate mechanical stresses and growth conditions, but this region does not initiate longitudinal growth.

d: The medullary cavity elongates once bone trabeculae have formed from cartilage within the epiphyseal disk.

e: The primary ossification center identifies the locus within a region of cartilage at which the process of cartilage-to-bone conversion begins and from which the process radiates.

2. Answer b:

Although the dominant action of vitamin D is to stimulate osteoclasts that have functioning vitamin D receptors and to stimulate bone resorption, it does paradoxically and simultaneously stimulate osteoblasts.

a: Estrogen has two effects relevant to osteoblast function

1) It acts directly on nuclear receptors in osteoblasts to inhibit formation of osteoclast-differentiating factor (s).

2) it enhances formation of biologically active vitamin D and thereby tends to stimulate osteoid deposition by osteoblasts.

Therefore, lack of estrogen, as might be observed in menopause, will decrease osteoid deposition by osteoblasts and enhance osteoclast activity. The enhanced osteoclast activity is mostly due to increased secretion of osteoclast differentiating factor(s) from osteoblasts and peripheral monocytes.

c: Excess levels of glucocorticoids depress collagen synthesis and osteoblast activity and at the same time enhance bone resorption. Hence elevated levels of glucocorticoids are associated with both decreased bone growth and net loss of bone. The mechanism may be, in part, a stimulatory effect of glucocorticoids on PTH levels (in some settings) and a potentiating effect of glucocorticoids on PTH action. The dominant action of PTH is to enhance differentiation and activity of osteoclasts.

d: Hyperthyroidism is accompanied by excess bone loss and can be associated with increased plasma levels of Ca^{++} and PO_4^{3-}. Their urinary excretion may increase as a result.

3. **Answer a:**

PTH is synthesized in and secreted from chief cells in the parathyroid glands in response to low levels of ionized calcium in the extracellular fluid. This is an unusual phenomenon in that most secretory processes are associated with increased influx of Ca^{++}. In the parathyroid, on the other hand, increased Ca^{++} influx inhibits PTH secretion. PTH secretion is promoted by increases in cytosolic cAMP such as would be observed after activation of adenylate cyclase by β-adrenoceptors or H_2 histamine receptors.

b and d: There is some evidence that vitamin D can inhibit PTH secretion but no evidence that PTH secretion is stimulated by vitamin D. Receptors for 1,25-$(OH)_2D_3$ vitamin D have been identified in parathyroid tissue and 1,25-$(OH)_2D_3$ vitamin D is given intravenously to suppress PTH secretion in patients with chronic renal failure. The mechanism of this PTH suppression may be direct via activation of vitamin D receptors in the parathyroid or indirect via vitamin D-mediated hypercalcemia.

c: Hypercalcemia inhibits PTH secretion.

e: Even though phosphorus is among the most abundant constituents of most tissues and is involved in almost all metabolic processes, it does not appear to be a primary stimulus in a process of feedback regulation. The major control over phosphorus economy is exerted at the level of the kidney.

4. **Answer b:**

The primary function of PTH is to increase extracellular [Ca^{++}]. Three mechanisms are involved:
1) Increased bone resorption
2) Increased Ca^{++} reabsorption from the distal convoluted tubule
3) Vitamin D-mediated increase of intestinal Ca^{++}absorption because PTH promotes formation of 1,25-$(OH)_2D_3$ vitamin D.

Promotion of osteoclast differentiation by PTH leads to histologically specific diminution of bone mass. Elevation of serum [Ca^{++}] causes deposition of calcium salts in the renal parenchyma. Renal stones are usually composed of calcium oxalate or calcium phosphate.

a: PTH stimulates the action of 1 α-hydroxylase and therefore promotes formation of 1,25-$(OH)_2D_3$ vitamin D.

c: High levels of PTH do promote bone resorption as well phosphate absorption from the GI tract. The tendency of these two effects to cause hyperphosphatemia is counteracted by the very efficient renal clearance mechanism. Increased tubular load and a PTH-mediated decrease in tubular reabsorption of PO_4^{3-} together ensure normal serum phosphate levels in hyperparathyroidism.

d: PTH promotes bone resorption and elevation of serum [Ca^{++}].

5. The nomenclature for biologically active vitamin D is confusing and must often be derived from

the context in which an expression is used. The most common synonyms are: Vitamin D; 1,25-$(OH)_2D_3$ vitamin D, 1,25-dihydroxy Vitamin D3, 1,25-$(OH)_2D_3$, 1,25-$(OH)_2D$, 1-25 dihydroxycholecalciferol and calcitriol.

Answer a:

Biologically active vitamin D, 1,25-$(OH)_2D_3$, increases renal reabsorption of phosphate. The mechanism is believed to be changes in membrane lipid composition. PTH, on the other hand, depresses renal phosphate reabsorption. The major site of PTH action is the proximal convoluted tubule, but some inhibition of distal tubular phosphate reabsorption occurs as well.

b: Vitamin D and PTH both increase renal Ca^{++} reabsorption.

c: Vitamin D increases intestinal Ca^{++} absorption. PTH has little direct effect on intestinal Ca^{++} transport. However, it stimulates 1 α-hydroxylase and thereby promotes formation of biologically active vitamin D.

d: Both PTH and vitamin D elevate plasma [Ca^{++}]. PTH does it by stimulating differentiation of osteoclast precursors into active, bone-resorbing osteoclasts. Vitamin D, on the other hand, has the dominant effect of promoting [Ca^{++}] absorption from the intestinal lumen to the extracellular fluid.

6. Answer c:

Vitamin D deficiency would be expected to impede intestinal absorption, leading to hypocalcemia. Low serum [Ca^{++}] is a potent stimulus for PTH secretion.

a: The major actions of calcitonin are to inhibit osteoclasts and to inhibit renal Ca^{++} reabsorption. Both actions would tend to decrease plasma calcium levels. However, the effects of calcitonin are weak and easily compensated by PTH and vitamin D. Therefore it is unlikely that calcitonin deficiency could cause significant hypocalcemia.

b: Hyperparathyroidism would primarily stimulate bone resorption and cause increased plasma [Ca^{++}].

d: It is not possible to have high PTH levels in a hypoparathyroid state.

7. Answer b:

Alkaline phosphatase is a marker for osteoblasts and its cellular levels correlate with the ability of bone to mineralize. Its precise role in bone mineralization is not fully known but may involve neutralization of pyrophosphate, a potent inhibitor of mineralization. Mineralization begins with precipitation of calcium salts from their soluble forms and is followed by local accumulation of calcium phosphate in the form of brushite ($CaHPO_4 \cdot H_2O$). As mineralization progresses, the solid phase becomes hydroxyapatite [$Ca_5(PO_4)_3OH$]. At first it has a low ratio of calcium to phosphate. With maturation the ratio increases and the precipitate becomes increasingly crystalline.

a: Crystallization requires the specific microenvironment that is created at specific sites of the collagen fibers and in the presence of osteoblasts. Once nucleation has occurred, existing plasma concentrations of Ca^{++} and HPO_4^{3-} can sustain the processes of mineralization.

c: Bone mass reaches a maximum in the second or third decades of life and then declines gradually. This indicates that overall osteoclast activity is slightly greater than osteoblast activity.

d: In the adult human less than 50% of dietary Ca^{++} is absorbed.

e: In the western hemisphere about 80% of kidney stones are made of calcium oxalate or calcium phosphate (in the form of brushite or hydroxyapatite). Stone formation requires initiation and growth of stable nuclei. This requires both a high serum level of the appropriate salts and a malfunction in the mechanisms that normally inhibit stone formation. Urine contains potent inhibitors of nucleation and growth of calcium oxalate and calcium phosphate stones. One of these is pyrophosphate, affecting particularly calcium phosphate. It is a strong inhibitor of stone formation even at high urine phosphate levels.

8. **Answer a:**

PTH does stimulate conversion of 25-hydroxy vitamin D to 1,25 dihydroxy vitamin D_3 in the kidney.

b: Calcitonin inhibits osteoclasts, not osteoblasts. It also inhibits renal Ca^{++} reabsorption. This would tend to decrease plasma [Ca^{++}]. However, calcitonin normally has little effect on plasma [Ca^{++}] or [PO_4^{3-}].

c: Renal tubular Ca^{++} reabsorption occurs via passive mechanisms, driven by gradients in charge and concentration. 1,25-$(OH)_2D_3$ directly promotes Ca^{++} reabsorption from the nephron, but is not required for this reabsorption. The mechanism of enhanced Ca^{++} reabsorption is via a receptor-coupled effect of vitamin D in distal convoluted tubular cells.

d: PTH synthesis and secretion show little change with changes in plasma [PO_4^{3-}].

e: Hypoparathyroidism causes reduced levels of 1 α-hydroxylase and therefore causes reduced formation of 1,25-$(OH)_2D_3$. The major target organ for this biologically active form of vitamin D is the intestine where it causes increased reabsorption of Ca^{++}. Hypoparathyroidism is therefore associated with decreased reabsorption of Ca^{++} from the gut.

9. **Answer c:**

When bone is first formed it consists of a network of trabeculae. Compact bone is sufficiently old to have been filled in completely with mineralized matrix. The only exception is the lacunae and their interconnecting canaliculi.

a: Compact bone is made up of parallel bony columns. They are named osteons.

b: Compact bone does contain a network of canals (haversian canals and Volkmann's canals) that house blood vessels and nerves.

d: Compact bone does contain haversian canals. They are the canals forming the center of each osteon.

10. **Answer b:**

The hormone 1,25-dihydroxy vitamin D3 does not exist in the gut, and it is not actively transported anywhere.

a: The statement is true. Conversion of 25-hydroxy vitamin D3 to 1,25-dihydroxy vitamin D3 requires the enzyme 1 α-hydroxylase, which is located largely in mitochondria of epithelial cells in renal proximal convoluted tubule and even there in only small amounts.

c: Vitamin D3 that is not bound to receptors or metabolized is stored in adipocytes.

d: The statement is correct.

Chapter 11: Integrative Physiology

1. **Answer c:**

In the absence of food intake insulin levels would be expected to be low so that lipolysis would proceed unhindered in adipose tissue. A further stimulus to lipolysis is catecholamines. Their circulating levels are likely to be high in this stressful situation. Catecholamines promote lipolysis via β_1 adrenoceptors. They exert their effect on adipose tissue mostly by reaching it via the circulation. White fat is not innervated.

a: Cortisol levels would be expected to be elevated in this stressful setting.

b: In the absence of food intake plasma glucose levels will tend to decrease and glucagon levels will rise. Glucagon stimulates glycogenolysis and gluconeogenesis. Both processes cause increased glucose output from the liver.

d: When only water is consumed, but at least some Na^+ is lost via obligatory urine excretion, then there is a tendency for plasma Na^+ levels to decrease. This provides a stimulus for increased aldosterone production. Its purpose is to conserve Na^+.

e: In the absence of food intake skeletal muscle provides glucose (via glycogenolysis) and amino acids (via proteolysis). Cortisol is a major stimulus for increased proteolysis.

2. **Answer c:**

The major cardiovascular role of the parasympathetic nervous system relates to its inhibitory influence on heart rate. During exercise at the stated level of performance, cardiac parasympathetic activity will be suppressed in order to elevate heart rate. Therefore removing it altogether will have no significant effect on cardiovascular function.

a: Skeletal muscle function is controlled by the somatic nervous system, not by the autonomic nervous system.
b: There will be no significant effect on heart rate because cardiac parasympathetic function is already suppressed at this level of exercise performance.
d: Erection is caused mainly by release of vasoactive intestinal peptide (VIP) from a separate set of nerves. Removal of parasympathetic activity would not lead to an erection. Even if it did, the motion of the penis would be from toes towards the head. A more likely event would therefore be that the toes are pulled up rather than the chest being pulled down.

3. **Answer b:**
The sympathetic nervous system has several important functions during exercise. They include elevation of heart rate, elevation of cardiac performance, elevation of plasma epinephrine levels, and constriction of vascular beds that are not essential during exercise performance. Removing sympathetic nervous activity would have significant effects on cardiac output and the level of exercise performance.

a: The sympathetic nervous system constricts vascular beds that are not metabolically active. This has the effect of directing cardiac output preferentially towards those tissues that are metabolically active. Such tissues are dilated by the local vasodilator action of metabolic products like lactic acid.
c: The statement is true.
d: Norepinephrine is the postganglionic neurotransmitter of the sympathetic nervous system. Its plasma level is elevated at times of high sympathetic activity, like exercise. Removal of sympathetic activity would decrease plasma levels of norepinephrine.

4. **Answer e:**
H^+ is elevated above normal levels. Hence there is a state of acedemia. Since P_{CO_2} is at the normal level (40 mm Hg), but $[HCO_3^-]$ is below normal, suspect metabolic acidosis. Is the respiratory compensation appropriate? If $\Delta P_{CO_2} = 1.2\Delta[HCO_3^-]$ the compensation is appropriate. $\Delta[HCO_3^-] = 24 - 14 = 10$. Hence, if respiratory compensation were appropriate the P_{CO_2} should be decreased 12 mm Hg below normal. In this patient P_{CO_2} is 40 mm Hg, higher than is appropriate. Therefore there also is a state of respiratory acidosis.

a and b: Simple respiratory acidosis would be accompanied by elevation in HCO_3^-. The reason is that the acid load would be buffered primarily by intracellular buffering in the short term and by increased renal excretion of ammonium (NH_4^+). The former raises $[HCO_3^-]$ because of the chloride shift in erythrocytes. The latter raises $[HCO_3^-]$ because NH_4^+ excretion generates new HCO_3^-.
c: It recognizes the presence of the primary problem but overlooks the additional respiratory complication.
d: Metabolic alkalosis exists when the plasma level of HCO_3^- is higher than appropriate. This is not the case here.

5. **Answer d:**
Although $[H^+]$ is normal, the $[HCO_3^-]$ is depressed, indicating the presence of metabolic acidosis. Intracellular buffering of the acid load would have resulted in translocation of K^+ out of the cell, raising extracellular $[K^+]$.

a and c: It is highly unlikely that sufficient quantities of dietary intake or intravenous infusion could be administered to elevate extracellular $[K^+]$ in a healthy individual. It would be possible, however, to have elevated $[K^+]$, even without dietary or intravenous loading, in a state of insulin deficiency.
b: There is no physiologic mechanism for the regulation of the unmeasured anion gap (UAG). The UAG is the difference

between $[Na^+]$ + $[K^+]$ and $[Cl^-]$ + $[HCO_3^-]$. It is the ionic footprint left by species of anion other than Cl^- or HCO_3^-. It is often called "the poor man's lactate reading". In this case UAG = 39. Normal UAG is [(140 + 5) – (100 +24)] = 21. Elevation of the UAG in this case is further evidence for the presence of metabolic acidosis even though there is not a state of acidemia.

e: Repolarization of excitable tissue at the end of an action potential does involve the translocation of K^+ from intracellular to extracellular space. This may result in localized hyperkalemia near a very active nerve, but it is highly unlikely that enough K^+ could be translocated by nervous activity alone to increase extracellular K^+ concentration 60% above normal values.

6. **Answer b:**

The existing $[H^+]$ is higher than normal by 7 nm/L. Therefore a state of acedemia exists. It is respiratory acidosis because P_{CO_2} is higher than normal. It is not metabolic acidosis because $[HCO_3^-]$ is not lower than normal. Is it acute or chronic respiratory acidosis? Acute if $\Delta[H^+] = 0.77\Delta P_{CO_2}$; chronic if $\Delta[H^+] = 0.32\Delta P_{CO_2}$. From the values given, $0.77\Delta P_{CO_2} = 0.77(63 - 40) = 17.7$ and $0.32\Delta P_{CO_2} = 0.32(63 - 40) = 7.4$. The data indicate a state of chronic respiratory acidosis.

Is the metabolic compensation appropriate? Yes, if $\Delta[HCO_3^-]/\Delta P_{CO_2} = 3.5/10$. The readings indicate that $\Delta[HCO_3^-]/\Delta P_{CO_2} = 8/23 = 0.35$. Metabolic compensation is appropriate, and there is no mixed disturbance present.

a: The condition has been of sufficient duration to allow generation of HCO_3^- for buffering of extracellular H^+, making the ratio $\Delta H^+/\Delta P_{CO_2}$ sufficiently low to rule out acute respiratory acidosis.

c: Metabolic acidosis does not exist because $[HCO_3^-]$ is not lower than appropriate.

d: $[HCO_3^-]$ is not higher than appropriate. It could be argued that in the absence of a patient history the primary problem is metabolic alkalosis (elevated $[HCO_3^-]$) combined with respiratory acidosis. Apart from a history, a measurement of urine $[Cl^-]$ would settle the question. Metabolic alkalosis can persist only if there is an intense need to reabsorb Na^+ in the nephron, over and above that which can be reabsorbed along with Cl^-. In this setting, after all Cl^- has been reabsorbed, additional Na^+ reabsorption would take place along with HCO_3^- and urine $[Cl^-]$ would be near zero.

7. **Answer e:**

Secretion of estradiol by the maturing follicle maintains a sufficiently high level of this hormone for a sufficient length of time to permit a temporarily positive feedback of estradiol on LH secretion. This causes the LH "surge" and causes ovulation, it does not prevent it.

a: It correctly describes the hormone levels during the luteal phase (the time between ovulation and menses) and the feedback effects of estradiol and progesterone during that time.

b, c, and d: These are each a correct statement describing hormonal events of the menstrual cycle.

8. **Answer b:**

Testosterone is an anabolic hormone. It increases the synthesis and decreases the breakdown of proteins, thereby increasing the rate of growth.

a: Testosterone increases basal metabolic rate. The reason is believed to be the increase in muscle mass relative to body surface area.

c: Testosterone stimulates the final stages of spermatogenesis.

d: Testosterone inhibits LH secretion in part by inhibiting GnRH secretion and in part by inhibiting pituitary gonadotrophs.

e: Testosterone is a vital determinant of secondary sex characteristics such as size of internal and external genitalia, distribution of body hair, size of larynx, length and thickness of vocal cords, body configuration, psycho-social attitude, and behavior.

9. **Answer d:**

The meiotic division by which each daughter cell receives the haploid number of 23 chromosomes does not occur until the final stages of spermatogenesis. Thus the early spermatid is the first stage in which the haploid number is present.

a, b, and c: Spermatogonium (germ cell) and both early (c) and late spermatocyte (b) represent earlier stages. The full diploid number is still present in them.

e: Leydig cells contain the full diploid set of chromosomes.

10. **Answer a:**

Sertoli cells do not contain LH receptors. Such receptors are found in the membranes of Leydig cells.

b: Sertoli cells do secrete the fluid that bathes the interior of the seminiferous tubule.

c: Sertoli cells do synthesize and secrete the peptide, inhibin.

d: Sertoli cells do provide the environment and nourishment for maturing sperm.

e: The statement that Sertoli cells secrete androgen-binding protein is a correct statement.

11. **Answer d:**

Excitement can be brought on by psychogenic factors as well as by physical factors such as visual, olfactory, or tactile sensations.

a: There are changes in regional blood flow. They cause such changes as skin mottling, erection of nipples or penis, or enlargement of the clitoris.

b: Continuation of excitement to the next phase, plateau requires reinforcement. If that is absent then return to a nonsexual state can occur quickly.

c: Cardiovascular and respiratory changes are evident. They manifest themselves as changes in heart rate, blood pressure, and ventilation.

APPENDIX

*Glossary of Abbreviations**

ACTH	Adrenocorticotropic Hormone
CRH	Corticotropin-Releasing Hormone
DA	Dopamine
DIT	Di-iodothyronine
END	Endorphins
FSH	Follicle-Stimulating Hormone
GH	Growth Hormone
GHRH	Growth Hormone-Releasing Hormone
GnRH	Gonadotropin-Releasing Hormone
IGF	Insulin-like Growth Factor
IP_3	Inositol Trisphosphate
LH	Leuteinizing Hormone
MIT	Mono iodothyronine
MSH	Melanocyte-Stimulating Hormone
PIF	Prolactin-Inhibiting Factors
POMC	Proopiomelanocortin
PRF	Prolactin-Releasing Factor
SA	Serum Albumin
SS	Somatostatin
T_3	Tri-iodothyronine
T_4	Thyroxine
TRH	Thyrotropin-Releasing Hormone
TSH	Thyroid-Stimulating Hormone
TYR	Tyrosine
TBG	Thyroid-binding Globulin
TBPA	Thyroid-binding Pre Albumin
TGL	Thyroglobulin
VIP	Vasoactive Intestinal Peptide

*Used in Chapter 9.

INDEX

Note: Page numbers in italics indicate figures; page numbers followed by *t* indicate tables, page numbers followed by *n* indicate footnotes.

Mosby's Review's Series

How to install this program—Windows users

1. Place the disk in Drive A: (or B:)
2. From Program Manager, select File, then Run, then enter:
   A:SETUP (or B:SETUP if your disk drive is B:)
3. Follow the instructions on screen.

How to run this program—Windows users

Open the MOSBY Program Group and select the ACE program.

How to install this program—Macintosh users

1. Create a new folder on your hard disk called MOSBY, then open it.
   If you already have a MOSBY folder, you may use any other name.
2. Insert the disk into your floppy drive and open it.
3. Drag all items from the disk to the new folder.
4. If more than one disk is included, perform steps 2 and 3 for each disk.

How to run this program—Macintosh users

Open the MOSBY folder and select the ACE program.

For complete instructions on using the program, please read the "How to use this Program" file.